Research Methods for the Biosciences

research methods for the biosciences

D. Holmes

P. Moody

D. Dine

OXFORD
UNIVERSITY PRESS

Great Clarendon Street, Oxford OX2 6DP

Oxford University Press is a department of the University of Oxford.
It furthers the University's objective of excellence in research, scholarship,
and education by publishing worldwide in

Oxford New York

Auckland Cape Town Dar es Salaam Hong Kong Karachi
Kuala Lumpur Madrid Melbourne Mexico City Nairobi
New Delhi Shanghai Taipei Toronto

With offices in

Argentina Austria Brazil Chile Czech Republic France Greece
Guatemala Hungary Italy Japan Poland Portugal Singapore
South Korea Switzerland Thailand Turkey Ukraine Vietnam

Oxford is a registered trade mark of Oxford University Press
in the UK and in certain other countries

Published in the United States
by Oxford University Press Inc., New York

First published 2006

British Library Cataloguing in Publication Data
Data available

Library of Congress Cataloging in Publication Data
Data available

Typeset by Newgen Imaging Systems (P) Ltd., Chennai, India
Printed in Great Britain
on acid-free paper by
Ashford Colour Press, Gosport, Hamsphire

ISBN 0–19–927692–7 978–0–19–927692–9

10 9 8 7 6 5 4 3 2 1

To Gill, and all those who teach and inspire us.

Preface

Getting the most out of this book

We write this section with some uncertainty since we rarely read a preface in a book and wonder if anyone is going to read what we've written here. Nonetheless, there are several important things we need to explain to you to enable you to get the best out of this book and its Online Resource Centre, so we'd encourage you to take a few moments to read on.

Who should read this book?

This book is primarily written for undergraduates who wish to develop their understanding of designing, carrying out, and reporting research. We anticipate that graduates, although familiar with most of this material, will find this book and its Online Resource Centre to be a useful reference too. We have therefore used examples almost entirely drawn from real undergraduate and some graduate research projects. We are indebted to all our students who have allowed us to use their ideas and data in this book.

Learning features

In our experience we have found that students best understand the statistics element of this book if they first use a calculator to work out the calculation. Therefore, in this book we have arranged the statistical information in boxes with general details and a worked example for you to follow.

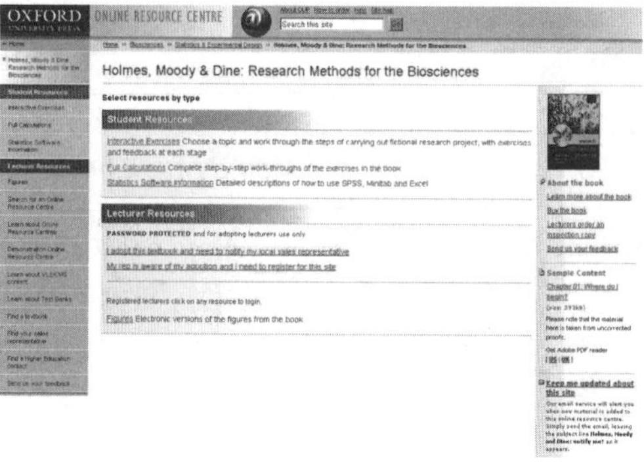

We then include in the Online Resource Centre an explanation of how to use SPSS, Excel, and Minitab to carry out the same calculation.

In the calculations, we've included in this book we have rounded all values, usually to five decimal places. However, we carried out all the calculations using **all** decimal places (as you should). This means that some of our sums do not appear to quite add up. Any minor differences in the calculations should be the result of this rounding of values.

We know that most people will dip in to this book and so we have included a **glossary** of most terms that you need to be familiar with. Key terms are shown in coloured type the first time they appear; definitions of these key terms then appear in the glossary at the end of the book.

Online Resource Centre

Research Methods for the Biosciences is more than just this printed book. The *Research Methods for the Biosciences* **Online Resource Centre** features extensive online materials to help you really get to grips with the skills you need to carry out research work.

The **student area** of the Online Resource Centre includes:

- full details of all calculations in this book: every step in each calculation is shown so you can see exactly how we reach the answers shown in the book
- walk-through explanations of how to use SPSS, Excel, and Minitab to carry out these calculations
- interactive tasks for you to work through to test your understanding of the topics in this book, and hone your research methods skills.

You can access these resources at www.oxfordtextbooks.co.uk/orc/holmes

You'll see the Online Resource Centre icon throughout the book. This icon tells you that the part of the text you are reading has online materials to accompany it.

For the lecturer

The **lecturer area** of the Online Resource Centre provides additional materials to make the book easier to teach from. These materials include:

- **figures from the book**, available to download for use in lecture slides
- all the content from the student area as a **VLE cartridge,** so these materials can be readily uploaded into your own online teaching pages.

Simply go to www.oxfordtextbooks.co.uk/orc/holmes and register as a user of the book to gain free access to these materials.

Getting started

You may be coming to this book with very little in the way of training in maths. If you do not know how to calculate this sum $(4 - 3)^2/2$, or if you do not recognize the symbols $<$ or $>$, then we suggest you first look at **Appendix c** and the additional examples on the Online Resource Centre.

You may wish to analyse data you have gathered from an experiment carried out in your course. For this we suggest you start with **Appendix b**.

You may wish to prepare a critique of published research. The chapter that considers this is **Chapter 2**, with cross-referencing to earlier and some later chapters.

If you wish to design a research project and you are familiar with terms such as variable, parametric, aim, hypothesis, etc., then go to **Appendix a** and **Chapter 2**. If you are not familiar with these terms still refer to **Appendix a**, and then continue from **Chapter 1**.

And finally . . .

Just in case OUP ask us to produce a second edition, we would like to hear about any errors (we hope there are none) and any suggestions you have for improvements, so that a second edition can be even better than the first. You can contact us by using the 'Send us your feedback' option in the Online Resource Centre.

Acknowledgements

We have used many of the bright ideas and results that our undergraduates and graduates have produced to ensure that this book is relevant to you. We especially wish to thank Helen Bagley, Alys Black, Roz Chalk, Elaine Chape, Suzanne Charlton, Jeremy Cox, Stephanie Ellis, Shaan Gabriel, Mikaela Haslem, Robin Holbrook, Becky Lee, Dan Price, Anita Rattu, Olivia Renshaw, Michelle Robertson, Angie Roxburgh, Emma Georgiou, Stephanie Stallard, Gemma Sykes, Viv Tolley, and Robert Vernon.

During the preparation of this book we have benefited considerably from the excellent, thorough comments provided by our reviewers: Dr Martin Cox, Coventry University; Professor Graeme Ruxton, University of Glasgow; and Dr Richard Small, Liverpool John Moores University. We are most grateful to Minitab.com for providing us with the Minitab software. Most of the figures have been produced by Rosemary Holmes, with John Holmes managing the KID program. The wonderful illustrations are by Jenny Joseph.

And finally our thanks go to Jonathan Crowe and OUP for giving us this opportunity to review our thinking about research methods and to explain our understanding of this topic to you.

Contents

4 An introduction to hypothesis testing

5 Hypothesis testing: Do my data fit an expected ratio?

6 Hypothesis testing: Associations and relationships

7 Hypothesis testing: Do my samples come from the same population? Parametric data

8 Hypothesis testing: Do my samples come from the same population? Non-parametric data

9 Research, the law, and you

Boxes that outline the methods for the statistical tests included in this book

Standard deviation and variance

Parametric and non-parametric data

Chi-squared and G tests

Correlation and regression

z and *t* tests

Parametric ANOVAs

Mann–Whitney U test and Wilcoxon's matched pairs test

Non-parametric ANOVAs, including the Kruskal–Wallis test

1

Where do I begin?

You may be using this book because you are starting to plan a piece of independent research, you have been carrying out an experiment designed by someone else, or you are reading about published research. This book is written for those of you who need to have an understanding of experimental design and as part of this need to develop your statistical skills. We begin in this chapter by introducing you to some of the basic principles and terms that you will come across in your research. The topics covered in this chapter are supported by further examples in our Online Resource Centre. If you are about to embark on a piece of research and need help choosing a research topic, we have included some suggestions in Appendix a. If you are working through this chapter, it should take about 1 hour to complete all the exercises.

1.1. Data

When carrying out an investigation you will generate data (singular – datum). Data consist of a number of observations or measurements on the topic you are investigating. For example, 103 students registered on a particular science course and each week the number attending a scheduled session was recorded (Table 1.1.). The numbers of students attending each week are the data.

1.2. Types of investigations

Your data are collected when you carry out research. Research can be broadly defined as the process by which you try to find out more about a particular topic. In this book we focus mainly on primary practical

Table 1.1. The number of students attending each week on a particular course

Week number for the course	1	2	3	4	5	6 (Assessed practical)	7	8	9 (Exam)
Number of students attending the class	100	95	93	80	55	100	52	51	100

research: that is, the research you do yourself in a laboratory or in the field. However, the ideas covered here are also relevant to preparing literature reviews or critiques of published research papers. Primary research falls roughly into two categories which we call Observational and Experimental. Each of these has particular features that you need to be aware of when planning your research.

1.2.1. **An observational investigation**

An observational investigation is one that:

- does not generally start out with a question (hypothesis, 1.3.3.)
- sets out to examine an aim and objectives (1.3.)
- does generate data that describe a specific system
- may lead to the construction of hypotheses after the investigation
- is often used in preliminary research.

For example, most surveys set out to find out 'what is there'. These can include surveys of species, surveys of people's opinions, etc. None of these necessarily set out to test a question (hypothesis, 1.3.3.) and if they don't then they are 'observational' investigations. The survey of the numbers of students attending a course (Table 1.1.) is an observational investigation. An ecological survey may also be an observational investigation.

i. An observational investigation does not generally test hypotheses

In most undergraduate studies the investigations you carry out in practicals are designed to test a specific question or hypothesis (1.3.3.). As a result most students tend to think that all research tests hypotheses and feel as though the work is incomplete or they are somehow missing the point if there are no hypotheses to test. Some investigations, however, start without any hypotheses. It is still perfectly valid and valuable research.

ii. An observational investigation sets out to examine an aim and objectives

Even though an observational investigation does not generally set out to test hypotheses, the work is not directionless but should be guided by an aim and objectives.

iii. An observational investigation does generate data that describe a specific system

If, for example, you did carry out a survey of people's opinions, it is clear that data will be collected. Observational investigations generate data that will need to be evaluated. Since you do not usually test hypotheses, then you will not need the hypothesis-testing statistics (Chapters 4–8). However, you may wish to summarize your data (Chapter 3) and you may decide to use tables and/or figures to communicate your findings (Chapter 10).

iv. An observational investigation may lead to the construction of hypotheses after the investigation

One of the most common outcomes from carrying out an observational investigation is that having collected and evaluated your data you are then left with a series of questions or hypotheses. To test each pair of hypotheses (1.3.3.) you would carry out an experiment (1.2.2.).

v. An observational investigation is often used in preliminary research

Observational investigations are defined by the aim and objectives, but are usually carried out in systems where very little, if anything, is known. They usually precede experiments and are therefore often preliminary – the first type of investigation carried out on a system.

1.2.2. An experimental investigation

An experimental investigation is one that:

- does start out with a question (hypotheses, 1.3.3.)
- sets out to examine an aim and objectives (1.3.)
- also generates data that describe a specific system
- may lead to the construction of more hypotheses after the investigation
- usually follows observational investigations.

EXAMPLE 1.1. Student attendance and assessments

From Table 1.1. it appears that student attendance tends to fall off during a term, but increases when there is an obvious direct relationship to an assessment. A lecturer investigated this further by collecting data from all courses in a degree programme for 5 years and compared attendance with dates for assessment. He hypothesized that attendance differed between days perceived to be directly related to assessment and days perceived to be indirectly related to assessment. This then was an experimental investigation.

i. An experimental investigation does test hypotheses

The preliminary observational investigation shown in Table 1.1. has generated hypotheses (Example 1.1.). These are investigated by designing and carrying out an experiment.

ii. An experimental investigation sets out to examine an aim and objectives

All research needs to be focused and, as in observational research, experiments also need to be structured by an aim and objectives.

iii. An experimental investigation also generates data that describe a specific system

As in the observational investigation, data are collected from the investigation. The data can be used to test the hypotheses (Chapters 4–8), they can be summarized (Chapter 3), and if the results are to be communicated they may be presented in tables and/or figures (Chapter 10).

v. An experimental investigation may lead to the construction of more hypotheses

Most experiments will answer the questions they have set out to test, but leave the researcher with yet more questions to ask. For example, student attendance may well be linked to assessment points, but is this association found in other universities?

vi. An experimental investigation usually follows observational investigations

To carry out an experiment you need to know some basic information about the system you are investigating (2.2.2.). This means that most experiments follow observational investigations. The observational investigations provide some of the background information that is used to design your experiment. In addition to this, you may also pilot or trial your experiment. This is a test run of your experiment to see if there are any unforeseen difficulties. For example, if you wish to use a questionnaire to obtain information a pilot study will allow you to check that your questions are understood by your volunteers and to assess the size of sample you will need (2.3.).

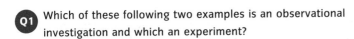

Q1 Which of these following two examples is an observational investigation and which an experiment?

a. An undergraduate wished to examine the trigger factors in people who suffered from migraine. She asked volunteers to complete a questionnaire and list the factors that they felt triggered their migraines.

b. An undergraduate wished to examine people's opinions in relation to human cloning. She compared students taking a Biology course with students taking a Health Studies course. She asked volunteers to complete a questionnaire.

A1

1. Observational. The student is surveying people's experiences; she is not starting out with any hypotheses.
2. Experimental. Although this is a similar type of investigation the student here is starting out with a question: Is there a difference of opinions between the Biology students and the Health Studies students?

1.3. Aims, objectives, hypotheses

When designing or reading about research you often come across the terms aim, objective, and hypothesis (plural – hypotheses). It is a common failing to confuse the 'aim' with the 'objective', which can lead to a lack of direction in the research and research report. All investigations require at least one aim and at least one objective. You therefore need to understand the distinction between these terms and why this difference is important.

EXAMPLE 1.2. The time given to paid employment during term-time by students studying on degree courses

A wide-ranging research project was carried out by the Higher Education Academy to assess the time spent on paid employment during term-time by students studying on degree courses and the impact this has on student learning. This example will be used to illustrate all the following terms and show their relationships to each other.

1.3.1. The aim

Most research is developed through a series of investigations. All the investigations are carried out within a certain area of interest. This area of interest is the aim. In undergraduate projects it is difficult to appreciate the need for an aim, since most investigations of a topic involve a single experiment.

The aim: is a generalized statement about the topic you are investigating.

For Example 1.2. the aim might be to evaluate the time given to paid employment during term-time by students studying on degree courses and the impact this has on student learning. The aim in this example is very broad and will require several investigations to cover all aspects of this topic. Each experiment or group of experiments is therefore described by an objective.

1.3.2. The objective

To thoroughly investigate the topic described by an aim will require one or more investigations. Each investigation will usually examine a small aspect of the overall topic. Each investigation will have a particular design and record particular data. It is therefore important to have objectives which provide additional information about each step of the investigation. This enables the researcher to keep track of their work and anyone reading their reports can see how the separate investigations relate to each other and to the aim.

An objective: will describe a specific investigation where data will be collected.

Given the aim we outlined for Example 1.2. the objectives might include:

- Objective 1. To determine the time given to paid employment during term-time by students taking an English, Law, or Psychology degree course in 20 universities in England.
- Objective 2. To investigate the association between time spent on paid employment during term-time and student performance for all students entering Science degree courses at age 18 years and with three A levels.

1.3.3. Hypotheses

Experiments set out to test a specific question. This question is written in a format called a hypothesis. Chapter 4 introduces you to hypothesis testing and explains in more detail how to write hypotheses and why this format is used.

A hypothesis: is the formal phrasing of each objective and includes details relating to the experiment and the way in which the data will be tested statistically.

For Example 1.2., objective 1, the hypothesis would be:

> **H_0:** There is no significant difference in the mean time (hours) spent on paid employment during term-time by undergraduates studying English, Law, or Psychology.
>
> **H_1:** There is a significant difference in the mean time (hours) spent on paid employment during term-time by undergraduates studying English, Law, or Psychology.

From the example we include here you can see that hypotheses come in pairs. These two hypotheses are called the null hypothesis (H_0) and the alternate hypothesis (H_1). The word 'significant' has a specific meaning in statistics that we will discuss later (Chapter 4). For the moment, you can think of it as being used in the 'everyday' sense of 'big enough to be important'. In Chapters 2 and 4 we discuss the design of investigations. There are more examples of aims, objectives, and hypotheses in these chapters. Exercises for you to test your understanding are included in the Online Resource Centre.

1.4. Items and observations

In an investigation you will collect measurements about or from the subject of your research. These are your observations. The individual or sample from which the measurements are taken is usually called the item. We use this term throughout the book; however, in our Online Resource Centre we show you how to use the statistical software SPSS to carry out calculations. In this software the term 'case' is used in place of the word 'item'.

Item: One representative of the subject you wish to measure.

Observation: A single measurement taken from one item.

For example, in the study of time spent on paid employment during term-time by undergraduates studying English, Law, or Psychology an ITEM is a particular student and an OBSERVATION is the time spent by that student on paid employment during term-time. All these observations are your DATA.

 **Look at objective 2 (1.3.2.). Which is the observation and which the item?**

 In this example there are two measurements. The first is the time spent on paid employment. The second is the students' performance. Therefore, your data will consist of two sets of observations (time and performance). The item is each student.

1.5. Populations

There are several different definitions for the word population. In ecology the term usually means the individuals that can be observed within an identifiably discrete group; for example, the number of individuals in a wood, river, or city. In genetics a population usually means the grouping in which all the individuals of a particular species interbreed. When you are using statistics to analyse your data, the term population is used in yet another way.

Population: All the individual items that are the subject of your research. For example, in the study of time spent on paid employment during term-time by undergraduates in higher education the statistical population is the entire body of students in higher education.

The statistical population reflects your aim. Sometimes a statistical population will be the same as an ecological population. For example, if your aim was to investigate the characteristics of individuals of one species within a wood, then the ecological population and the statistical population are the same. However, if your aim was to examine the characteristics of a particular species worldwide then the statistical population will be all individuals of that species on the planet. The ecological population will still be the group of individuals of that species in each wood.

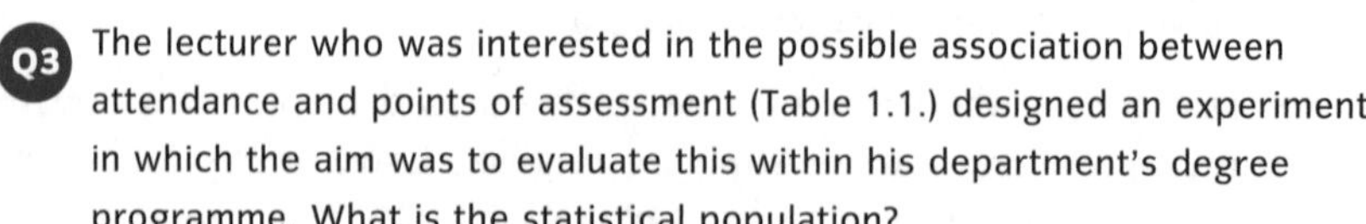

Q3 **The lecturer who was interested in the possible association between attendance and points of assessment (Table 1.1.) designed an experiment in which the aim was to evaluate this within his department's degree programme. What is the statistical population?**

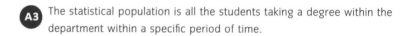

A3 The statistical population is all the students taking a degree within the department within a specific period of time.

1.6. Sample

It is often not possible to study the whole statistical population because of practical and economic constraints. The testing procedure may also be destructive and you may therefore wish to limit your investigation. For example, if you wanted to carry out taste testing on cheese and onion crisps it would not help the factory managers if you tasted all the crisps. Usually a subset of the population is examined. This is the sample.

Sample: A number of observations recorded from a subset of items in the population.

For example, in the investigation into the time spent on paid employment during term-time by undergraduates in higher education a SAMPLE would be those studying English, Law, or Psychology at 20 universities in England (Example 1.2.).

But is this a representative sample?

1.6.1. Representative samples

If it is not possible to measure the whole population then a representative sample is required. A representative sample should be obtained by a clearly defined and consistently applied method. The observations recorded by this method should closely match the result that would be obtained if every item in the population was measured.

In Example 1.2. a sample could be 25 students studying Life Sciences in a small rural university. The relatively small sample size, subject specificity, and university location may mean that this group of students is not typical of the majority, i.e. they are not representative. To resolve this, the investigators may decide to record the time spent on paid employment by these 25 students each year for 3 years. They will then have 75 observations. But clearly the three observations for each student are not **independent** of each other: the three measures from each student are likely to be more similar than if three different students were observed. This will also not be a representative sample.

What would you do?

1.6.2. How do you obtain a representative sample?

There are a number of sampling methods that may be used. Most of these are designed for particular circumstances and to produce a representative sample. We describe the common methods here.

i. Random sampling

Here each item must have an equal chance of being sampled each time. This method avoids conscious and unconscious bias. On average random sampling does produce a representative sample and is therefore in common usage.

How do you randomly sample? In ecology a common error is to believe that throwing quadrats over your shoulder is going to provide you with a random sample. This is untrue: you will only sample within a small area around yourself and will tend to throw forwards or backwards because this is how arms work! A better approach is to use a random number table to identify an item within your population. In Example 1.2., rather than studying 25 students on one degree course at one university a more representative sample would be obtained if students were selected through the Universities and Colleges Admissions Service (UCAS) based on the use of a random number table to generate applicants' numbers irrespective of their course of study or university.

ii. Systematic/periodic sampling

This is a regularized sampling method, where an item at regular intervals is chosen for observation. This method can be suitable, for example, to examine the general distribution of plants across a footpath using a line transect where data are collected at regular intervals. This method is also commonly used when studying the distribution of flora and fauna along a shore from cliff to sea.

iii. Stratified random sampling

This is a combination of the two methods outlined above, where random sampling occurs at regular intervals. So an area may be divided into sections and random sampling occurs within each section, the numbers sampled reflecting the relative size of the section. An example of this would be where soil samples are collected from a contaminated site and where it is important that all parts of the site are represented within the sample. The site can be divided into a number of discrete areas and random samples are taken within each area.

iv. Homogeneous stands

In some work you may need to be selective and identify areas for your investigation that are homogeneous, i.e. similar. In the UK this approach is most common when collecting data about a plant community and assigning a National Vegetation Classification (e.g. Rodwell, 1991). Full details about this approach, including suitable quadrat sizes for the canopy, ground, and field layer, are given within this series of books.

Q4 Anthills are dispersed throughout a grassland. You wish to study the percentage cover of specific plant species on the anthills. What sampling strategy would you use?

There may be several practical and effective answers. Here is ours. Take numbered plant pot labels with you and place one in each anthill. Number a corresponding set of tickets and place in an opaque container. In planning your investigation you will already have determined how many anthills need to be sampled (2.2.8.). Draw out this number of tickets and these will give you your random sample of anthills.

1.7. Populations, parameters, statistics, and samples

Later in this book we introduce you to mathematical methods that you may apply to your data: both the methods used to summarize data (Chapter 3) and those used to test hypotheses (Chapters 4–8). Where the answers from these calculations relate to a population they are usually called parameters. Where they relate to a sample they are called statistics. Most research, of necessity, collects observations from samples and therefore all our chapters focus on designing investigations and evaluating data from samples. Even so, there are frequent occasions when you will come across terms that relate to a statistical population and a sample.

1.7.1. Mathematical notation for populations and samples

In statistics we use mathematical notation. For those not familiar with this notation we have included further explanation in Appendix c. Different symbols are used if we are thinking about populations or samples. The

Table 1.2. Symbols used to indicate population parameters and sample statistics

Term	Population	Sample
Mean	μ	$\bar{x}$
Variance	σ^2	s^2
Standard deviation	σ	s

most common symbols are shown in Table 1.2. Different symbols are used because there are important differences in the meanings and methods involved when calculating parameters of populations or statistics of samples. More information, including an explanation of the terms mean, variance, and standard deviation, is given in Chapters 3 and 4.

1.7.2. Calculations

Some mathematical tests exist in two forms, one suitable when all population values are known and one when a sample is being studied. One value you may often use is called a variance. We look at these equations in Chapter 3 onwards. The variance you calculate may be a population parameter and in this instance the formula for the variance from a normal distribution (3.5.3.) can be written as:

$$\sigma^2 = \frac{\Sigma(x - \mu)}{N}$$

If the variance you are working out is for a sample then you may use the sample notation:

$$s^2 = \frac{\Sigma(x - \bar{x})}{n - 1}$$

(The meaning of the symbols Σ, $\bar{x}$, x, and n is explained in Appendix c.)

Calculators and computer software invariably have the facility to calculate both population parameters and sample statistics, and it is therefore important to ensure that you are using the correct version. Calculators and computer software often indicate the version being used either by the symbols shown in Table 1.2. or by the symbols n (population) or $n - 1$ (sample).

1.8. Treatments

This is a word that has two valid meanings and both are used in this book. The examples we use here appear in Chapters 10 and 5 respectively.

Treatment (definition 1): When carrying out an experiment your items are exposed to a particular environment that is manipulated by you, the investigator.

EXAMPLE 10.1. The response of tobacco explants to auxin

In an undergraduate investigation into the effect of different concentrations of auxin on the growth of tobacco in tissue culture the relative increase in diameter (%) of leaf explants was recorded after 2 weeks.

In this example the tobacco explants were *treated* by exposure to different concentrations of auxin.

Treatment (definition 2): A statistical term in which any samples or groups of observations that are being compared are called treatments.

EXAMPLE 5.3. Shell colour in *Cepea nemoralis* in coastal and hedgerow habitats

An investigation was carried out into the frequency of banding and colour patterns in snail shells in two different habitats (a coastal region and a hedgerow).

In this example, although the investigator has not imposed these habitats on the snails, when the data are analysed each sample is called a TREATMENT. So the coastal region is one 'treatment' and the hedgerow is a second 'treatment'.

The word treatment is often found coupled with other terms; hence you will come across treatment variable (e.g. Chapter 5), non-treatment variable (Chapter 2), and treatment effects (Chapter 8). In all these contexts it is the broader definition of 'treatment' (definition 2) that is used.

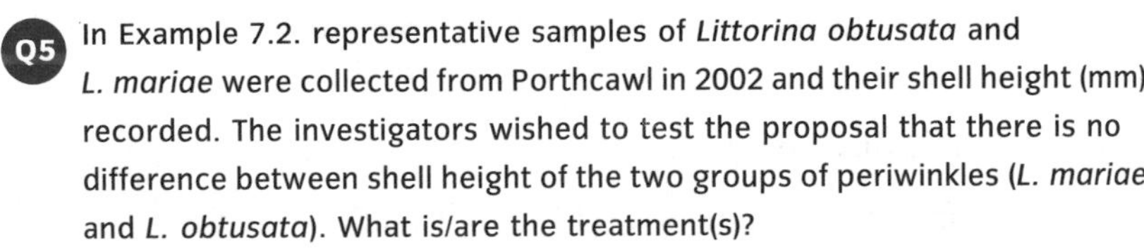

Q5 In Example 7.2. representative samples of *Littorina obtusata* and *L. mariae* were collected from Porthcawl in 2002 and their shell height (mm) recorded. The investigators wished to test the proposal that there is no difference between shell height of the two groups of periwinkles (*L. mariae* and *L. obtusata*). What is/are the treatment(s)?

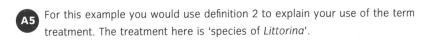

A5 For this example you would use definition 2 to explain your use of the term treatment. The treatment here is 'species of *Littorina*'.

1.9. Variation and variables

In any investigation you will record observations. Usually these observations do not all have the same value – they vary. A set of observations

is said to show variation and the characteristic that is being measured is called a variable. In Example 10.1. the variable is the relative increase in diameter (%) of leaf explants recorded after 2 weeks.

Variation: When the observations within your data do not all have the same value.

Variable: The characteristic that is being measured from each item.

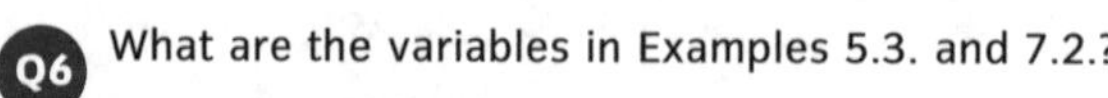

Q6 **What are the variables in Examples 5.3. and 7.2.?**

A6 Example 5.3. The number of individuals with particular patterns on their shells. Example 7.2. The variable is shell height (mm).

Variation between observations can be due to many factors. For example, the variation in human height is due to genetics, diet, and disease. If you wanted to investigate the effect of diet on human height, how can you decide how much of the variation is due to diet and how much variation is due to other factors? As an investigator you want to be able to 'partition' the variation in your data:

variation in your data = variation due to the factor of interest
+ variation due to other factors

In our example this would be:

variation in human height = variation in height due to diet
+ variation in height due to all other factors

Using the second definition of the term treatment (1.8.) this relationship can be written in more statistical terms as:

variation in your data = treatment variation
+ non-treatment variation

In our example the treatment is 'diet' and all the other factors that contribute to human height would be non-treatment variables which result in non-treatment variation. In the rest of this book we develop these ideas and show how experimental design and data analysis can be used as powerful tools to allow you to examine the effect of the treatment within your research. In Chapter 2 we begin by considering good practice in experimental design, and in Chapter 4 we focus on hypothesis testing and data analysis in relation to treatment and non-treatment variation. In Chapters 5, 6, 7, and 8 we return to this topic when we review experimental designs such as the Latin square.

Summary of Chapter 1

- When designing an investigation it will be either observational or experimental and both will generate data (1.1., 1.2., and developed in Chapter 2).
- To provide a focus for your research you will have an aim and objectives. These are not the same thing and when designing your investigation you will need to distinguish between them (1.3. and developed in Chapter 2).
- Some investigations start out with a question (hypothesis). When you use statistics to test the hypothesis then the hypothesis is phrased in a specific way (1.3. and developed in Chapter 4).
- Within your investigation you will record observations for particular items. You may collect observations for every item in a population or you may sample (1.4., 1.5., 1.6., and developed in Chapter 2).
- When sampling, your intention is usually to obtain a representative sample that reflects what is happening in the population. Statistical tests, including those programmed in calculators and computer software, use different symbols when referring to populations and samples (1.5., 1.6., 1.7., and developed in Chapter 2).
- The term 'treatment' is used when designing and evaluating experiments. Both definitions are used in this book (1.8. and developed in Chapter 2).
- In most investigations the observations do not all have the same value – they vary. This variation is central to examining the effect of treatments (1.9. and developed in Chapter 3).
- The Online Resource Centre includes interactive exercises that test your understanding of this chapter with other topics, particularly those considered in Chapters 2 and 4.

2

Experimental design

The best way to learn about the design of investigations, particularly those with hypotheses, is to look at what other people have done and to have a go yourself. In this chapter we take you through both these approaches. The aim is to show you that you are already very able to understand the strengths and weaknesses of an experimental design and identify the ways in which the design might be improved. We also provide information about the process of designing experiments and the things you need to consider.

Designing an experiment is an iterative exercise. It requires an understanding of certain design principles, but also an understanding of data and how they are to be analysed. This means that, although in this chapter we cover all the steps you need to consider when designing an investigation, you may not fully appreciate the importance of some of our comments until you have become more familiar with the content of the other chapters in this book. One of the most common problems in designing research that is a direct consequence of this need for a rounded approach is that many people ignore the importance of considering how the data will be analysed before they begin their investigation: they only come to this when the data are collected and they then find out that the data cannot be analysed. To avoid this you must accept that this chapter, although providing an overview of experimental design, does not stand alone and must be used with subsequent chapters. If you work through this chapter it should take about 2 hours to complete all the exercises. There are additional examples for you to consider in the Online Resource Centre.

2.1. Evaluating published research

The following example is based on a short research paper that appeared some time ago in a prestigious journal. Read it through and then answer the questions at the end.

EXAMPLE 2.1. Social drinking and morning-after breath alcohol levels

The investigator wished to study the level of morning-after breath alcohol concentrations after an evening of social drinking and to see how these levels were affected by the levels of habitual intake. An important element of this investigation was that real social drinking behaviour was examined rather than effects observed in a laboratory trial.

Fifty-eight men aged aged 20–50 years took part in the study. Habitual alcohol intake was estimated by a 4-week prospective drinking diary or accounts of the frequency of drinking and quantity per session in a typical month. To examine morning-after breath alcohol concentrations, the participants were asked to count the number of drinks they consumed at an evening social event where they anticipated drinking heavily. Other people present were asked to verify the amount drunk and the time that drinking stopped. All the men ate shortly before or during drinking. The morning-after alcohol concentrations were then recorded in each person's home 7–8 hours after drinking had stopped (Wright, 1997).

If possible, discuss this example of an experimental design with your friends and combine your ideas. Think about the following three questions:

What is/are the **aim**(s) and **objectives**?

What are the strengths of this experimental design?

What are the weaknesses of this experimental design?

2.1.1. What are the aim(s) and objective(s)?

The aim describes the broad area being examined by the researcher and the objectives break this down to reflect the various investigations that were carried out (1.3.). We identify the aims and objectives for the study described in Example 2.1. as follows.

Aim: The influence of habitual intake on morning-after breath alcohol concentrations in men.

There are two objectives:

1. To quantify the habitual intake of alcohol in 58 men.
2. To assess morning-after breath alcohol levels in 58 men.

Compare our aim and objectives with your own. How, and in what way, did they differ?

2.1.2. Strengths of the experimental design

It is always so easy to be critical, and sometimes overcritical, of research. It is therefore best if you start by considering what is good about the experimental design before considering the experiment's limitations. Which strengths have you listed for this experiment? We identified the following strengths:

- The experiment was carried out in the environment in which this activity usually takes place.
- Attempts were made to keep the effect of the experiment on people's behaviour to a minimum.
- Fifty-eight is a reasonable sample size (2.2.8.).

2.1.3. Weaknesses of the experimental design

There are three types of weaknesses that you may identify in any report of an investigation.

i. Faults

When you are considering published research the first type of weakness is a fault that could be avoided through a change in the design of the investigation without changing the aim. In any publication or report these faults should be acknowledged and should be taken into account when evaluating the results (Chapter 10). If you design and carry out a piece of research, then careful preparation and planning should ensure that there are few if any faults in your work.

ii. Limitations

The second type of weakness (a limitation) can arise in an investigation, but cannot be overcome without changing the aim. These limitations should also be taken into account when the results are evaluated (Chapter 10). If you are evaluating published research then your task is to identify these limitations, since you clearly cannot change the aim. If you were designing an investigation and were aware before you started of significant limitations in your design, then you could consider changing your aim to enable you to carry out an investigation without these weaknesses. You should also show that you are aware of any limitations in your design, when reporting on your research.

iii. Communication

The third weakness is one of communication. Omissions and lack of clarity can make it difficult to fully comprehend the design of an investigation. The format of the paper required by the journal can sometimes exacerbate this.

If you look at Example 2.1. you will quickly appreciate that there is plenty of scope for discussion here about what is a fault in the experimental design, what is a limitation and what is due to omissions in the journal article. We think the following fit in these three categories. What did you list?

a. Fault in the design The age of the men varied considerably and it is not known how this may affect the results. There is a need to demonstrate that this sample is truly representative of the average male drinking population and if it were then this range in age is a strength more than a weakness. Alternatively, using statistics to confirm that 'age' as a factor did not influence the results would resolve this concern.

It seems unlikely that the drinker and friends will remember accurately how much was drunk and over what time period. This may be a factor that affects the heavy drinkers more. The author does acknowledge this weakness in the article. Could a bartender not be recruited as an independent recorder?

Two methods for establishing habitual intake were used. Will the drinkers be honest about how much they habitually drink? There is a need to test these methods of recording in advance to see how reliable these forms of reporting are and which is the better method, and then to be consistent in their use.

b. Limitation The times of eating varied and may have affected alcohol metabolism. If eating was controlled so that certain items were eaten at certain times during the drinking session this may resolve the concern over the impact of eating on alcohol metabolism, but would almost certainly interfere with the normal behaviour of the participants.

The rate at which the alcohol was consumed was not recorded. Quickly drinking three pints and then nursing the fourth may result in a different rate of alcohol metabolism than a steady drinking pattern. It is difficult to see how this can be either controlled or recorded without affecting the participant's behaviour.

c. Communication The published report did not indicate the lengths of the drinking period or what was eaten by the participants during their drinking session. It was not clear how the two methods for establishing habitual intake were used. Were they alternative methods or used in conjunction with each other?

What weaknesses have you identified? How do they compare with those we have included? Do you disagree with any parts of our evaluation? Why?

By thinking about this investigation we have introduced you to some of the important elements you need to consider when designing an investigation. To start with these include knowing your aim and having clear objectives. There were two **treatment variables**: 'the amount drunk habitually' and 'the level of breath alcohol when measured on the "morning after"'. You then identified a number of other factors that were either limitations or faults in the design of the investigation. All these factors that are not treatments are called **non-treatment variables** and can result in **sampling error** (2.2.6. and Chapter 4).

2.2. Have a go!

In our Online Resource Centre we have a range of different topics for you to have a go at designing investigations. They cover a broad range of subject areas, so there should be one that is closely related to your area of study. Alternatively, think about the following topic: 'The effect of wind strength on maximum seed dispersal in the common creeping thistle *Cirsium arvense*'. Have a go at designing an investigation using the checklist in this section. There are many acceptable experiments that could be devised. We have provided you with one version to illustrate each of the design steps.

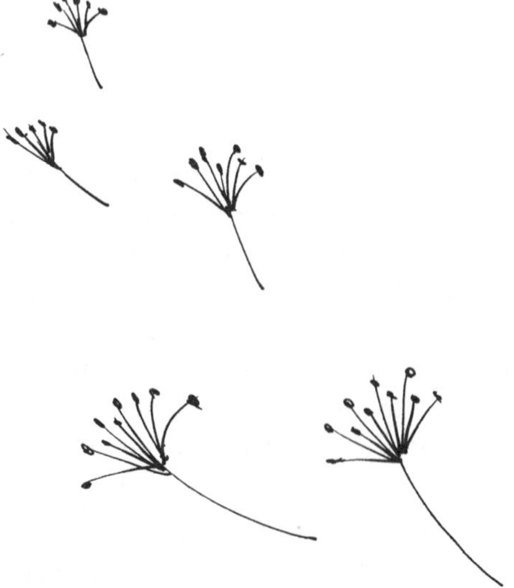

When designing an investigation you will find that you need to be prepared to go through the process several times, first drafting your design and then redrafting as all the elements fall into place. So many factors need to be considered that as you adjust one you will need to go back and re-think all the others again. We have organized this process under the following 12 subheadings, 2.2.1.–2.2.12. You should consider all these before finalizing your design.

2.2.1. Type of investigation

In Chapter 1 we explained that there are two types of primary research: **observational** and **experimental** (1.2.). Usually only experiments have **hypotheses** to test. The first step then is to decide whether or not you have hypotheses.

To examine the effect of wind strength on maximum seed dispersal in *Cirsium arvense* we will test a question (hypothesis), which is: Does the wind speed affect the maximum distance travelled by the seed? Therefore, we will be designing an experiment.

2.2.2. Background information

Background information can come from a variety of sources, most often from literature reviews and asking other researchers. Do not rely on abstracts, such as those easily available on the Internet. This will not provide you with sufficient detail to be useful when designing an investigation. You may also carry out a preliminary investigation, such as an observational investigation or a pilot study. This can help you to identify the appropriate scope of a treatment or the timescale that the experiment may need to run. For example, in an investigation of bacteria on different types of chopping boards you will need to find out before carrying out your experiment the numbers of bacterial colonies you expect in your samples so that you can use appropriate levels of dilution. If you are interested in the effect of music on the heart rate of dogs, you need to know how long the dogs need to hear music before their heart rate shows an effect. If you are interested in the leaching of nutrients into an adjacent stream from fertilizer applied to a meadow, you would need to know the cation exchange capacity of the soil.

Background information is essential for a number of reasons and you should take the time to compile adequate background information before you finalize your experimental design.

i. The justification

When you come to report on your investigation you will need to explain the scientific reasons for carrying out a research project. Just because you can record or measure something does not make it appropriate research. For example, you may be able to record the effect of adding whisky to soil on the growth rate of daisies. But why investigate this? Do you have a sensible reason?

ii. The practicalities

Many topics may be affected by seasonal factors, such as climate, timing of growth in plants, or behaviour in animals. For example, you may wish to collect geological samples from the top of Snowdon, Wales. When will it be safe for you to do this? Will this fit in with the academic year or other time constraints? Alternatively, if you were intending to examine the effect of temperature on the behaviour of *Oniscus asellus* (woodlice), what temperatures do they normally experience in their habitat? What temperature is likely to kill them?

Further comments on practicalities are included in Appendix a: How to choose a research project.

iii. The context

In some investigations there is a need to have some baseline data or information that enables you to place your study within its context of other research. You will achieve this by designing preliminary or pilot investigations or you may obtain such data from another source, such as the records from a previous investigation. For example, to investigate the effect of flooding on the species richness of a particular meadow you would need to know what was present before the flood.

What background information would you need to develop an investigation into the effects of wind speed on maximum seed dispersal in *Cirsium arvense*? We suggest the following:

To provide a justification

There may be several reasons for carrying out this investigation. They might include a wish to understand this plant better, a need to control seed dispersal, or an interest in flight mechanics.

Practical implications

This investigation requires 'live' material; consequently there are many practicalities to consider. These include: when are the seeds produced; what is a normal range of wind speeds experienced; what equipment would we need to examine wind speed; can *Cirsium arvense* be grown in pots; is it known from other studies what the natural dispersal of thistle seed is; what factors are known to affect seed dispersal in general and in *Cirsium arvense* in particular?

The context

Information collected the previous year during the seed dispersal period at the site can be used to identify the range and average wind speed experienced by this species. This can be confirmed by measuring wind speed during the current season.

2.2.3. Aim and objectives

Having completed your literature review and collected other background information you should now be able to write a draft aim and objective(s). Aims and objectives are discussed in Chapter 1 (1.3.).

For the investigation into the effects of wind speed on maximum seed dispersal in *Cirsium arvense* the aim and objective could be:

Aim: To investigate the effect of wind speed on maximum seed dispersal in *Cirsium arvense*.

Objective: To examine the maximum distance travelled by the seed of *Cirsium arvense* when placed in a wind tunnel and exposed to wind speeds reflecting the range and average wind speed in the natural population.

2.2.4. Population and sampling

The term **population** has different meanings depending on whether you are using the term, for example, in ecology, genetics, or statistics (1.5.). When starting to plan your investigation you may be using any one of these definitions. You need to be clear, therefore, as to what is your statistical population. However, it is rarely possible or even of any value to measure every individual in your population. You will usually sample. **Samples** need to be representative and you will therefore need a suitable sampling strategy (1.6.).

Having identified the statistical population and considered a draft sampling strategy, you will then need to decide how many **items** you will need in your sample. Your choice depends on the type of data you will collect (3.1.) and how you will analyse this (Chapters 4–8). This is not an easy task. We therefore spend more time on this topic in 2.2.8. and 2.3.5.

Population

Our aim indicates our interest in maximum seed dispersal in the species *Cirsium arvense*. All individuals of this species then form our statistical population.

Sample

In an investigation on seed dispersal you cannot realistically examine every plant in the population; therefore, you will need a representative seed sample. For our sampling strategy we have decided to use whole plants, as this will then mean that the seeds are held at the same height as they would be in the field and within an inflorescence, as they would normally be.

Therefore, in our design one ecological population of *Cirsium arvense* was chosen at random and all the apparently mature flowering plants with an apparently undisturbed inflorescence were given a number. Fifteen numbers were chosen at random using random number tables and these plants were dug up. Care was taken not to dislodge any seeds. (The rationale for choosing this sample size is outlined in the following sections.)

Q1 **Do you think that our design so far will provide us with a representative sample?**

A1 It is difficult to say with any confidence that this will be the case, since a sample of plants from a single ecological population may not reflect this species across its range. We must consider revising our experiment, and there are two ways forwards at this stage. We can either extend the sampling into a number of ecological populations over a broad range of the species distribution or we can revise the aim and objective so that they reflect the limited nature of this study. We will take the second course. Therefore, the revised aim and objective are:

Aim: To investigate the effect of wind speed on maximum distance seed from *Cirsium arvense* is dispersed at Windmill Hill, Worcester.

Objective: To examine the maximum distance travelled by the seed of *Cirsium arvense* sampled from an ecological population in Worcester when placed in a wind tunnel and exposed to wind speeds reflecting the range and average wind speed in the natural population.

The statistical population is now all the individuals of *Cirsium arvense* at Windmill Hill.

2.2.5. Controls

A control is an experimental baseline against which any effects of the treatment(s) may be compared. A control should be an integral part of the investigation and must be carried out at the same time as the treatments under investigation.

In laboratory experiments it is usually relatively easy to construct a control. In field experiments, if a control is necessary then you will need to select an area which is as similar as possible to the one exposed to the treatment(s). In complex experiments where you are investigating the effects of more than one treatment you may require more than one control.

A control may not always be necessary. For example, in an investigation into the effect of temperature on the behaviour of *Oniscus asellus* (woodlice) you may select temperatures of −5, 0, 5, 10, 15, and 20 °C as your treatment range. There is no one temperature that could be called a control.

The investigation into the effects of wind speed on seed dispersal in *Cirsium arvense* is similar to the *Oniscus asellus* example in that there are no controls.

EXAMPLE 2.2. The antibacterial properties of triclosan and tea-tree oil

An undergraduate took swabs from the hands of volunteers working in the medical profession. These were the 'before' measures. She then washed the volunteers' hands. One hand was washed with triclosan and one with tea-tree oil. Having allowed these washed hands to air dry, the student took a second swab from each hand. These gave the 'after' results. At the laboratory the swabs were used to inoculate plates. As the bacterial count could be high, serial dilutions were also made from the swabs and plates were then inoculated. After 2 days incubation the numbers of bacterial colonies on the plates were counted.

In the design described in Example 2.2., are any controls needed? If so what?

Yes, controls are needed to check that the bacteria are coming only from the participants' hands. The controls in this case need to be a series of plates inoculated with

- nothing
- triclosan
- tea-tree oil
- an unused swab
- the solution used for serial dilution.

2.2.6. Variables

When you record a set of measurements in an experiment you usually see variation in these measurements. This variation may be due to the factors you are investigating, i.e. the treatments (1.8.) or other causes, i.e. non-treatment variation (1.9.). Non-treatment variation is commonly known as sampling error (4.1.3.). If the variation in your data caused by non-treatment effects is considerable, it can mask the effect of the treatment(s). Therefore, it is important to be clear about what is/are your treatment(s); identify the possible causes of non-treatment variation; minimize the effect of non-treatment variation; and if possible mathematically separate out (partition) the **variation** in your data due to the treatment and the variation due to non-treatment factors.

i. What is your treatment?

There are so many terms that relate to the factors we investigate, such as '**variable**', 'treatment' and other terms that relate to elements in our investigations, such as 'sample', that it can become confusing as to which is which. This step may seem pedantic, but if you note down your treatment it does mean you are clear about what you are, and by inference what you are not, investigating.

In our *Cirsium arvense* investigation the treatment is wind speed.

---

ii. Identify the possible causes of non-treatment variation

This has to be your next step as you cannot minimize something if you are not aware that it might affect your experimental system. Background or preliminary investigations, common sense, and reading help in identifying causes of non-treatment variation.

There are many possible causes of non-treatment variation in this investigation. These include: the effects of digging up the plants; the maturity of the seeds and inflorescence; the height of the inflorescence; the number of seeds in the inflorescence; the temperature and humidity of the wind tunnel; and the location of the plants during testing.

---

iii. Minimizing the effect of non-treatment variation

In your investigation you will be collecting observations from a number of items (1.4.). This data will almost certainly show variation (1.9.). Often your items are grouped in some way, such as different samples. If you wish to gain an understanding of the degree to which the variation in your data is due to your treatments, you need to ensure that all other non-treatment effects are equalized in some way across all the items in all the groups. This can be achieved in two ways: 'Equalize the effect' and 'Randomization'. Where possible, both approaches should be incorporated into your design.

a. Equalize the effect One approach to minimize the effect of non-treatment variables is to ensure that the causes of non-treatment variation are made as constant as possible across your experimental system. By smoothing the effect across all treatments the difference between treatments will still be detectable.

To minimize the non-treatment effects in the experiment on seed dispersal, the temperature and humidity of the test area was consistent throughout the experiment. Each plant was placed at the same point in the wind tunnel, the inflorescence being placed above a mark on the floor.

---

b. Randomization A common practice that, on average, will act to reduce the impact of non-treatment variables is that of randomization. The nature of the randomization will depend on your investigation and the non-treatment variables. Most often randomization occurs in relation to treatment, location, or time of day; for example, the inoculated plates from Example 2.2. were arranged at random within the incubator to minimize the effects of any micro-variation in temperature within the incubator. Randomization also therefore has an equalizing effect. Usually the non-treatment effects after randomization will be the same for all treatments, and so the variation due to the treatments will be identifiable in your data.

Three wind speeds (low, medium and high) were selected to reflect the range of wind speeds experienced by the thistles in their natural population. The 15 plants were assigned at random, 5 to each of the 3 treatments. In this way any variation between the plants in terms of their maturity, height of inflorescence, and size of inflorescence should be distributed equally across each of the 3 treatments.

---

iv. Mathematically separate out (partition) the variation in your data

Another approach to dealing with non-treatment variation is to design your investigation in such a way that you can mathematically obtain an estimate of the variation in your data that is due to non-treatment effects. The simplest method to achieve this is by using replicates (2.2.7.).

2.2.7. **Replication**

Replication means that you have more than one of something. It can be more than one item being exposed to a treatment or more than one group of items exposed to one treatment or more than one item in a sample.

i. When should you use replicates?

Replicates should be used whenever possible for two reasons: first, to increase the reliability of your estimate of the population parameters; and second, to allow mathematical estimates of non-treatment variation.

a. To increase the reliability of your estimate of the population parameters In Example 2.2. the student could carry out her investigation on one person, where one hand is washed in triclosan and one in tea-tree oil. If the aim is to investigate the effectiveness of these two hand-washes on people working in a medical profession, then this is clearly a very small sample and any values from the single individual are not likely to closely match the population parameters. It would be better to increase your sample size and to do it in the same way for both treatments. In fact, 25 people took part in this investigation. Each hand washed in triclosan was a replicate within the treatment 'triclosan'. Each hand washed with tea-tree oil was a replicate within the treatment 'tea-tree oil'.

b. To allow mathematical estimates of non-treatment variation Using replicates may allow you to use certain statistical tests, such as an analysis of variance (Chapter 7). These tests estimate the variation in your data due to the non-treatment effects and the variation due to your treatment(s) and make allowances for the non-treatment variation when testing hypotheses (Chapter 4). An understanding of the level of variation within your population may be critical. For example, in the study on the effectiveness of tea-tree oil the results showed that in some cases tea-tree oil reduced bacterial counts, in some cases the counts were increased, and in some cases counts were too low or too high to be measured effectively by this technique. In a practical setting such as this, the variation demonstrated between people (replicates) is crucial to gauging the real effectiveness of a bactericide.

In the thistle experiment 5 plants were exposed to each of the 3 wind speeds. Each plant in any one group was a replicate. The 15 plants were each given a number. A random number table was used to determine which plant was to be tested first. The plant had already been assigned to a treatment so this was the first wind speed tested. This was repeated for each plant so that the order in which the plants were tested and therefore each replicate and each treatment were tested was randomized. Each plant was exposed to the given wind regimen for 10 min. After this time the distance (mm) travelled from the mark on the floor to the seed furthest from the plant was recorded.

---

ii. Features of replicates

Replicates must be an integral part of the experimental design and must all be established and examined within the single experiment and not subsequently. If this is not the case then they are pseudo-replicates. Replicates must also be **'independent'** of one another. This means that you must not use the same animals, plots, etc. For example, if you were investigating the response of plants to a watering regimen you might measure the growth of several leaves on the same plant exposed to one treatment and several leaves on another plant exposed to a different watering regimen. Clearly the leaves on any one plant are not independent of each other: they share the same genotype and microclimate. A more appropriate method for replication would be to expose many plants assigned at random to each treatment and either randomly or using a stratified method sample one leaf from each plant (2.2.4.)

Another example of pseudo-replication frequently arises when items are grouped in the same Petri dish, pot, etc. For example, in an investigation into the effect of fertilizer on the growth of marigolds there were 10 plant pots each containing 4 marigolds. Five of these pots were treated with a new fertilizer and the other 5 pots were treated with the standard plant food. After 6 weeks the dry mass of the shoots and roots from each plant as measured. A common mistake is to treat each plant as a replicate and therefore believe that there are 20 replicates for each treatment, but the 4 plants in a pot are not independent of each other. (If you do design an experiment like this you should refer to 7.9. as the 4 plants are 'nested' within the 5 pots.)

iii. How many replicates?

One of the hardest things to establish is how many replicates are needed in any particular investigation. An indication of the number of replicates comes from requirements of the particular type of statistical test to be used and the extent of the non-treatment variation. It may be necessary to carry out a preliminary investigation to identify the degree of non-treatment variation before choosing the number of replicates for the main investigation. We look at this thorny issue again in relation to sample sizes in 2.2.8. and 2.3.5.

2.2.8. **Statistics**

This critical stage of experimental design is the one most frequently left out of the planning stages, and in our experience is the most frequent cause of disappointing research. You will see that we feel very strongly about this as we keep drawing your attention to it. This step is very much a forwards and backwards process. Your draft design is used to determine which is the most appropriate statistical test to use, but this choice of test may then influence factors such as how many observations you should have in each sample or how many replicates. There are four elements covered under this topic: choosing your statistical tests; reconsidering your experimental design; finalizing your aims and objectives; and writing your hypotheses – these take you through both the 'forward' and the 'backwards' steps. Understanding this step in the process of designing research requires an understanding of some of the concepts we do not cover until later in this book. Therefore you will need to jump forward in places to read these sections where indicated. Choosing your statistical test is also summarized in Appendix b.

i. Choosing your statistical tests and reconsidering your experimental design

In our first step in designing an investigation you were asked to decide whether your research was an observational investigation or an experiment that was going to test hypotheses (1.3. and 2.1.). If your investigation is observational then you will not need to use the hypothesis-testing statistics (Chapters 5–8). You may need to present your data in a table or figure, or summarize them (Chapters 3 and 10). If you are designing an experiment then you MUST take into account in the planning stages how you intend to test your hypotheses. You may otherwise generate data that cannot be analysed, which is a waste of your time.

It is simplest if we illustrate how you go about this using the experiment we have been designing in this chapter.

a. *Which type of hypotheses am I testing?* The next step is to decide which type of hypotheses you are testing. There are three types:

1. Do the data match an **expected** ratio?
2. Is there an **association** between two or more variables?
3. Do samples come from the same or different populations?

More details are given in Chapter 4 and you will need to be familiar with these details to enable you to make the correct decision.

In our experiment we have no reason at the outset to expect the data to match a particular mathematical outcome. We are only examining one variable or treatment (the effect of wind speed). However, we do want to know if all the samples behave in the same way, i.e. is the maximum distance travelled by a seed the same for all wind speeds? The hypothesis for our experiment therefore falls into the third group.

---

b. *What type of data will I be collecting?* Types of data are described in Chapter 3. Read about these terms and then think about the units you will be recording. Are they mm, pH, percentages, etc.?

In this experiment we will be recording the distance travelled by the furthest seed (in mm). Length measurements like this are measured on a **continuous** scale: they are numbers and therefore are **quantitative**, and these numbers can be arranged in a specific order so they are **rankable** (3.1.)

It is difficult until you have carried out your experiment to be certain whether the data will be **parametric**. Background reading about seed dispersal indicates that the data are unlikely to be parametric. So this is what we assume in the planning stage, but we must then confirm this after the experiment has been carried out.

---

c. *Which statistical test should I use and how many replicates, samples or observations do I need?* Chapters 7 and 8 give details about the most common statistical tests that are appropriate for testing hypotheses of the type we have in our experiment. In Chapter 7 all the tests relate to data that is parametric. In Chapter 8 all the tests relate to data sets that are **non-parametric**. We have already decided that it is most likely that our data will be non-parametric, so we should refer to Chapter 8.

The first section in Chapter 8 sets out to guide you to the most appropriate statistical test. The following is an extract from Chapter 8.

Chapter 8. How to choose the correct test

Each test has several requirements that must be met and these details are given at the start of each section. The following guide takes you to the most likely test for your data. It is assumed that you have non-parametric data.

You have one treatment variable. You are going to compare two samples. The data are **unmatched**. You have 20 observations or less in each sample.	Mann–Whitney U test (8.1.)
You have one treatment variable. You are going to compare two samples. The data are unmatched. The data are measured on a continuous scale and you have more than 30 observations in each sample.	*z* test for unmatched data (Chapter 7 (7.1.))
You have one treatment **variable**. You are going to compare two samples. The data are **unmatched**. You have more than 20 **observations** in each sample.	Sokal & Rohlf, 1981
You have one treatment variable. You are going to compare two samples. The data is **matched**. You have fewer than 30 pairs of observations.	Wilcoxon's rank paired test (8.2.)
You have one treatment variable. You are going to compare two samples. The data are **matched**. You have more than 30 pairs of observations.	*z* test for matched data (Chapter 7 (7.2))
You have one treatment variable. You are going to compare two or more samples. You wish to test **general** and **specific** hypotheses.	One-way ANOVA (Kruskal–Wallis test) (8.3. and 8.4.)
You have more than one treatment variable. You are going to compare two or more samples. You wish to test general and specific hypotheses. You will be using a calculator.	Two-way non parametric ANOVA (8.5. and 8.6.)
You have more than one treatment variable. You are going to compare two or more samples. You wish to test general hypotheses. You want to use a computer.	Scheirer–Ray–Hare test (8.7.)

From this table it is clear that for our experiment with one variable (wind speed) and three samples (low, medium, and high wind speeds) we should use the Kruskal–Wallis test.

The criteria for using a Kruskal–Wallis test are outlined in 8.3. We need to check these, as we hope that they will confirm our choice of test and help us choose sample sizes. To use a Kruskal–Wallis test you:

1. Wish to test for differences in population **medians**.
2. Have one treatment variable and three or more samples.
3. Have data that are non-parametric, but can be ranked.
4. Do not need equal sample sizes.
5. Must, if there are only three samples, have more than five **observations** per sample.

In our experiment we do have one variable (wind speed) and three samples (wind speed low, medium, and high). We do wish to test for a difference between population medians. The data are probably going to be non-parametric and can be ranked. From criteria 4 and 5 we can see that although in our design we have equal sample sizes (a seed from each of five plants), this is not critical. We can also see that since we have only three samples we must have more than five observations per sample. In our current design we only had five plants per sample. So although most of the criteria for using this test are met, the size of sample proposed in our current design is not large enough for us to be able to use the Kruskal–Wallis test. We must either redesign our experiment or seek out another statistical test.

Many statistical tests of hypotheses have minimum and maximum guides to the numbers of observations in the sample. These may be determined by the range over which the test is most effective; or in large samples it may be that you will obtain no additional useful information from having more observations. Clearly, by checking your design in the manner we have described you can select the most appropriate statistical test, and as part of this process you have also checked on details such as how many observations you need in each sample. If your draft design does not meet the criteria, you still have time to amend it before you carry out the experiment.

In our current draft design the sample size we had chosen arbitrarily (2.2.4.) is not going to be adequate. Therefore, we will increase the sample size to 30 and assign 10 plants at random to each treatment.

In experimental design it is this step that is probably the hardest. But do not shy away from it. There are lots of examples in the Online Resource Centre to give you practice in choosing the most appropriate statistical test and so enable you to boost your confidence.

ii. Finalizing your aim and objectives

The steps above often lead to a change in your objective(s) and sometimes in your aim. Therefore, you should now finalize them, taking into account the planning stages you have completed.

Aim: To investigate the effect of wind speed on maximum distance seeds are dispersed in *Cirsium arvense* at Windmill Hill.

Objective: To examine the maximum distance travelled by seed from 30 *Cirsium arvense* sampled from one ecological population in Worcester when placed in a wind tunnel and exposed to low, medium or high wind speeds which reflect the range and average wind speed in the natural population.

---

iii. Writing your hypotheses

We introduced you to hypotheses in Chapter 1 (1.3.). If you are designing an experiment and have got this far in your planning, you should be able to write your hypotheses. More information about writing hypotheses is given in Chapter 4 (4.1.2.).

Our example is an experimental investigation, so there will be hypotheses. We believe that the data that we are collecting will be non-parametric, so our hypothesis testing will compare population medians (4.1.2.).

H_0: There is no significant difference in the median maximum distance (mm) seed from *Cirsium arvense* are dispersed when exposed to low, medium, or high wind speeds (m/s) in a wind tunnel.

H_1: There is a significant difference in the median maximum distance (mm) seed from *Cirsium arvense* are dispersed when exposed to low, medium, or high wind speeds (m/s) in a wind tunnel.

2.2.9. **Influencing outcomes**

Some particular methodologies may themselves influence the outcome of an investigation. If this is likely, then you should try to estimate the extent of this effect, and take it into account when designing your experiment and when you interpret the results. One of the most obvious examples of this can occur when studying animals, where your own presence and behaviour and the presence of experimental apparatus may influence the outcome. The contamination of samples from your own body or equipment is another example. The problems that can arise from not wishing to influence the investigation are clearly seen in Example 2.1.

In the experiment we have been designing in this chapter it is difficult to see how we might influence the outcome. This step in our planning is probably not applicable in this instance.

2.2.10. Assumptions and bias

Despite all your careful planning you may still have to make certain assumptions about the experimental system. Make a note of these assumptions as you become aware of them. These assumptions may be testable as a separate experiment. You should always show that you are aware of these assumptions and possible causes of bias when you interpret your results and communicate your findings. In research that involves animals, including humans, one common cause of bias is that of 'self-selection'. Some animals can become 'trap-happy' and choose to be caught, usually because traps contain a source of food. Clearly, such a subset of animals may not represent the population under investigation. Similarly, humans who volunteer to take part in a study may not be representative. This can be exacerbated if a reward is offered as an incentive (9.5.2).

Look through the information you have noted down when 'having a go'. How many assumptions and possible causes of bias can you identify? For our design these include that:

- The random method employed when collecting plants from the wild did result in a representative sample.
- No bias was introduced by the loss of any seeds from the inflorescence during the sampling.
- The random allocation of plants to the treatments minimized the effect of non-treatment variables.
- The wind speeds used in the experiment were representative of those experienced by the species at Windmill Hill, Worcester.
- The order of testing the plants within the wind tunnel had no effect on the distance travelled by the seeds.
- The seeds were blown along in a manner that is similar to that in the field, including seeds that were blown along the ground.

2.2.11. Repeatability

If an effect is real then it is reasonable to expect that a repeated investigation, carried out using the same method, will identify the same trend. A scientific finding that has been confirmed in this manner is considered to be 'sound'. Clearly there may be circumstances that do not allow for a repeated experiment. For example, an investigation into the effects of hurricane Ivan on the distribution of manatee along the Florida coast could not be repeated.

One common failing is to confuse a repeated experiment with a replicate. If an experiment is repeated at a different time this is not a replicate. The data so obtained cannot be combined with the data from the earlier experiment without using certain mathematical steps to prove that it can be pooled. We give one example of how this might be done in Chapter 5 (5.2.).

In our example the experiment could be repeated with another 30 plants taken from the same population in a similar manner.

2.2.12. **Back to the beginning**

As we explained at the beginning of this chapter, when designing an investigation you usually have to develop a draft and then refine it, as at each step any change you make may affect steps you have already been through. So at this point look through your design again using our checklist and make sure your design is as good as it can be. The experimental design we came up with is summarized below. How does it compare with your design?

Topic: The effect of wind strength on maximum seed dispersal in *Cirsium arvense*.

Aim: To investigate the effect of wind speed on maximum distance seed from *Cirsium arvense* are dispersed, at Windmill Hill, Worcester.

Objective: To examine the maximum distance travelled by the seed of 30 *Cirsium arvense* sampled from one ecological population in Worcestershire when placed in a wind tunnel and exposed to low, medium, and high wind speeds which reflect the range and average wind speed in the natural population.

Hypotheses:

H_0: There is no significant difference in the median maximum distance (mm) seed from *Cirsium arvense* are dispersed when exposed to low, medium, or high wind speeds (m/s) in a wind tunnel.

H_1: There is a significant difference in the median maximum distance (mm) seed from *Cirsium arvense* are dispersed when exposed to low, medium, or high wind speeds (m/s) in a wind tunnel.

Method

All the apparently mature flowering plants with an apparently undisturbed inflorescence within a specific ecological population at Windmil Hill in Worcestershire were given a number. Thirty numbers were chosen at random using a random number table and these plants were dug up. Care was taken not to dislodge any seeds. Each plant was potted up in 12 cm pots in John Innes no. 1 compost.

Three wind speeds (low, medium, high) that reflected the range of wind speeds experienced by these plants in the natural population were selected. Each plant was given a number. The first 10 numbers chosen at random were assigned to the low wind speed treatment, the second 10 numbers chosen at random were assigned to the medium wind speed treatment and the remainder were assigned to the high wind speed treatment. In this way any variation between the plants in terms of their maturity, inflorescence size, and height should be distributed equally across each of the three treatments.

This numbering system was also used to determine the order in which the plants were tested. Again, using a random number table a plant was selected and then exposed to its predetermined treatment. A second plant and treatment were selected in the same way until all plants had been tested. This ensured that the order in which the three treatments were tested was also randomized. To minimize the non-treatment effects in the experiment the temperature and humidity of the test area were consistent throughout the experiment. Each plant was placed at the same point in the wind tunnel. Each plant was exposed to the given wind regimen for 10 min and the distance from the centre of the pot to the seed that had travelled the furthest was measured (mm). The hypotheses for this experiment will be tested using a Kruskal–Wallis test.

2.3. Questionnaires, focus groups, and interviews

We have considered the essential steps needed to design most investigations in 2.1. and 2.2. However, some information may be gathered using methods such as unstructured and structured interviews, questionnaires, and focus groups. If you are planning to use one of these methods there are additional points that need to be considered.

Questionnaires, focus groups, and interviews may allow you to collect both quantitative (numerical) data and **qualitative** (descriptive) information. Qualitative approaches are based on the notion that reality varies for different people in different contexts; therefore, you cannot use a single scale against which you make measurements. An example of this is people's perception of pain. Qualitative information can be summarized and evaluated, but the tools used to do this are outside the scope of this book. In this section we therefore focus on how these methods may be used effectively to obtain numerical or quantitative data and how these data may be analysed.

2.3.1. **What is a questionnaire, interview, or focus group?**

These three methods for obtaining information differ in terms of the degree of interaction between participants and the interaction between the researcher and the participant.

i. Questionnaires

Questionnaires are, as the term implies, collections of questions given to all participants. The participant writes down their answer and returns the questionnaire to the researcher. Questionnaires are usually anonymous and confidential (Chapter 9). They require very little contact between the researcher and volunteers and usually there is no interaction between participants. A questionnaire is most useful if you wish to generate quantitative data.

ii. Focus groups

A focus group is a discussion-based interview involving more than two people. A theme or focus is provided by the researcher, who also directs the discussion. These discussions are usually recorded as audio or video tapes and evaluated later. Most of the output from a focus group will be a record of who said what when. Therefore, less-quantitative data are usually generated by this method.

iii. Interviews

An interview is a meeting between the researcher and a participant. Interviews can be structured, in that the interviewer asks the same series of questions to all the interviewees, or unstructured, where the interviewee is left to make their own comments about a topic. Recording the results from an interview can be done at the time if the researcher is able to take notes. More often the interviews are recorded as audio or video images to be evaluated later. Interviews can be used in a similar manner to questionnaires and may also be useful if you wish to generate numerical data.

2.3.2. **Open and closed questions**

There are two types of questions: open and closed. A closed question is one where you give the participant a limited number of answers and they have to choose one or more or these options. An open question is where the answers are not prescribed.

For example:

Closed question: Are you warm at present? YES/NO (delete as applicable)

Open question: How warm do you feel at present?

i. Closed questions

Closed questions may be used in questionnaires or structured interviews. They have an advantage over open questions in that the answers can be collated and the frequency of respondents giving a particular answer can be recorded. For example, if we asked 20 people if they felt warm at present it may be that 18/20 replied YES and 2/20 replied NO. These data can be presented and summarized using the methods outlined in Chapters 3 and 10. If you are comparing two or more groups of people, then a chi-squared test can be used to compare the relative distribution of the answers between the groups. (We discuss chi-squared tests in Chapter 5.) For example, if we ask students studying in two rooms if they are warm at present the outcome from this question can be summarized (Table 2.1.).

ii. Open questions

Open questions may be used in all the methods we are considering in this section, including structured interviews and questionnaires. Open questions lead to descriptive answers or comments. The chief advantage with open questions is that you do not restrict the information you gather. However, this information is much harder to summarize. One approach is to record the number of times a keyword appears within the answers. These frequencies can then be evaluated in the same way as data from closed questions. For example, we asked 20 students how warm they felt at present. Looking through the answers it was clear that several keywords, such as hot, cold, and warm, were consistently used. By looking at each answer it is possible to categorize each student's response (Table 2.2.).

Table 2.1. Responses by two groups of students to the question: Are you warm at present?

	Answer to question	
	YES	NO
Room 114	18	2
Room 130	14	6

Table 2.2. Responses from 20 undergraduates to the question: 'How warm are you at present?' categorized by one keyword in each answer

	Keywords used to answer question		
	Cold	Warm	Hot
Number of students	2	14	4

EXAMPLE 2.3. How useful is this book?

A cohort of 40 students were asked to complete a questionnaire about their experience of using this book. Twenty of these students were studying on a degree course in Microbiology and 20 in Forensic Science. The aim of the investigation was to compare the experience of the two groups of students. The questions were:

1. Have you found this book helpful? YES NO
2. If YES, why?
3. How helpful have you found this book? *Circle one value. 1 (least) 10 (most)*
   1 2 3 4 5 6 7 8 9 10

Q3 In Example 2.3., which of these three questions are open and which closed?

A3 Question 2 is open. The respondent is left to comment freely in response to the question.

Questions 1 and 3 are prescribed and the respondent has to chose from a limited number of options. These are closed questions.

2.3.3. Sensible questions

Whether you use open or closed questions or a combination of both, all questions need to be phrased so that the meaning is clear. Here is a checklist with examples that you can use to check your own questions against.

Fault in the phrasing of your question	Examples of poorly structured questions	Improved structure to questions
Vague	How did you get here?	Give the mode of transport used to travel here, e.g. on foot, bus, train, car, bike, motorbike, etc.
Too few options in a closed question	How regularly do you go swimming: every day/never	How regularly do you go swimming? Every day/A few times a week/Once a week/Once a month/A few times a year/Never
Leading	What is your opinion of the terrible and frightening developments in molecular biology?	What is your opinion of the developments made in molecular biology in the last 5 years?
Double negatives	Do you not believe that your degree course is not adequate in preparing you for employment?	Do you believe that your degree course is adequate in preparing you for employment?

Fault in the phrasing of your question	Examples of poorly structured questions	Improved structure to questions
Jargon	Are you a member of the HEA?	Are you a member of the Higher Education Academy?
Too many topics	Are you in favour of fox-hunting and the control of deer populations through shooting?	a. Are you in favour of fox hunting? b. Are you in favour of the control of deer populations through shooting?
Unrealistic	How old were you when you read your first word?	(Delete such questions)
Status questions	Are you employed/unemployed? ('Employed' is the first answer, implying that this has a higher status.)	(These questions are very difficult to improve but you need to be aware of the potential impact that the order of the answers may have on a volunteer. The participant may be more likely to lie and/or be offended (9.5.1.).)
Insufficient details for analysis	How old are you? 10–20, 21–70	How old are you? 10–19, 20–29, 30–39, 40–49, 50–59, 60–69, 70 or over
Overlap in answers	How old are you? 18–20, 20–22, 22–24	How old are you? 18–19, 20–21, 22–23, 24–25

If your method for collecting information is structured as in a structured interview or questionnaire, you should carry out a pilot study before you begin. This is an invaluable way of checking that the questions really do provide you with the answers you are expecting. In an unstructured interview or a focus group you will have the opportunity to clarify any questions during the interview/discussion.

2.3.4. Your participants

For any investigation that involves people you need to make sure your approach is suitable for your 'audience' (Chapter 9). There are many elements in this and not all will apply to all the methods you might use. Here are some prompts:

- Language. Use appropriate language. If necessary define terms, provide explanations of terms or ideas without influencing responses.
- Accessibility. Consider the layout of a questionnaire. Is the format and font suitable, for example, for dyslexic participants, for children . . . ? Do you need to produce a Braille copy? If you are providing a questionnaire online, can it be read by the standard

screen-reader software? Is the room you are using accessible for all participants? Do you need a translator?

- Environments. If the location for carrying out this research is important as in the interviews and focus groups, is the room free from distractions?
- Time. How long will the interview, questionnaire, or discussion take? Do your participants have that much time?

2.3.5. **Sample sizes**

For most investigations you will not involve the whole population; instead you will sample (1.6.). Apart from deciding on your sampling strategy (1.6.2.), you will also need to decide on the size of your sample. The size of your sample will be determined in part by you satisfying yourself that your sampling strategy will generate a representative sample. In addition, sample size is determined by the way in which the data will be analysed. We consider this last point in particular for closed (i) and open (ii) questions.

i. Closed questions

If you wish to compare two or more groups and you are using closed questions, then your sample size is dependent on two criteria: first is your sample representative (1.6.1.), and second is your sample size appropriate for the statistical test you intend to use?

As we discussed earlier (2.2.8.), if you are carrying out an experiment then you will wish to test hypotheses. To achieve this you need to identify the statistical test you will use before you carry out your investigation, as it is usually this that determines the size of your sample. We illustrate this process by looking at question 1 in Example 2.3. The statistical test most often used to compare two or more sets of answers to a question is the chi-squared test (Chapter 5). Which chi-squared test you use will depend on the number of answers in the closed questions and the number of samples you intend to collect. The following illustrates the process where you have two or more samples. The same steps can be applied when you intend to gather data from only one sample.

The responses of 40 students in Example 2.3. to question 1 have been collated (Table 2.3.).These two groups of students can only choose one of two possible answers: yes or no. To compare the answers from these two groups of students you would therefore use a chi-squared test for association with Yates's correction (5.4.2.). All statistical tests have a set of criteria that should be met by the data that are to be analysed. If the criteria are not met then the use of the test is invalidated. The criteria

Table 2.3. A comparison between Microbiology and Forensic Science students in their response to Example 2.3. question 1: Have you found this book helpful?

	Number of respondents for a particular answer	
	YES	NO
Microbiology students	15	5
Forensic Science students	10	10

for using a chi-squared test for association with Yates's, correction are that you:

1. Wish to test for an association between two treatment variables.
2. Have data that are organized into two discrete categories for each variable.
3. Have data that are counts or frequencies and are not percentages or proportions.
4. Have observations that are independent of each other.
5. Have expected values that are more than 5.

Since we do not expect you to have necessarily read Chapters 3, 4, and 5 at this point, you will have to believe us when we say that all these criteria are met. The criterion that is most important when determining sample size is the last one. Expected numbers are produced as part of the chi-squared calculation. We show you in Table 2.4. how they are calculated for this example. The expected numbers for this set of data are all greater than 5. So for the results from this question it appears that our sample size was adequate (i.e. expected values are all greater than 5).

Since you cannot calculate expected values until you have carried out the investigation and collected your data, how then can you determine what sample size to use? The answer is to try out your closed questions on a group of people similar to those who will take part in your study. That way you can judge if you are likely to get an appropriate distribution of answers.

But what happens when we analyse the answers from question 3? This is also a closed question but there are ten possible answers, and as Table 2.5. shows us, not surprisingly our data are more 'spread out' compared with question 1 (Table 2.3.).

Since there are more than two possible answers, we use a chi-squared test for association (without a Yates's correction) to test our hypotheses (5.3.).

Table 2.4. Chi-squared test for association comparing Microbiology and Forensic Science students in their response to Example 2.3. question 1: 'Have you found this book helpful?'

	Number of respondents for a particular answer		
	YES	NO	TOTAL
Microbiology students observed numbers	15	5	20
Microbiology students expected numbers	$25/40 \times 20 = 12.5$	$15/40 \times 20 = 7.5$	
Forensic Science students observed numbers	10	10	20
Forensic Science students expected numbers	$25/40 \times 20 = 12.5$	7.5 $15/40 \times 20 = 7.5$	
TOTAL OBSERVED	25	15	40

The criteria for using this test are that you:

1. Wish to test for an association between two treatment variables.
2. Have data that are organized into more than two categories for at least one of the variables and into two or more categories for the second variable.
3. Have data that are counts or frequencies and are not percentages or proportions.
4. Have observations that are independent of each other.
5. Have expected values that are greater than 5.

Again, you may have to take our word for it that the criteria 1–5 are met. Criterion 5 is again the one most relevant to determining our sample size. We have calculated the expected values for these data (Table 2.6.) and unlike question 1 (Table 2.4.), all the expected values are less than 5. This means that although this sample size of 40 was adequate to allow analysis of question 1 where there were only two possible answers, it is not large enough to allow us to analyse question 2, where there are 10 possible answers. In fact you would need more than 100 participants in each group

Table 2.5. A comparison between Microbiology and Forensic Science students in their response to Example 2.3, question 3: 'How helpful have you found this book?'

	Number of respondents for a particular answer									
	1	2	3	4	5	6	7	8	9	10
Microbiology students	1	0	1	2	2	3	3	4	3	2
Forensic Science students	0	0	0	0	7	5	4	3	1	0

Table 2.6. Calculation of expected values for numbers of respondents studying Microbiology or Forensic science when answering Example 2.3., question 3: 'How helpful have you found this book?'

	Number of respondents for a particular answer									
	1	2	3	4	5	6	7	8	9	10
Microbiology students	1	0	1	2	2	3	3	4	3	2
Microbiology students expected numbers	0.5	0	0.5	1	4.5	4	3.5	3.5	2	1
Forensic Science students observed numbers	0	0	0	0	7	5	4	3	1	0
Forensic Science students observed numbers	0.5	0	0.5	1	4.5	4	3.5	3.5	2	1

to generate expected values that were large enough for the answers given by the two groups of students to be compared.

Therefore, what had been an adequate sample size for question 1 is not adequate for question 3. In general, the more 'possible' answers you include for your closed questions, the larger the sample size needed. Clearly, you need to balance the number of possible answers so that you obtain useful information against the size of sample you will then need to survey to generate data that can be analysed.

If you unfortunately find that having carried out an investigation where you planned to use either of the chi-squared tests for association (5.3. and 5.4.) you do have expected values less than 5 then you should refer to section 5.6.

ii. Open questions

In this book we only consider how you may extract quantitative information from open questions and analyse this. If you wish to evaluate the qualitative elements of your information, you will need to look to other texts (e.g. Robson, 2002).

Imagine some answers that may have been given in response to question 2 in Example 2.3. These might include:

Person 1. 'The glossary was useful.'

Person 2. 'The explanations are clear, the glossary and boxes are helpful and I like the illustrations.'

Person 3. 'Having a glossary meant I could check unfamiliar terms.'

If you have a set of open answers like this it is possible to compile a list of words that appear regularly. Each persons answer can then be checked for the presence of these keywords.

Record sheet for the evaluation of question 2. I have found this book helpful because:

	Clear explanation	Glossary	Boxes	Illustrations
Person 1		x		
Person 2	x	x	x	x
Person 3		x	x	x

What can you do with the data? You could total the number of times a keyword or phrase is mentioned. But if one person mentions four keywords and others only one, then the person who includes four keywords is over-represented in your sample and these observations are said to be not **independent** of each other. An alternative approach would be to record the first keyword in each answer. Although this resolves the problem of independence, clearly this approach does not then reflect all the information you have. Trying to obtain quantitative data from qualitative answers can be indicative at best.

One of the chi-squared tests for association may be appropriate to analyse this type of data (Chapter 5). Clearly you again have a problem when trying to determine sample size, as you will not know in advance how many keywords or phrases you will identify from your respondents' answers. Again the solution is to run a pilot test on a similar group of people. The general relationship between numbers of possible answers and sample size applies here to the number of keywords and phrases you identify. The more keywords and phrases there are then in general the larger the sample size required.

iii. Achieving the required sample size

One of the great drawbacks of using these methods to gather information is obtaining a large enough sample size simply because the members of your population are not sufficiently motivated or do not have enough time to take part. You need to bear this in mind when writing your aim and objectives. Will the population identified by the aim provide sufficient numbers of participants to allow you to test any hypotheses you have? But you must work ethically in your research, so any investigation that includes an element of coercion is not acceptable. Forms of coercion can include asking your family and friends, asking everyone in a lecture or tutorial group, or offering a reward. More details are included in Chapter 9.

2.4. Research and the law

In this chapter we have considered a number of different types of investigations and the factors that you need to think about as you design the research. There are several areas which we have touched on in part that are so important that we have devoted a separate chapter to them (Chapter 9). These are where the law relates to you as a researcher and to your research. The four areas we consider are health and safety, sampling and access, animal welfare, and working with humans.

2.5. Managing research

To be effective as a researcher, either as an undergraduate or a graduate, you must develop good management skills. In research, management falls into at least three areas: managing time, managing space, and managing data.

2.5.1. Time management

Time management is critical. You may have steps in a method which need to be carried out for certain periods of time, or you may need to collect samples or complete preparations all on one day. In addition, most research is carried out within a time frame determined either by your course or by funding. It is therefore important to consider, before you start, how long each part of your investigation is likely to take and to draw up a timetable. A timetable should help you identify critical points in the execution of your investigation or even experimental designs that just cannot be carried out in the time available. The time constraints imposed by seasonality, and so on, must also be recognized in the planning, as should external factors such as assessment points for other courses. In Example 2.2. the student collected the samples on one morning, but it then took all afternoon and into the late evening to inoculate the plates. This student's planning was exemplary and she knew from her experience with a pilot experiment how long the work would take. She was therefore able to plan her time effectively and carry out the project, having made proper arrangements in advance to work late.

From our experience there are two common errors that arise in relation to time management because of inexperience. The first of these is developing complex experimental designs. It is very tempting to try to investigate lots of factors all within one experiment. This can be counterproductive for two reasons. The factors can interact and you may not be able to untangle which factors had which effect. Trying to do so can be very time consuming and is usually not very successful. Complex designs often need complex statistics. This is fine as long as you have thought this all through before you start and then allowed in your planning for the time it will take to analyse the data correctly. It is usually better to have more objectives to examine an aim than one complex objective. Second, research takes longer than you think. A general rule of thumb is to plan your investigation and then double the time you have allowed for each part. If you are required to write a thesis or report you should also allow yourself twice as much time as you think it should take. This realistic estimate of time should be built into your timetable.

2.5.2. **Space management**

In research you are often working in a confined space, such as a laboratory, and in company with other scientists. This means that your management of your space is very important, allowing you to complete your research without interfering with other peoples' work. You must be considerate of other users of the areas around where you are working: we consider this further in 9.2. and 9.5.

2.5.3. **Data management**

When designing and carrying out research, your ability to keep and manage comprehensive records is paramount. For most of us records are kept in two ways: in a transportable hard-copy format, such as a laboratory or field notebook, and as electronic files.

i. Laboratory or field notebooks

The best approach to recording information in such notebooks is to record more detail rather than less. You should date each day's work, give full details of the method used that day, which may for fieldwork include a site description and details of the weather, and a record of the results you obtained. You may also need to record contacts' details, but should be wary of any records that may compromise volunteers' confidentiality (9.5.). Finally, add notes about your ideas, why you are taking the research in a particular direction, what the data appear to show you, and how this may relate to other people's studies. All this information will be invaluable, both in tutorials and when communicating your results in a more formal way (Chapter 10). Having more detail rather than less is particularly important for graduates, who may not write formal reports for several months and therefore will be less able to rely on memory to compensate for unclear notes.

EXAMPLE 2.4. The growth rate of rye seedlings

Researchers at an agricultural research station were interested in the natural variation in the rate of growth of rye. The first measurement was made when the rye was at the seedling stage and the total height (mm) was recorded for 15 seedlings. In her notebook the researcher noted down the following:

12.0, 12.0, 11.5, 18.0, 14.0, 11.0, 14.5, 11.5, 10.0, 10.0, 19.5, 19.0, 21.0, 15.5, 14.5

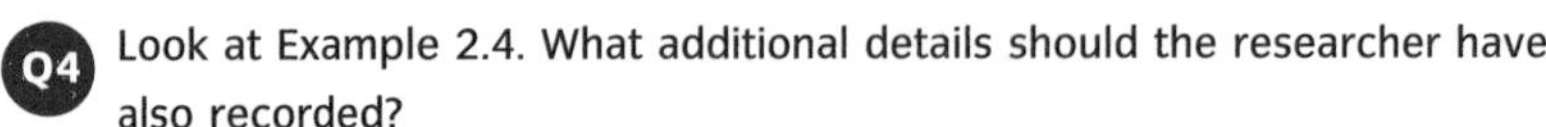

Q4 **Look at Example 2.4. What additional details should the researcher have also recorded?**

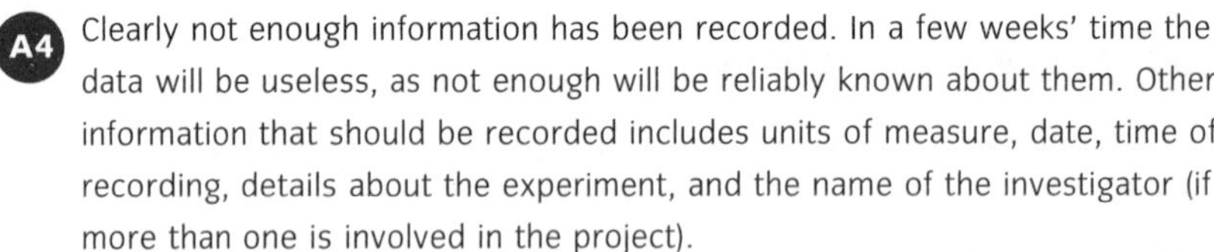

A4 Clearly not enough information has been recorded. In a few weeks' time the data will be useless, as not enough will be reliably known about them. Other information that should be recorded includes units of measure, date, time of recording, details about the experiment, and the name of the investigator (if more than one is involved in the project).

In Chapter 9 we consider health and safety legislation. One outcome from this is that you will need to produce a risk assessment. This should also be pasted firmly into your field or laboratory notebook. There is no point in finding out how to deal with an emergency if you do not have the relevant information with you when it is needed!

ii. Electronic records

There are two useful practices here that can save enormous heartache from lost or apparently missing files. The first of these is keeping track and the second is backing-up. As your research project develops and you alter electronic files, you need to know which version of a file you are working with. The simplest method to achieve this is to include a table at the top of each file as illustrated (Table 2.7.). Using a table like this will prompt you to make sure you keep track of the date the file was last altered and your thoughts as you develop your work. When saved, files do indicate the date last altered, but we have found this 'table' system to be more effective as a file-management tool.

Backing-up electronic files is essential and you should get in the habit of doing this at least every day, so a file-management system (e.g. using a table as in Table 2.7.) is invaluable. Data storage and collection are subject to several laws, including the Human Rights Act 1995 and the Data Protection Act 1998 (9.5.4. and 9.5.5.): you must therefore make sure that you are careful about how all records are stored, where, and for what period of time.

Table 2.7. Information that helps you keep track of electronic files

Title of project	
Last revised	
Aim	
Objectives	
Check for accuracy	
Check for inadvertent plagiarism	
Checked supervisor's feedback	
COMMENTS	

Summary of Chapter 2

- By working through the first section in this chapter, you will realise that you already know about experimental design and are able to identify the strengths and weaknesses of a design (2.1.).
- We then consider the 12 steps that take you through designing an experiment including: the type of investigation (2.2.1.); the importance of background information (2.2.2.); aims and objectives (2.2.3.); populations and sampling (2.2.4.); controls (2.2.5.); variables (2.2.6.); replication (2.2.7.); statistics (2.2.8.); influencing outcomes (2.2.9.); assumptions and bias (2.2.10.), repeatability (2.2.11.), and a final review (2.2.12.). Some sections (e.g. 2.2.8.) of necessity draw on subsequent chapters (especially Chapters 4–9).
- There are additional points that need to be considered when designing investigations in which you may use questionnaires, focus groups, or interviews to obtain quantitative data (2.3.), including the need to consider the type of question and how this may determine how the data are analysed and the sample size.
- You are reminded of the importance of compliance with UK law, including ethics, and health and safety (2.3.4., 2.4., 2.5.3., and developed in Chapter 9).
- Tools for managing research in relation to the planning stages, time management, space considerations, and data management are discussed in this chapter (2.5.) and developed in Chapters 9 and 10 and Appendix a.
- The Online Resource Centre includes interactive exercises that test your understanding of this chapter along with other topics, particularly those considered in Chapters 4–9.

3 What to do with raw data

So far we have looked at the general terms that you may encounter when reading about or carrying out research relating to experimental design (Chapter 1) and the steps you need to follow when designing an investigation or evaluating other people's research (Chapter 2). In this chapter we consider what to do with the data you have collected from your investigations; these ideas are developed in Chapters 4–8.

As we explained in Chapter 2 (2.2.8. and 2.3.5.) the first time to think about the data that your research may generate is whilst you are designing your investigation. This is critical. Most research uses statistics as a tool to help identify the trends in the data. But each statistical test has certain requirements that need to be met. For example, to use the Mann–Whitney U test (8.1.) you need to have between 5 and 30 observations in each sample. If you have not decided on which statistical tests to use before you start your investigation, your sample size may be too small and your data cannot therefore be analysed. This is a waste of your time and reflects badly on you as a scientist as it is clear you have not planned your work properly in the first place. To choose the correct test and hence determine sample size you first need to understand terms such as 'parametric' and 'qualitative'. These are explained in this chapter (3.1. and 3.8.). You also need to understand about 'distributions' (3.2.) and 'transforming data' (3.9.). We provide an overview of how to choose the correct statistical test in Appendix b.

The second time you need to think about your raw data is after you have completed your investigation. You will need to identify the trends in your data and to communicate these to other people. When you come to communicate your findings you may wish to use a figure (10.8.2.), a table (10.8.1.), or summary statistics (3.3.–3.7.). In learning how to summarize your data you will also be introduced to some of the central steps in statistics, which are the calculation of a sums of squares, variance, and standard deviation (BOX 3.1.). More examples on all topics are included in the Online Resource Centre. If you work through the exercises in this chapter it will take about 2 hours.

3.1. Types of data

When carrying out an investigation you will generate data as a series of **observations** measured on a particular scale. For example, if you are investigating the change in human body temperature in relation to exercise, the scale of measurement used here is degrees Celsius (°C). There are several terms that are used frequently to describe the scales of measurements used when collecting data. You need to become familiar with these terms: they are essential in helping you decide how to design your investigation and how to communicate your findings. These terms are: qualitative, quantitative, discrete, continuous, rankable, nominal, ordinal, interval, and derived variables.

Qualitative refers to information that is not numerical but descriptive. In 2.3. we refer to methods such as focus groups and interviews which may gather qualitative responses, opinions, and thoughts expressed in words. The term 'qualitative' can also be used when collecting numerical data, but where the scale of measurement is qualitative. In this case qualitative data will be numerical observations assigned to named, descriptive categories that are mutually exclusive and non-numerical, for example, the number of *Lotus corniculatus* (birdsfoot trefoil) with a yellow or a red keel. The categories yellow and red are qualitative. These scales of measurement are always discrete.

Quantitative refers to the use of numbers. A quantitative scale of measurement is one where the observations are assigned to ordered numerical categories; for example, the height of adult males (m) is measured on a continuous quantitative scale.

Discrete measurements fall into a series of distinct, mutually exclusive categories and the number of categories is limited. For example, the Royal Horticultural Society's (RHS) scale for recording petal colours (e.g. red, pink, white) is a qualitative discrete scale of measurement. The number of eggs in a clutch is a quantitative discrete scale of measurement. In this context you cannot have half an egg.

Continuous scales are ones where observations do not fall into a series of distinct categories and may take any value within the scale of measurement. For example, height (m) can be measured from 0 m upwards with no limit. Some scales, such as percentages and pH, are also continuous, but only within prescribed boundaries. For example, the percentage scale is restricted to 0–100%.

Rankable. In some scales of measurement the categories can be ranked (put in order). These can include qualitative, quantitative, discrete, and continuous, and therefore ordinal and interval scales. For example, if a

Table 3.1. Height (cm) of 87 male students

	Frequency classes for height of male students								
Height	150.0–154.9	155.0–159.9	160.0–164.9	165.0–169.9	170.0–174.9	175.0–179.9	180.0–184.9	185.0–189.9	190.0–194.9
Frequency	3	4	12	15	19	15	12	4	3

questionnaire included a question 'How warm are you?' the answers (very hot, hot, warm, cool, cold), although qualitative, can be ranked. Similarly, a numerical but discrete scale, such as the number of eggs in a clutch, can be ranked (e.g. 0, 1, 2, 3, etc.). Continuous data are ranked in two ways: either by numerical order or by class order. For example, if the heights of five men were recorded, the observations could be ranked 166.0 cm, 166.5 cm, 168.0 cm, 168.2 cm, 170.0 cm. Where you have many observations recorded on a continuous scale, these can be arranged into classes (Table 3.1.). These classes can also be ranked. For some statistical tests observations are ranked and then 'assigned a rank order'. We explain this process in 3.8.2.

Nominal scales of measurement fall into discrete categories. The number and nature of the categories can be either an inherent property or what is being measured or imposed by the investigator, and should ensure that every observation in the data set should be able to be classified. The categories have no specified order. For example, the RHS scale used for recording petal colours is an artificial device that provides a method for categorizing flower colours; the flower colours have no particular order and cannot therefore be ranked. If a closed question is included in a questionnaire (2.3.2.) where the prescribed answers are 'yes', 'no' or 'don't know', these mutually exclusive categories are an inherent property of what is being measured but there is no rationale for ranking them: they are therefore nominal.

Ordinal scales of measurement are similar to nominal scales in that they are also based on discrete, mutually exclusive categories. However, in this case the categories can be ranked. The categories can be qualitative or quantitative. In ecology a common scale that is used when examining species abundance is the ACFOR scale, which includes categories 'abundant', 'common', 'frequent', 'occasional', and 'rare'. These are a qualitative measure but have an inherent order and so can be ranked. The number of eggs in a clutch also falls into discrete categories. These quantitative categories can also be ranked and are therefore ordinal. If volunteers were asked to rate their responses to a question or statement (see Example 2.3., question 3) the results would be ordinal.

Interval data are measured on a continuous and rankable scale, and unlike data measured on an ordinal scale it is possible to measure the difference between each observation. Examples of interval scales include temperature (°C), distance (m), and mass (g).

Derived variables are scales of measurements that are the result of a calculation; they are therefore quantitative and usually continuous. The four types of derived variables are ratios, proportions, percentages, and rates. Some of these scales, such as percentages, are constrained within certain limits (e.g. 0–100%); others, such as a rate (km/h), are not.

You will usually find that more than one term can be used to describe any one measurement. For example, if an investigation examines the height of students within a higher education institution the scale of measurement will be centimetres. This is a quantitative, continuous, rankable, interval scale.

Q1 Which terms best describe these data:

(i) The number of prickles on holly leaves
(ii) Percentage (%)
(iii) pH
(iv) grams.

A1 (i) Quantitative, discrete, rankable, ordinal
(ii) Quantitative, continuous, rankable, derived variable
(iii) Quantitative, continuous, rankable, derived variable
(iv) Quantitative, continuous, rankable, interval.

The terms we have considered so far all relate to the scale on which the measurements are made. Understanding which type of scale is being used is important in helping you choose the type of statistical test you should use when you design your experiment and analyse your data (Appendix b), and which figure or table you could use when communicating your findings (10.8.). There are two more terms (parametric and non-parametric) which are also critical. These terms relate to your observations and the characteristics of the distribution of your data. To be able to tell if your data are parametric or non-parametric you have to first understand distributions (3.2.) and how to calculate a mean and variance (3.4., 3.5.). We therefore consider the terms parametric and non-parametric in 3.8. and 3.9. and explain how to tell if your data are parametric in BOX 3.2.

3.2. Distributions of data

In investigations examining a single **treatment variable** the data may be plotted for example as a **bar chart** or **histogram** (e.g. Fig. 3.1.). These figures can be seen to have a particular shape or distribution. Some distributions are well known, such as the normal (3.2.1.), binomial (3.2.2.), Poisson (3.2.3.), and exponential (3.2.4.) distributions. The shapes of these distributions have been described mathematically and these equations have been used to demonstrate other relationships, including the tendency of a distribution to have a central point (e.g. a mean, 3.4.) and the spread of the data around the central point, such as the variance (3.5.).

3.2.1. Normal distribution

This is a very important distribution which can be best explained using an example.

In a study of the height of 87 male students, the raw data are first organized into a frequency table (Table 3.1.) (If you are not familiar with frequency tables, see Chapter 10.)

As the data are measured on an interval scale the data can be plotted as a histogram (Fig. 3.1.). If you collected more observations, then this

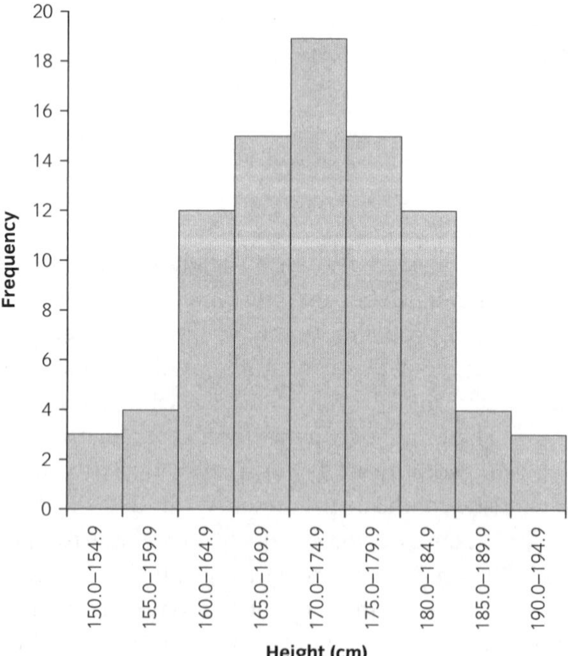

Fig. 3.1. Height of 87 male students.

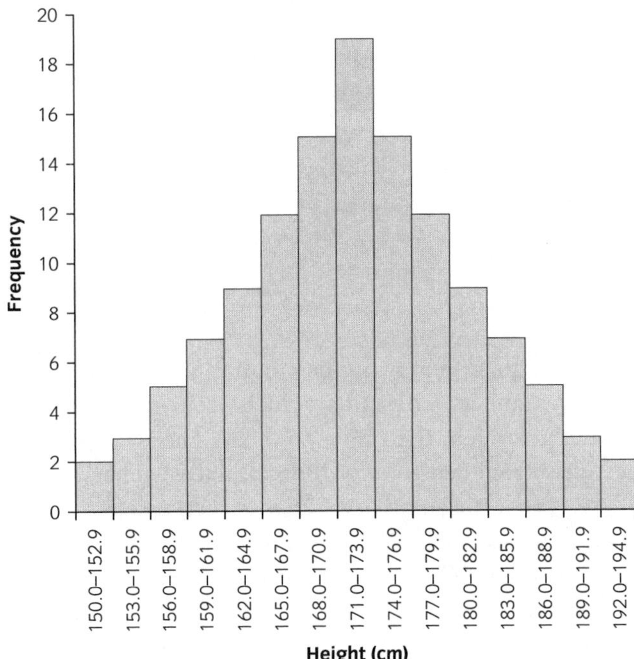

Fig. 3.2. Height of 124 male students.

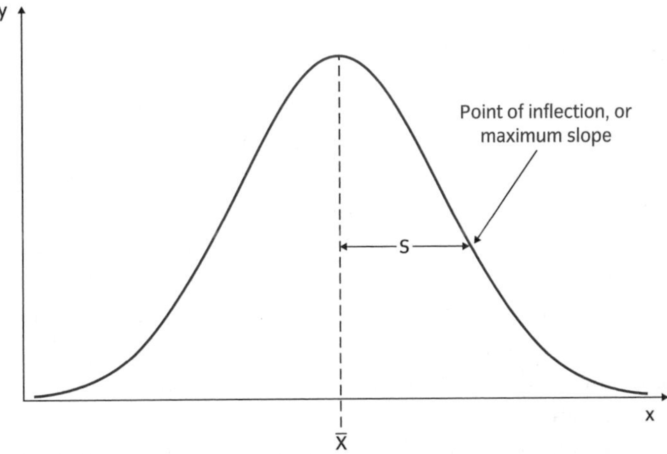

Fig. 3.3. Normal distribution with mean ($\bar{x}$) and standard deviation (s).

increased sample size would allow you to reduce the sizes of the classes you are using and the distribution looks much smoother when plotted (Fig. 3.2.). If you could extend this study still further the distribution becomes even smoother, symmetrical, and 'bell-shaped' (Fig. 3.3.) with a single central peak (unimodal). This is typical of a normal distribution.

If your data is normally distributed like this, then there is a mathematical relationship between the x values and y values. This mathematical description is called the Gaussian equation, where:

$$y = \frac{1}{\sqrt{(2\pi s^2)}} e^{-h}$$

and

$$h = \frac{(x - \bar{x})^2}{2s^2}$$

The terms in this equation may not be familiar to you. The symbols e and π are particular numbers (constants) which occur so often in maths that they have been given letters to refer to them. Their values are 2.72 and 3.14 respectively (rounded to two decimal places). The $\bar{x}$ (mean) and s^2 (variance) are two of the summary statistics that we consider in 3.4. and 3.5. x is any one observation in your **sample** and y is the y value calculated for any given x value. This equation is considered again in Chapter 5 (5.1.3.) where we show you how to test whether your data can be described by the Gaussian equation, and Appendix c where we explain the symbols in this equation in more detail. A normal distribution has several features that are widely exploited in statistics and we refer back to this distribution throughout the rest of the book.

3.2.2. Binomial distribution

The second distribution we consider is one you would expect to obtain if each item examined can have either one or another state. For example, a seed could be germinated or not germinated. Another, although non-biological, example arises when you toss a coin. If you toss a coin the probability of obtaining 'heads' is 1/2. The alternative outcome would be getting a 'tail' and the chance of this on each toss is also 1/2. You may toss the coin three times. In this case the chance of getting a 'head' three times in a row will be $1/2 \times 1/2 \times 1/2 = (1/2)^3 = 0.125$. This can be extended to more throws and any combination of heads and tails. It is possible to write a mathematical equation that allows you to work out the probability for any particular number of throws and any particular numbers of heads and tails:

$$y = \frac{n!}{x!(n - x)!} \times p^x \times q^{(n-x)}$$

This equation describes the mathematical relationship between y (the probability of obtaining a particular number of heads in a given number of throws) and x (a selected number of heads). If you are not familiar with

Table 3.2. The probability of obtaining a certain number of heads when tossing a coin 10 times

Number of heads in 10 throws (x)	Probability (y)
0	0.001
1	0.010
2	0.044
3	0.117
4	0.205
5	0.246
6	0.205
7	0.117
8	0.044
9	0.010
10	0.000

the terms we have used in this equation (e.g. !) we give more details and some worked examples in Appendix c.

We know that when tossing a coin the probability of getting 'heads' (p) is 1/2 and the probability of getting a tails (q) in each throw is also 1/2. If we tossed a coin 10 times (n) we can use this information to work out the probability of obtaining any number of heads within those 10 throws of the coin (Table 3.2.). The probability of obtaining 4 heads and 6 tails is 0.205; the probability of obtaining all heads is very small, only 0.001.

When the data are plotted (Fig. 3.4.) you can see that this distribution also has a symmetrical shape with a single (unimodal) high point. If you collect data that have this shape and that can be described by this mathematical equation, then your data are said to be binomial and they have a binomial distribution. We show you how to check if your data have a particular distribution in 5.1.3.

3.2.3. Poisson distribution

This is a distribution often found where events are randomly distributed in time or space, such as the distribution of cells within a liquid culture or the dispersal of pollen or seed by wind. This distribution is unimodal, but is often very asymmetrical with a protracted tail either to the right (a positive skew) or to the left (a negative skew). Figure 3.5. illustrates four

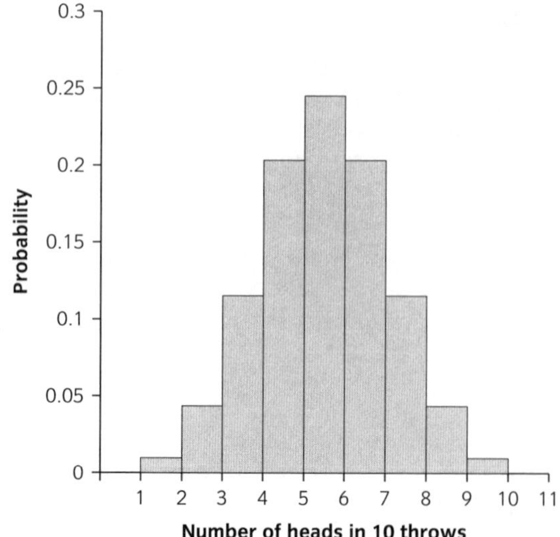

Fig. 3.4. Binomial distribution.

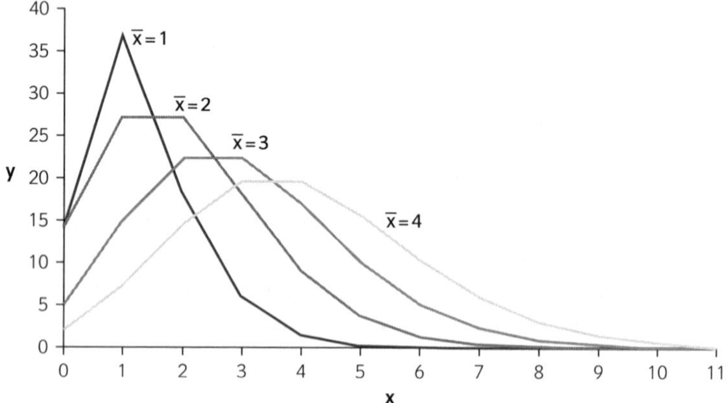

Fig. 3.5. Poisson distribution.

Poisson distributions which vary in their degree of skewness. The Poisson distribution can be described by the equation:

$$y = \mathrm{e}^{-\bar{x}} \times \frac{\bar{x}^x}{x!}$$

As we explained when we looked at the Gaussian equation, e is a constant with the approximate value of 2.72 and $\bar{x}$ is the mean of the sample. x is the number of individuals in the sample and y is the value for a given x. One Poisson distribution is shown in Table 3.3. and Fig. 3.5., where $\bar{x} \approx 2.0$ m.

We explain in 5.1.3. and 3.5.3. how you may check your data to confirm that they have a Poisson distribution.

Table 3.3. The distance seeds are dispersed from the canopy edge of one *Taxus baccata* (yew) tree

Distance (m)	Number of seeds
0	13
1	27
2	27
3	18
4	9
5	4
6	1
7	0
8	0
9	0
10	0

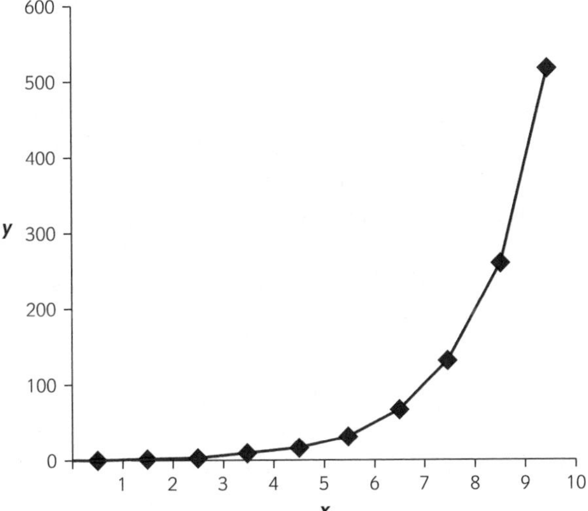

Fig. 3.6. Exponential distribution.

3.2.4. Exponential distribution

This distribution can be found in bacterial liquid cultures during their peak growth where the numbers of individuals keep doubling, or if you recorded the increase in the number of copies of DNA during a polymerase chain reaction (PCR) (Fig. 3.6.). The mathematical description of an exponential distribution uses mathematical notation relating to integration, which is a mathematical technique beyond the scope of this book. You should refer to other statistical texts for further details.

3.3. Summary statistics

In this section we consider summary statistics. These have two places in research. In the first summary statistics are used when communicating your results, to avoid the inclusion of extensive tables of raw data and to allow you to present the main trends in your data more succinctly. The second role of summary statistics is that many of the calculations, particularly the mean and variance, are terms that appear in many statistical calculations and it is therefore important to appreciate types of summary statistics, their relative strengths, and how they are calculated.

Summaries of data focus on two things: the central point of the distribution and the variation around this central point. The calculations of the central point can be derived from the mathematical relationship, such as the Gaussian equation or Poisson equation, and in these instances this central point is called the mean. Where the distribution is not known then other measures, such as the median and mode, can be used. A similar distinction occurs in relation to estimates of the spread of the data about the central point. Some distributions are not symmetrical about the central point and this deviation from symmetry can be indicated by using measures of skewness (3.4.4.) The shape of the peak can also vary and a useful measure of this kurtosis is also considered in 3.4.4.

The summary statistics that we consider in this section can be used in two ways: either you have data where there are no **replicates** or you have data with replicates. We indicate the appropriate summary statistics for data with no replicates in Table 3.4. If the data are nominal or ordinal or grouped interval data with no replicates, then you summarize across all groups. In Example 3.1. (Table 3.6.) we show you how to summarize the results for year 2004 and year 2005 separately.

Table 3.4. Appropriate summary statistics for different types of data

	Nominal	Ordinal	Interval – unknown distribution	Interval – known distribution
Central tendency	Mode	Mode Median	Mode Median	Mean
Spread of data (Variation)		Range Interquartile range Percentiles Confidence limits	Range Interquartile range Percentiles Confidence limits	Standard deviation Coefficient of variation Confidence limits

Table 3.5. The number of seeds per umbel in four random samples of *Allium schoenoprasum* in 2004

	Number of seeds per umbel								
Sample	13	14	15	16	17	18	19	20	21
a	1	1	3	0	1	0	2	1	0
b	2	0	3	1	2	0	3	2	4
c	0	3	1	3	1	3	1	0	2
d	2	2	1	2	3	0	0	2	0

If there were replicates in your design (e.g. Table 3.5.) then you may wish to summarize within each category. If you summarize within a category, for example we may wish to summarize within the category 13 seeds per umbel (Table 3.5.), then your data within that category will either be ordinal or interval. Table 3.4. indicates the most appropriate summary statistics for ordinal or interval data.

3.4. Estimates of the central tendency

There are three common methods used to measure central tendency: the mode, median, and mean. The mean should not be confused with the average, which is an estimate of the mean applied to data whose distribution is not known (interval or ordinal). Though commonly used, this is an incorrect calculation and may identify an incorrect central point.

EXAMPLE 3.1. The number of seeds per umbel in *Allium schoenoprasum*

A single *Allium schoenoprasum* (chive) may produce one inflorescence with many flowers held in an umbel. Each inflorescence may produce many seeds. We recorded the number of seeds produced per umbel over two consecutive years (Table 3.6), examining nine umbels in 2004 and eight in 2005.

Table 3.6. The number of seeds per umbel in *Allium schoenoprasum* in 2004 and 2005

Year	Number of seeds per umbel								
2004	13	14	15	15	15	17	19	19	20
2005	13	14	15	15	17	19	19	20	–

3.4.1. Mode

The mode is the category that contains the greatest number of observations. If the data are nominal or ordinal the categories are already specified. For example, the scale of measurement used in Example 3.1. is ordinal with categories that can be ranked. When summarized (Table 3.7.) it is clear that the mode is 15 seeds in 2004 (i.e. unimodal). In 2005 there are two modes: 15 seeds and 19 seeds (i.e. bimodal).

If the data are measured on an interval scale but the underlying distribution is not known, then you may use the mode as a measure of central tendency but will first need to organize your data into a frequency table. The modal class is the class with the highest frequency and the mode is the mid-point of this category. Clearly, since you have imposed these categories on the data there is a degree of artificiality in relation to the mode; grouping the data into other size classes may generate a different mode. You should be aware of this both when choosing the classes for the frequency table and when interpreting your results. (For further comments about choosing classes see 10.8.1.)

3.4.2. Median

When data can be ranked (i.e. ordinal or interval data), a simple measure of the central tendency is to take the 'middle' value. This is the median.

Table 3.7. The number of umbels of *Allium schoenoprasum* producing a certain number of seeds (data recorded in 2004 and 2005)

	Number of seeds							
	13	14	15	16	17	18	19	20
Number of umbels recorded in 2004	1	1	3	0	1	0	2	1
Number of umbels recorded in 2005	1	1	2	0	1	0	2	1

The median is dependent on the number of observations in the data set (n). When n is an odd number then the median is the middle value. When n is an even number, then the median is calculated as half the sum of the two middle values. In Table 3.6. for year 2004 $n = 9$, the data are already in numerical order and the middle value is 15 seeds per umbel. In 2005 $n = 8$, here the median is $(15 + 17)/2 = 16$ seeds per umbel.

3.4.3. **Mean**

The mean is used for interval data where the distribution is known. Here we give the methods for calculating a mean for a normal distribution (described by the Gaussian equation), a binomial distribution, and a Poisson distribution. For other distributions you will need to refer to other texts and/or computer software.

For any one distribution the sample mean ($\bar{x}$) and **population** mean (μ) would be calculated in the same way. The sample mean is used as an estimate of the population mean when not all items in the population have been measured. A value called the confidence interval can be calculated to demonstrate the area around a sample mean in which the population mean will probably fall (3.7.).

i. Normal distribution

When calculating a mean for a normal distribution, most often you will have data that are not organized in classes. However, if you have data that are organized into classes a mean may still be calculated using a different method that provides a reasonable estimate of the value.

a. Your data are not grouped in classes In normally distributed data the mean is calculated as the sum (Σ) of all observations (x) divided by the number of observations in the data set (n). If you had collected measurements from all items in the population then the calculation is the same, but you would refer to N rather than n and the mean would be referred to as μ (1.7.). If you are not familiar with these terms you may also wish to refer to Appendix c. The calculation of the mean for a sample is written as:

$$\bar{x} = \frac{\sum x}{n}$$

EXAMPLE 3.2. Length (mm) of two-spot ladybirds (*Adalia bipunctata*)

An investigator was interested in the length of two-spot ladybirds (*Adalia bipunctata*). In an observational investigation she measured the length (mm) of 50 ladybirds collected at random from a garden (Table 3.8.).

Table 3.8. The length (mm) of 50 *Adalia bipunctata* sampled in a garden

Length of *Adalia bipunctata* (x)									
1	5	2	5	7	8	3	6	7	4
4	5	6	4	5	5	7	5	3	5
4	5	1	7	9	2	6	5	6	3
3	6	8	6	4	6	6	8	5	6
7	4	8	9	5	4	3	4	2	5

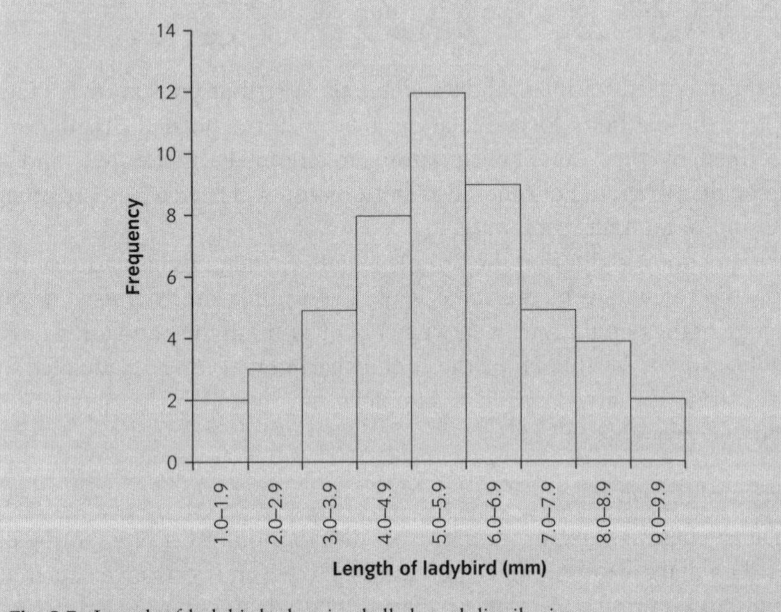

Fig. 3.7. Length of ladybirds showing bell-shaped distribution.

We demonstrate in BOX 3.2. that these data are normally distributed and we can therefore calculate a mean assuming that the data are described by a Gaussian equation. For this example if you add all the x values together (i.e. $\Sigma x = 1 + 4 + 4 + 3 \ldots\ldots + 3 + 6 + 5$); this equals 254, and there are 50 observations. Therefore, the sample mean is:

$$\bar{x} = \frac{254}{50} = 5.08\,\text{mm}$$

Table 3.9. Frequency table of length of two-spot *Adalia bipunctata* (ladybirds) showing how to calculate a mean for grouped data

	Size classes for length (mm) of *Adalia bipunctata* (ladybird)								
	1.0–1.9	**2.0–2.9**	**3.0–3.9**	**4.0–4.9**	**5.0–5.9**	**6.0–6.9**	**7.0–7.9**	**8.0–8.9**	**9.0–9.9**
Mid-point of class (m)	1.45	2.45	3.45	4.45	5.45	6.45	7.45	8.45	9.45
Frequency (f)	2	3	5	8	12	9	5	4	2
$m \times f$	2.9	7.35	17.25	35.6	65.4	58.05	37.25	33.8	18.9

b. Your data are grouped into classes Some data may be from a normal distribution but may be grouped. In this case the mean can be estimated in a different way. For this calculation you need the mid-point for each group (m) and the frequency within that group (f). Then:

$$\bar{x} = \frac{\Sigma mf}{\Sigma f}$$

Table 3.9. is a frequency table for the data from Example 3.2., where the mid-point for each class and the mid-point multiplied by the frequency (mf) are given.

We know that there are 50 observations, so $\Sigma f = 50$. From the table we can add all the $m \times f$ values together $(\Sigma mf) = 276.5$. Therefore, $\bar{x} = 276.5/50 = 5.53$. This method will tend to overestimate the mean and therefore should only be used when you do not have the raw data.

Q2 Calculate the mean and mode using the data from Table 3.1.

$\Sigma mf = 15003.15$, $n = 87$, $\bar{x} = 171.4500$ cm

The mode = 172.45 cm

ii. Binomial distribution

In 3.2.2. we illustrated the idea of a binomial distribution in relation to tossing a coin and getting heads or tails. The number of heads you would expect to get would be the number of times you toss the coin (N) × 1/2. If you had six coins and you tossed these together then you would expect to get 6 × N × 1/2 'heads'. The mean for this binomial data will be this total number divided by the number of throws (N):

$$\bar{x} = \frac{\text{number of coins } (n) \times N \times 1/2}{N}$$

The two N values cancel each other out so the mean can be worked out simply as $\mu \approx \bar{x}; = np$, and so for six coins tossed together $\bar{x} = 6 \times 1/2 = 3$ 'heads'.

iii. Poisson distribution

The mean in a Poisson distribution relates both to the skew of the distribution and the variation in the data. The mean for a Poisson distribution is calculated as:

$$\mu \approx \bar{x} \approx \frac{\Sigma xy}{\Sigma y}$$

The terms Σxy and Σy appear in many statistical calculations and therefore we have shown you in detail how these terms are calculated. The steps in this calculation are shown in Table 3.10. in the first three columns. The final column (x^2y) is a step which occurs in the calculation of a variance (3.5.) for data with a Poisson distribution. Therefore, the mean for this example is:

$$\mu \approx \bar{x} \approx \frac{\Sigma xy}{\Sigma y} = \frac{197}{99} = 1.99\,\text{m} \approx 2.0\,\text{m}$$

3.4.4. Skew and kurtosis

There are two further measures that may usefully be used to describe features of the central tendency. These are skew and kurtosis. We have already seen an example of variation in the degree of skew in a distribution. In data with a Poisson distribution as the mean decreases the

Table 3.10. Calculating a mean and variance for data with a Poisson distribution: the distance seed dispersed from the canopy edge of one *Taxus baccata* (yew) tree

Distance (m) (x)	Number of seed (y)	xy	x^2y
0	13	$0 \times 13 = 0$	$0^2 \times 13 = 0$
1	27	$1 \times 27 = 27$	$1^2 \times 27 = 27$
2	27	$2 \times 27 = 54$ etc.	$2^2 \times 27 = 108$ etc.
3	18	54	162
4	9	36	144
5	4	20	100
6	1	6	36
7	0	0	0
	$\Sigma y = 99$	$\Sigma xy = 197$	$\Sigma x^2 y = 577$

distribution becomes less symmetrical (Fig. 3.5.). A measure of this degree of skew is the relationship between the mean, median, and mode. If a distribution is symmetrical, such as the normal distribution, then the mean should equal the mode, which should equal the median.

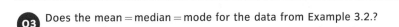

Q3 **Does the mean = median = mode for the data from Example 3.2.?**

A3 In 3.4.3.i. we calculated the mean $(\bar{x}) = 5.08$ mm

Median. Arrange all the values in numerical order. The mid-position lies halfway between two values of 5.0 mm. Therefore the median is 5.0 mm.

Mode. If you examine Table 3.9 it is clear that the most frequent value is 5.0 mm.

Therefore, the mean almost equals the median and the mode.

There are several other measures of skew of which the most common is:

$$\text{skew}(\gamma^3) = \frac{\sum(x - \bar{x})^3}{(n-1)s^3}$$

The value for skew for a perfectly symmetrical distribution should be zero If the distribution has a positive skew (Fig. 3.8) then the value of γ^3 will also be positive and for a negative skew (Fig 3.9.) the value of γ^3 will be negative. An assessment of skew is one of the tests for normality (BOX 3.2.) and may also be used as a descriptive statistic with which to compare distributions.

Kurtosis (γ^4) indicates how sharp the peak of the central point is. The shape of the central point can also indicate that the distribution is not symmetrical. A frequency distribution with a pointed narrow peak is called leptokurtic (Fig. 3.10.), a moderate peak is called mesokurtic,

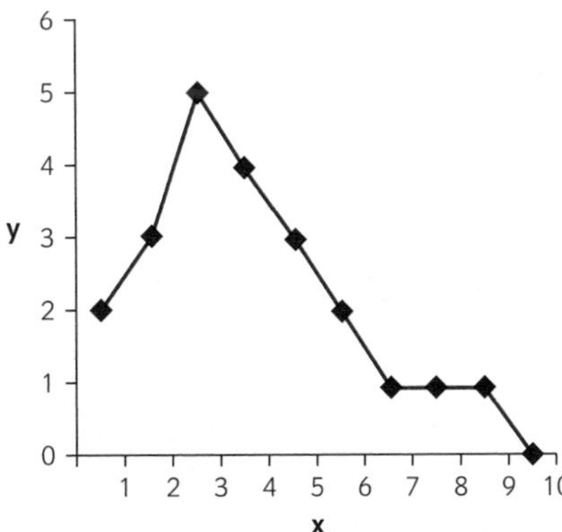

Fig. 3.8. Distribution skewed to right (positive skew).

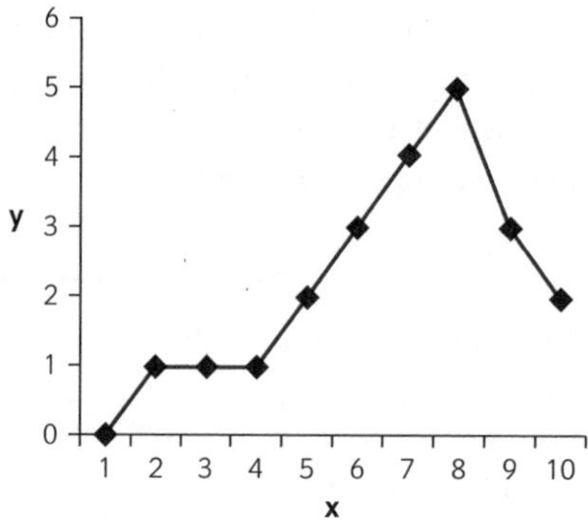

Fig. 3.9. Distribution skewed to left (negative skew).

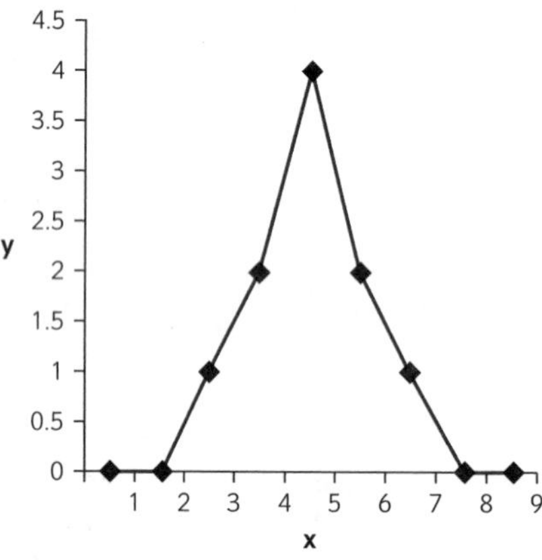

Fig. 3.10. Leptokurtic distribution.

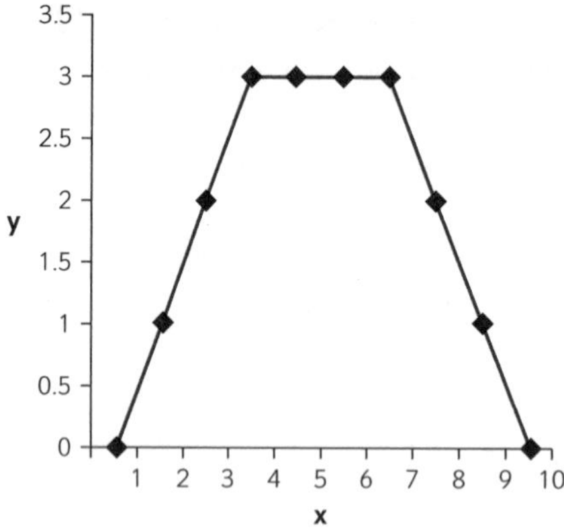

Fig. 3.11. Platykurtic distribution.

and a very flat peak is called platykurtic (Fig. 3.11.). Kurtosis can be measured as:

$$\text{kurtosis}(\gamma^4) = \frac{\sum(x - \bar{x})^4}{(n - 1)s^4}$$

Normally distributed data should have a mesokurtic appearance with a kurtosis of 3, a leptokurtic distribution will have a kurtosis of more than 3, and a platykurtic distribution will have a kurtosis value less than 3.

In Example 3.2. the skew $(\gamma^3) = 1.62$ and kurtosis $(\gamma^4) = 2.83$. This indicates that although these data can be described by all criteria outlined in BOX 3.2. as normally distributed, it does have a slight positive skew but the kurtosis indicates that the peak is mesokurtic.

These two methods for calculating skew and kurtosis assume that the mean and variance have been calculated using the equations outlined for a normal distribution. Therefore, for data with a distribution other than a normal distribution you may use these methods for indicating skew and kurtosis, having first calculated a mean and standard deviation using the normal equations. The seed dispersal around *Taxus baccata* is known to have a Poisson distribution (Table 3.10.). If the mean and standard deviation for this data are calculated using the normal equations then $\bar{x} = 2.29069$, $s = 1.12622$, $\gamma^3 = 2.79814$ and $\gamma^4 = 6.23915$. This supports the notion that this distribution is skewed to the right and is very leptokurtic.

3.5. Estimates of variation

The calculation of a measure of the central point can be used to compare different samples; however, apart from the consideration of skew and kurtosis, it does not tell you much about the shape of the distribution. For nominal data which cannot be ordered there is no appropriate method for indicating the variation in the data (Table 3.4.). For ordinal and interval data there are a number of measures that can be used to obtain an idea as to how the data is distributed around the central point. These together (the central point and estimate of variation) are used to provide a summary of your raw data.

3.5.1. Range

For ordinal and interval data where the underlying distribution is not known the range is the simplest indication of the spread of data around a central point. The range is the distance between the highest and lowest observations and is therefore calculated by subtracting the value of the lowest observation from the highest observation. For example, the range for the data illustrated in Table 3.6. for both 2004 and 2005 is 20 seeds – 13 seeds = 7 seeds per umbel. The disadvantage of this measure of variation is that it can be vastly altered by the presence of a single **outlier** and the range does not take into account the shape of the distribution between the two ends. The range as calculated does not make any allowance from sample size, yet commonly an increase in sample size will tend to increase the range.

3.5.2. Interquartile range and percentiles

The effect of outliers can be reduced by taking a range not from the smallest to the largest observation but at some specified point within the range. Most commonly the points used are the 25th centile (where 25% of the observations are smaller) and the 75th centile (where 25% of the observations are larger). This is known as the interquartile range. Alternatively, you may select another set of points, such as the 90th centile and 10th centile. These are known as percentiles. The interquartile range is the difference between the observation at the 25% point and the observation at the 75% point. In the data from Table 3.6. year 2004, the interquartile range is from 15 seeds to 19 seeds = 4 seeds per umbel. Only 50% of the observations fall within the interquartile range. Within a percentile range of 10–90%, 80% of observations are included. In both these ranges the impact of outliers is minimized, but neither system conveys much information about the shape of the distribution.

3.5.3. Variance

One of the most useful and frequently used measures of variation in data is the variance. This term may be calculated for interval data where the distribution is known. We illustrate how to calculate a variance for data with either a normal, binominal, or Poisson distribution. For other distributions you will need to find other sources of information. A variance for a sample (s^2) is one measure of the variation in the data. However, the variance is, as the symbol indicates, a squared value and the units of a variance are also squared. Therefore, it is common to take the square root of this measure of variation, the standard deviation (s), which has the same units as the original observations. If you have collected observations from every item in your population, then you should use the population equation to calculate the population variance (σ^2) and population standard deviation (σ). Unlike the mean, these calculations for the variance differ depending on whether you are working out the population parameter or the sample statistic.

i. Normal distribution

Normally distributed data can be evaluated using a particular set of statistics called parametric statistics. A common component of these statistics is the calculation of a variance and standard deviation. Therefore, we have included this calculation in a BOX for ease of cross-referencing from future chapters. The BOX is a format we use throughout the chapters on statistics and is based on two columns: one with the general information about the calculation and a second column with a specific worked example. In this case we use the data from Example 3.2., the length (mm) of the two-spot ladybird (*Adalia bipunctata*).

a. Your data are not grouped into classes The variance is calculated as the squared average deviation of an observation from the mean:

$$s^2 = \frac{\sum(x - \overline{x})^2}{n - 1}$$

It is possible to use this equation as it is. However, there is an alternative form of this equation that provides a very close approximation to the answer. This is the most commonly used equation, especially if you are working the variance out 'by hand'. We will take you through this method (BOX 3.1). Working out the variance by this method may give you a very slightly different answer to one calculated by a computer or calculator that is programmed with the other equation. Using the 'estimation' method is acceptable standard practice as the difference between answers is so small.

In many distributions, including the normal distribution and the Poisson distribution, there is a relationship between the mean and the standard deviation. For a normal distribution this is illustrated in Table 3.1. and Fig. 3.1. Eighty-seven men were measured. Most of them have an average or near average height. A few are much smaller or much taller. It is possible to use the mathematical description of the distribution to work out how many people you would expect to fall within a particular size class. To do this you use the standard deviation. The relationships between the number of people in a size class and the standard deviation can be seen in Fig. 4.1. The relationships we use most often in normally distributed data are approximately that:

68% of people should have heights in the size range $\overline{x} \pm 1s$

95% of people should have heights in the size range $\overline{x} \pm 1.96s$

99% of people should have heights in the size range $\overline{x} \pm 2.58s$

BOX 3.1. How to calculate a standard deviation and variance for normally distributed (parametric) data

GENERAL DETAILS	WORKED EXAMPLE
	Using the data from Example 3.2. the length of the two-spot ladybird (mm) (Table 3.8.).
a. First add all the observations in a sample together (Σx) and square the total $(\Sigma x)^2$. Divide this value by n: $\frac{(\Sigma x)^2}{n}$	a. $\Sigma x = 1+5+2\ldots+4+2+5 = 254$ $(\Sigma x)^2 = 64516$ $n = 50$ $\frac{(\Sigma x)^2}{n} = \frac{64516}{50} = 1290.32$
b. The next step is to square each of the observations (x^2) and add these squared values together (Σx^2).	b. $\Sigma x^2 = 1^2+5^2+2^2+\ldots+4^2+2^2+5^2$ $= 1474$
c. Subtract the value you worked out at step (a) from the value you worked out in step (b). The result is known as the **sums of squares** ($\mathrm{SS}(x)$).	c. Sums of squares $\mathrm{SS}(x) = 1474 - 1290.32 = 183.68$
d. Divide by $n-1$. This is the **sample variance** (s^2).	d. Variance $s^2 = \frac{183.68}{50-1} = 3.74857\,\mathrm{mm}^2$
e. The **sample standard deviation** (s) is the square root of the variance. $s = \sqrt{s^2}$	e. Standard deviation $s = \sqrt{3.74857} = 1.93612\,\mathrm{mm}$

This relationship is used to check if your data are normally distributed. A worked example showing this is included in BOX 3.2.

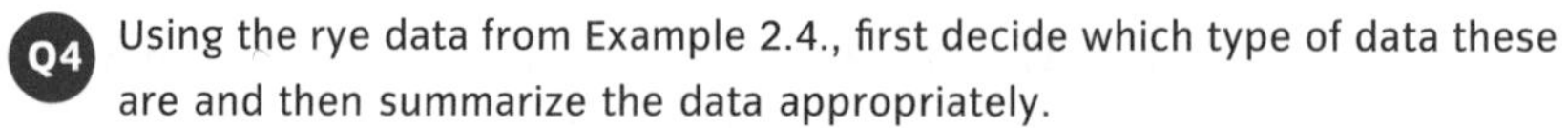

Q4 **Using the rye data from Example 2.4., first decide which type of data these are and then summarize the data appropriately.**

A4 The data are measured on an interval scale (cm), but there are few data and it will be difficult to establish the underlying distribution. The data have been organized into a frequency table in Chapter 10 (Table 10.4.). This indicates that the data are skewed and are unlikely to be normally distributed. Therefore, the data can be summarized using the median, mode, and a range.

The data can be written in numerical order:
10.0(×2) 11.0 11.5(×2) 12.0(×2) 14.0 14.5(×2) 15.5 18.0 19.5(×2) 21.0

There are 15 observations in total, therefore the median is the eighth observation which is 14.0 cm.

If the raw data are considered, there are four measures (14.5 cm, 12 cm, 11.5 cm and 10 cm) that occur twice and are therefore modes.

If the data are organized into a frequency table (Table 10.4.), the modal class is

10.0–11.9 cm, with mid-point 10.95 cm.

The range is between 10 and 20 cm = 10 cm

The interquartile range is 11.5 – 18.0 cm = 6.5 cm

b. Your data are grouped into classes In most research you will have access to the full data set, but there may be some occasions when you are only provided with data already summarized in a frequency table. In this case you may use the following to obtain an estimate of the variance and standard deviation assuming that the data are normally distributed.

The sample variance is derived by first calculating the Σx^2 from the mid-point of each class (m) and the frequency in each class (f) and where n is the total number of observations:

$$\Sigma x^2 = \frac{n\Sigma m^2 f - (\Sigma mf)^2}{n}$$

The variance is then:

$$s^2 = \frac{\Sigma x^2}{n-1}$$

When this approach is applied to the length of ladybirds (Table 3.9.), $\Sigma mf = 276.5$ (as calculated in 3.4.3.i.b.). Squaring each mid-point (m^2) and multiplying this by its frequency gives a series of values for $m^2 f$. Adding all these values together for this example, $\Sigma m^2 f = 1712.725$. Therefore, $\Sigma x^2 = 183.68$, $s^2 = 3.7486$, and $s = 1.93612$. These values for the variance and standard deviation are a very close match to those we calculated in BOX 3.1. using the same data but not grouped into classes.

Since the mean, variance, and standard deviation for normally distributed data are common components in statistics, calculators and computers invariably include software to allow these terms to be quickly calculated; it is worth becoming familiar with these tools. However, you must make sure that if you have a sample you use sample statistics not population parameters (1.7.).

ii. Binomial distribution In a binomial distribution the variance is determined by the number of observations (n) and the probability of obtaining each outcome (p, q). $\sigma^2 \approx s^2 = npq$. In our example in 3.2.2., $s^2 = 6 \times 1/2 \times 1/2 = 1.5$ 'heads'^{2}; the standard deviation (s) is the square root of the variance (s^2), which for our example is $s = 1.2247$ heads.

iii. Poisson distribution A measure of the variance in a Poisson distribution can be estimated by the following equation:

$$\sigma^2 \approx s^2 \approx \frac{\Sigma x^2 y}{\Sigma y} - \left[\frac{\Sigma yx}{\Sigma y}\right]^2$$

Using the terms $\Sigma x^2 y$, Σy and Σxy from Table 3.10. and 3.4.3.iii., the variance and standard deviation are:

$$s^2 \approx \frac{577}{99} - (197/99)^2 = 1.86858 \text{ and } s = 1.3669$$

In data with a Poisson distribution the mean should equal the variance. This is seen in Example 3.4. The estimate of the mean was 1.99 and the estimate of the variance was 1.88. These estimates are quite similar to each other. (More detailed proof of this relationship can be found in other statistical texts.) This means that if you believe you have data with a Poisson distribution you can initially confirm this by comparing the mean and variance. The definitive test is to 'fit' the Poisson equation to your data. We show you how to do this in principle in 5.1.3. and in practice in the Online Resource Centre.

3.6. Coefficient of variation

The standard deviation is the most commonly used measure of dispersion. If, however, you wish to compare variation in one set of data with that of another set of data then we use a relative (scale-less) measure of variation called the coefficient of variation (V). The coefficient of variation is especially useful when data you wish to compare are of different orders of magnitude or if they are measured on different scales. V is determined as a ratio between the mean and the standard deviation expressed as a percentage:

$$V = \frac{s}{\bar{x}} \times 100$$

The coefficient of variation can only be used for interval data with a known distribution and where a standard deviation can be calculated. In Chapter 7 we consider a survey carried out at Aberystwyth in 2002. Periwinkles from the mid and lower shore were sampled and the height of their shells (mm) recorded (Table 7.1.). The coefficient of variation for the periwinkles from the lower shore is $(1.89477/7.47) \times 100 = 25.36508\%$ and the coefficient of variation for the periwinkles on the mid shore is $(2.23011/5.46667) \times 100 = 40.079467\%$. This means that there is relatively more variation in the shell height (mm) of the periwinkles sampled from the mid shore than those sampled from the lower shore.

3.7. Confidence limits

In most research you do not collect observations from every **item** in your population but you collect observations from a sample. A useful measure, therefore, is to know how good a predictor of the population parameter is your sample statistic. We illustrate the principles behind these 'confidence limits' in relation to the means of samples, but confidence limits can be calculated for many population parameters including regression coefficients. For more details see Sokal & Rohlf (1981).

The confidence limits for a mean are derived from the standard error of the mean. The idea behind this is that for a given statistical population you may take many samples and these can all be described in terms of a mean. If you then used these means as your data you could calculate a standard deviation of these mean values, so indicating how much the means from the many samples varied. This standard deviation of the mean is more commonly known as the standard error of the mean. The term standard deviation is usually restricted to the standard deviation you calculate from the items in one sample. The standard error of the mean (SE) is:

$$\text{SE of mean} = \frac{s}{\sqrt{n}}$$

The distribution of these means will tend to be normal (even if the data in each sample is not normal), and so the standard error of the mean has the same relationship that we introduced in 3.5.3.i. in that there is a 95% probability that the population mean will fall within the range $\bar{x} \pm 1.96$ SE and there is a 99% probability that the population mean will fall within the range $\bar{x} \pm 2.58$ SE. Most often these confidence limits for means are included as vertical bars around mean values on figures (Fig. 10.4.).

If you have a small sample (less than 30) then the variation in the data due to **sampling error** will become relatively large and this can have an undue effect on the relationship between probability and the standard error of the mean. Therefore, a factor that recognizes the effect of the small sample size is included in the calculation of these confidence limits for small samples, so that the confidence limits are defined by the range $\bar{x} \pm (t \times \text{SE})$. The value for t can be found in the tables at the back of this book. We explain these tables in Chapter 4. To find a particular t value you need to choose the probability you wish to use in your confidence limits. A probability of 95% will be indicated by $p = 0.05$; a probability of 99% is indicated as $p = 0.01$. You also need to know the **degrees of freedom**. For these values of t the degrees of freedom are $n - 1$. Using your chosen p value and the degrees of freedom you can identify the value for t which you then use when calculating the confidence limits. Calculating confidence limits in this way is acceptable for interval data even from non-normal or unknown distributions.

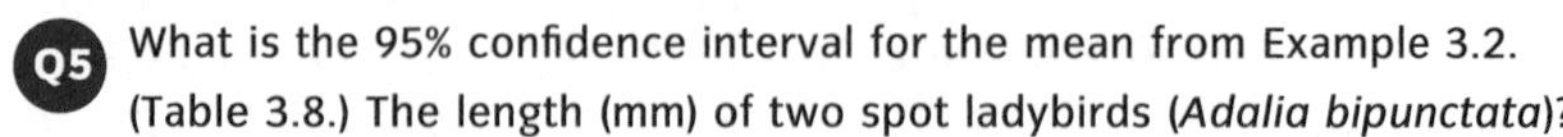

Q5 **What is the 95% confidence interval for the mean from Example 3.2. (Table 3.8.) The length (mm) of two spot ladybirds (*Adalia bipunctata*)?**

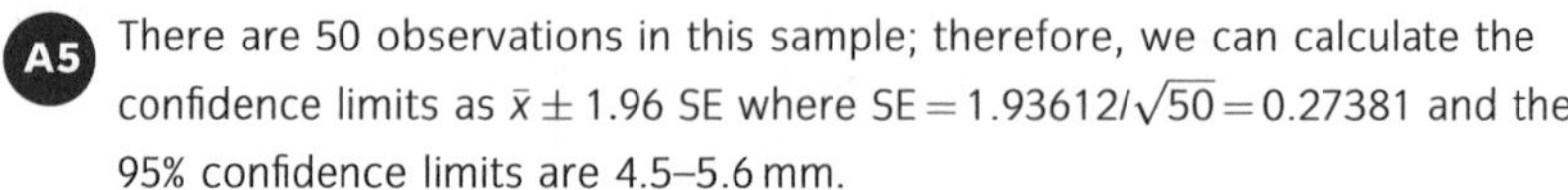

A5 There are 50 observations in this sample; therefore, we can calculate the confidence limits as $\bar{x} \pm 1.96$ SE where SE $= 1.93612/\sqrt{50} = 0.27381$ and the 95% confidence limits are 4.5–5.6 mm.

3.8. Parametric and non-parametric data

It is clear that where data have a particular distribution we are able to use this to inform how we manipulate the data. Therefore, for data with a known distribution, such as a normal distribution, we are able to calculate a mean, variance, and standard deviation. Data with a normal distribution should be symmetrical and the mean = median = mode. We have also introduced you to the relationship in normally distributed data between the number of observations in a particular region of the distribution, the probability, and the standard deviation, so that, for example, 95% of all observations should be found in the range $\bar{x} \pm 1.96s$. All this information has been utilized in a branch of statistics called parametric statistics, and normally distributed data are often called parametric data.

Parametric data are measured on an interval scale and so are quantitative, continuous, and rankable. The data have to have a normal distribution e.g. height of men (cm) (Table 3.1.). When testing hypotheses (Chapter 4) if you have parametric data you should use parametric statistics (Chapters 6 and 7).

Non-parametric data are data where either the distribution are not normal or more usually the distribution is unknown. When testing hypotheses and you have non-parametric data, you may either attempt to normalize the data by transforming them (3.9.) or use non-parametric statistics (Chapters 5, 6, and 8). Parametric data may also be analysed using non-parametric statistics, but not the other way round.

3.8.1. Are my data parametric?

Some data will have a normal distribution when plotted (3.2.) and can be described by a Gaussian equation. Data of this sort are called parametric. There is a whole branch of statistics that uses the Gaussian equation as its basis. These are parametric statistics (Chapters 6 and 7). Parametric statistics are more **powerful** and should always be used where possible. It is therefore important to be able to check if your data are normally distributed so that you can then use the parametric statistics.

We have already covered many of the key features of a normal distribution. These are summarized here for ease of cross-referencing from future chapters.

BOX 3.2. How to check if your data are normally distributed (parametric)

GENERAL DETAILS	WORKED EXAMPLE
To tell if your data are probably normally distributed and therefore if you can use parametric statistics you should check against the first four criteria at least.	Using the data from Example 3.2., Table 3.8. The length (mm) of two-spot *Adalia bipunctata* (ladybirds).
a. Are the data measured on an interval scale and therefore quantitative and continuous such as mm and grams? Data that are measured on an ordinal or nominal scale will not be parametric. Derived variables recorded on restricted scales such as percentages, where the scale of measurement only ranges between 0 and 100%, will also not be parametric. Some derived variables measured on a continuous scale such as rates (e.g. m/s) may be parametric.	a. Yes. The scale is mm, which is an interval scale.
b. Does the distribution appear to be a bell-shaped curve (e.g. Fig. 3.3.)? This criterion is often not very convincing as you may have few observations in your sample.	b. Using the frequency table (Table 3.9.), the data can be plotted as a histogram (Fig. 3.7.) There does appear to be a bell shape to the distribution.
c. Do about 68% of your observations fall within the range $\bar{x} \pm 1s$?	c. For this example the mean and standard deviation are: $\bar{x} = \frac{254}{50} = 5.08$ mm $s = 1.9361228$ mm (BOX 3.1.)

BOX 3.2. Continued

GENERAL DETAILS	WORKED EXAMPLE
	Therefore: $\bar{x} + s = 5.08 + 1.9361228 = 7.0161\,\text{mm}$ $\bar{x} - s = 5.08 - 1.9361228 = 3.1438\,\text{mm}$ If you examine Table 3.8., it is clear that 34/50 observations in this sample (68%) fall within the range 3.1–7.0 mm.
d. Does the mean = median = mode? This can be a difficult criterion to use, as it is not clear how much difference there can be between these values before they indicate that the data is not parametric.	d. The mean has already been calculated ($\bar{x} = 5.08\,\text{mm}$) Median Arrange all the values in numerical order. n = 50 so the median lies halfway between two values of 5.0 mm. Therefore the median is 5.0 mm. Mode If you examine Table 3.8. it is clear that the most frequent value is 5 mm. Therefore the mean, median, and mode are all very similar.
e. A normal distribution may be mathematically described by the Gaussian equation (3.5.). Using this equation and your own x values you can calculate the corresponding y values. These are an 'expected' data set and can be compared with your 'observed' data set by a x^2 goodness of fit test.	e. See 5.1.3. We conclude that the data are normally distributed and are therefore parametric.

In Q4. we calculated the summary statistics for data from Example 2.4. Use the first four criteria detailed in BOX 3.2. to decide whether these are parametric. If you would like to see the calculations in full, they are given in the Online Resource Centre.

a. Yes. The measures are on an interval scale.

b. No. When plotted, the data does not have a bell shape to the distribution.

c. No. The mean (14.26 cm) is approximately the same as the median (14.0 cm), but these differ from the mid-point of the modal class (10.95 cm).

d. No. Only 60% of observations fall within the range $\bar{x} \pm s$(17.9 – 10.6 cm).

Using these criteria the data (Example 2.4.) do not appear to be parametric. You could confirm this using criterion e in Box 3.2.

BOX 3.3. How to rank data for non-parametric statistics

1. If you have one sample

Sample 1: 1, 4, 5, 5, 6, 3, 4, 6, 5, 5, 7, 9

i. Arrange these values in numerical order

Sample 1: 1, 3, 4, 4, 5, 5, 5, 5, 6, 6, 7, 9

ii. Assign *possible* ranks

The first number in order is rank 1. The second number in rank order is 2, etc.

Sample 1: 1, 3, 4, 4, 5, 5, 5, 5, 6, 6, 7, 9

Ranks 1 2 3 4 5 6 7 8 9 10 11 12

iii. Where data have the same values, these observations all take the average rank.

e.g. In this set of data there are two 4s. In theory they should be ranked 3 and 4.

Sample 1: 1, 3, 4, 4, 5, 5, 5, 5, 6, 6, 7, 9

Ranks 1 2 3 4

But since the observations have the same value they should have the same rank.

The average rank for the 4s is therefore $\frac{3+4}{(2)}$ 3.5

The ranks are now

Sample 1: 1, 3, 4, 4, 5, 5, 5, 5, 6, 6, 7, 9

Ranks 1 2 3.5 3.5

This process is repeated for the four 5s and the two 6s.

$$\text{Rank for 5s} = \frac{5+6+7+8}{4} = 6.5$$

$$\text{Rank for 6s} = \frac{9+10}{2} = 9.5$$

The ranks are now

Sample 1: 1, 3, 4, 4, 5, 5, 5, 5, 6,

Ranks 1 2 3.5 3.5 6.5 6.5 6.5 6.5 9.5

Sample 6, 7 9

Ranks 9.5 11 12

2. If you have more than one sample

Many statistical tests require you to combine samples when assigning ranks.

e.g. Sample 1 1 3 5 3 2 2 4

Sample 2 1 1 4 6 7 8 7

i. Arrange all the observations in numerical order.

e.g. Sample 1 1 2 2 3 3 4 5

Sample 2 1 1 4 6 7 7 8

ii. Assign ranks. Where values are 'tied' assign the average rank.

e.g. Sample 1 1 2 2 3 3

Sample 2 1 1 4

Ranks 2 2 2 4.5 4.5 6.5 6.5 8.5

e.g. Sample 1 4 5

Sample 2 6 7 7 8

Ranks 8.5 10 11 12.5 12.5 14

iii. When the samples are then analysed they will have the following ranks:

Sample 1 1 2 2 3 3 4 5

Ranks 2 4.5 4.5 6.5 6.5 8.5 10

Sum of ranks $(\Sigma r_1) = 2 + 4.5 + 4.5 + 6.5 + 6.5 + 8.5 + 10 = 42.5$

Sample 2 1 1 4 6 7 7 8

Ranks 2 2 8.5 11 12.5 12.5 14

Sum of ranks $(\Sigma r_2) = 2 + 2 + 8.5 + 11 + 12.5 + 12.5 + 14 = 62.5$

These sum of ranks are often used in non-parametric statistical tests, e.g. Mann–Whitney U test (8.1.)

3.8.2. Ranking non-parametric data

When using parametric statistics (Chapters 6 and 7) you will need to know how to calculate a sums of squares, variance, and standard deviation, and we have explained these steps in BOX 3.1. If you are using non-parametric statistics you will usually need to know how to rank data. This is a central skill for most non-parametric statistical tests (Chapters 6 and 8). The data are organized in numerical order and each observation is given a 'ranking' depending on where in this order they fall. These 'ranking' values are then used in the calculation rather than the observations. Examine the data from Example 2.4. The growth rate of rye seedlings. These interval data can be organized into a numerical order. (You did this to answer Q6 when finding the median.) Ordinal data may also be ranked. For example, the qualitative ACFOR scale used in ecology can be ordered as shown: A – C – F – O – R. Nominal scales cannot be ranked. For example, if you graded petal colours as red, light red, pink, white, there is no reason for these colours to be organized in any particular order.

3.9. Transforming data

If your data are normally distributed (BOX 3.2.) then you are able to use the more powerful and generally more flexible parametric statistics to test your hypotheses (Chapter 4). Therefore, if you have data that are not normally distributed you should always consider transforming them. This process of transformation squeezes and/or stretches the scale you used when making your measurements so that the distribution takes on the appearance of a bell-shaped curve.

Transformations are only appropriate for ordinal or interval data. Selecting which type of transformation to use depends on first, whether you have observations for one treatment variable or more than one, and second, on the shape of the actual distribution seen in your original (untransformed) data. The presence of zeros in your data can have an undue influence in some transformations. Therefore, it is common practice, if you have zeros, to add one to all observations before transforming the data.

When transforming data you first take each observation and transform it (Table 3.11.). Next you should check to confirm that the data have been normalized (BOX 3.2.) before using the transformed values in your parametric calculations. When reporting the results from most tests of hypotheses for one variable you will not need to take further action in that the test statistic (Chapter 4) is usually independent of the original scale of measurement. You report the original untransformed data and the results from the analysis on the transformed data (Chapter 10).

Transforming two variables is not necessarily complex: however, interpreting the results from the analysis of transformed data following, for example, a regression analysis, can be complex and is outside the scope of this book. We therefore suggest that if you are considering the transformation of data where you have two treatment variables that you refer to other texts.

When you are considering transforming the data from one treatment variable the first step is to plot your data and examine the original distribution. If the data are not normally distributed then the distribution will have one or more of the following features: the data are very spread out and the distribution may be platykurtic (Fig. 3.8.); there is little variation and the distribution may appear to be leptokurtic (Fig. 3.9.); the distribution is skewed to the right (Fig. 3.10.); the distribution is skewed to the left (Fig. 3.11.). Table 3.11. indicates which transformation is most likely to be effective at normalizing your data. Most statistical calculators and computer software will carry out these transformations.

Table 3.11. How do I transform my data: one variable?

Type of distribution	Other features	Transformation
There is a lot of variation in the data so that the distribution appears to be spread out and may appear to be platykurtic.	The original measurements are on a scale where the values are greater than 10.	Take the $\log_{10}$ of each observation.
	Some of the original values are 0.	Add 1 to each observation and then take the $\log_{10}$ of each observation.
The variation is limited and the distribution is therefore narrow and may appear to be leptokurtic.	None of the original measurements are 0.	Take the square root of each observation.
	Some of the original values are 0.	Add 1 to each observation and then take the square root of each observation.
	The observations are recorded as percentages (%) calculated from counts with a common denominator, and most values lie between 0 and 20%.	Take the square root of each observation.
	The observations are recorded as percentages (%) calculated from counts with a common denominator and most values lie between 80 and 100%.	Subtract each observation from 100, and then take the square root of each of these values.
	Where the observations are recorded as % and there is a wide spread of values.	Divide each value by 100. Take the square root of each observation. Then find the angle whose sine equals each value or 'inverse sine' ($\sin^{-1}$).
	Where data is recorded as a proportion.	Take the square root of each observation. Then find the angle whose sine equals each value or 'inverse sine' ($\sin^{-1}$).

Table 3.11. Continued

Type of distribution	Other features	Transformation
The distribution is skewed to the right (positive)	There are a number of suitable transformations. The one to use depends on the degree of skewness. The following is in the order of little skew (try a) to considerable skew (try c or d). If you have zeros in your original data you should add 1 to each observation before transforming the data.	a. Take the square root of each observation. b. Take the natural logarithm (ln) of each observation. c. Divide 1 by each observation (reciprocal, 1/x). d. Take the square root of an observation and then divide 1 by this value (reciprocal square root $(1/\sqrt{x})$.
The distribution is skewed to the left (negative).	Again, there are a number of transformations which may be used depending on the degree of skew from least (try a.) to considerable (try b., etc.).	a. Square each observation. b. Cube each observation, etc.

Summary of Chapter 3

- When designing an investigation or choosing a statistical test you will need to be familiar with the terms used to describe scales of measurement. These terms are quantitative, qualitative, discrete, continuous, rankable, nominal, ordinal, interval, and derived variable (3.1.).
- Observations have a particular shape or distribution when plotted. Some of these distributions can be described mathematically including the normal distribution, Poisson distribution, and binomial distribution (3.2.).
- Distributions may be described in terms of the central point (mean, median, mode, skew, and kurtosis) and spread of data around the central point (range, interquartile range, percentile, variance, standard deviation). The methods used depend on whether the data are measured on a nominal, ordinal or, interval scale (3.3., 3.4., 3.5., 3.7.).
- The coefficient of variation can be used to compare the relative amount of variation in different data (3.6.).
- Data may be normally distributed and these data are also known as parametric data. If data are not normally distributed or the distribution is unknown then the data are non-parametric (3.8. and 3.9. and developed in Chapters 4–8).
- Non-parametric data may be normalized by transforming them. The choice of transformation depends on the original shape of the distribution and on how many variables need to be transformed (3.10.).
- When using parametric statistics you often need to know how to calculate the sums of squares, variance, and standard deviation (BOX 3.1.). If you are using non-parametric statistics you will need to know how to rank data (BOX 3.3.)
- The Online Resource Centre includes interactive exercises that test your understanding of this chapter with other topics, particularly those considered in Chapters 4–8.

4

An introduction to hypothesis testing

In this book so far we have thought about general principles relating to the design of investigations (Chapters 1 and 2) and introduced you to the ways in which you might start to deal with raw data (Chapter 3). What we have not done is shown you how to answer the questions your experiment may have set out to examine. This is what hypothesis testing is all about. There are generally five steps that you follow in all hypothesis testing:

1. What are the hypotheses to be tested?
2. Work out the calculated value of the test statistic.
3. Find the critical value of the test statistic.
4. The rule.
5. What does this mean in real terms?

These are the headings we use throughout our worked examples for all statistical tests in this book. We explain the underlying principles for each of these steps here and more examples are included in the Online Resource Centre. If you work through this chapter it should only take 1 hour to read and complete all the exercises.

4.1. What are the hypotheses to be tested?

In this chapter we are thinking about experiments that are designed to answer a specific question; for example, is there a difference in mineral content between organic and non-organic potatoes? In the flow chart overleaf you can see the two types of investigations we introduced in Chapter 1, usually only one of which (experiments) involves testing hypotheses.

Some investigations, often those producing baseline data, or a preliminary investigation, will not set out to test a hypothesis. For example, a survey of the flora and fauna of a wood is not 'asking' a specific

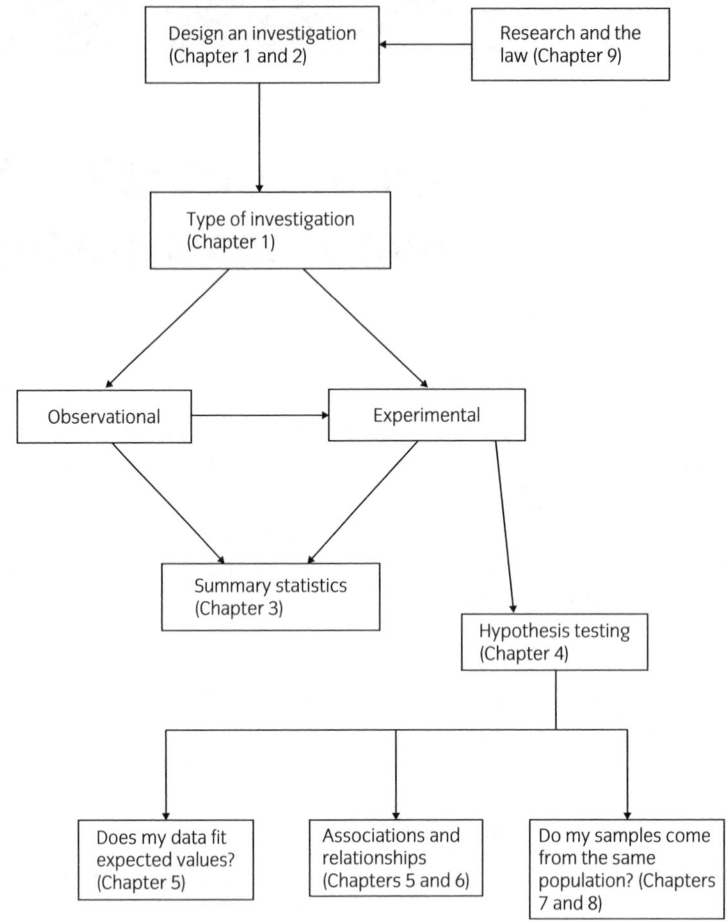

question: you will be finding out what is there. This type of investigation is observational and you are not likely to require the statistical tests used in hypothesis testing (Chapters 5, 6, 7, 8). If you are carrying out an experiment and do wish to test a hypothesis your first step is to decide which type of hypothesis is appropriate.

4.1.1. Types of hypotheses

In experimental investigations you will set out to ask a question. There are three types of questions you may ask in your experiment:

- Do the data fit expected values?
- Is there an association between two or more variables?
- Do samples come from the same or different populations?

i. Do the data fit expected values?

There are some occasions when we have reasons to make a mathematical prediction about our observations and we wish to compare our observations with our expectations. This can arise when:

- You can reasonably argue that all samples or observations should have the same value. For example, if you sampled the numbers of beetles falling into different-coloured pitfall traps, you might expect that there should be the same number of beetles collected in each trap.
- You have carried out a genetic cross or sampled in a population and have reason to expect a particular segregation ratio, population genetic ratio, or sex ratio, etc. For example, you may use the Hardy–Weinberg theorem to predict the allele frequencies in the F1 generation of a population of *Drosophila melanogaster* exposed to a known selection pressure.
- You wish to confirm that your data have a particular distribution, such as a normal or a Poisson distribution (3.2.).

The term expectation can be used in two ways, only one of which is correct. You may have a personal hunch ('expectation') about the likely outcome but this is not a mathematical prediction, also called an 'expectation'. When we talk about expectations it is the second of these that we mean.

Chapter 5 introduces you to the chi-squared tests and the G tests which can be used to test this type of hypothesis.

ii. Is there an association between two or more variables?

Two types of analyses may be carried out in relation to a possible association. These are:

a. The presence or absence of an association For example: Is there an association between blood group and eye colour in people living in north Wales? Is there an association between temperature and the growth rate of bacterial colonies in liquid culture?

When you wish to test for the presence of an association there are two groups of tests that may be useful: a chi-squared test or a G test (Chapter 5) and a correlation or regression (Chapter 6). There is some overlap between these two groups of tests, so you should look at both chapters before deciding which test is the most appropriate.

b. The nature of the association Some statistical analyses also allow an association to be described in a mathematical way: for example, in fitting a straight line to data and confirming the linear relationship between flower number and bulbil number in *Allium ampeloprasum* var. *porrum* (leek).

If you wish to examine the nature of an association in this way then go to Chapter 6 and look at regression analyses.

iii. Do samples come from the same or different populations?

The word '**population**' used here refers to a 'statistical' population. Usually you will not collect data from all **items** in the population. Instead, the population will be represented by a **sample**. Many experiments are designed so that you can compare samples from two or more populations. In some cases these will be samples from different ecological or genetic populations (1.5.). In other cases you may wish to compare different **treatments**. For example, you may wish to examine the effectiveness of different bactericide treatments on a number of bacteria. Or you may wish to compare phosphate concentrations in water samples collected above and below an outfall.

There are many tests that address this type of hypotheses. Your choice depends on the numbers of observations in each sample, the number of samples, whether the data are **parametric** or not, and the simplicity or complexity of the experimental design. Examples are shown below.

General criteria	**Parametric**	**Non-parametric**
One **variable** Two samples No **replicates**	Chapter 7 *t* test *z* test	Chapter 8 Mann–Whitney U test Chapter 7 (Sometimes the *z* test)
One variable Two samples No **replicates** **Matched** pairs of data	Chapter 7 *t* test and *z* test for matched pairs	Chapter 8 Wilcoxon's matched pairs test
One or more variables One or more samples With or without replicates	Chapter 7 Parametric analysis of variance (ANOVA) and Tukey's test	Chapter 8 Non-parametric analysis of variance (ANOVA), Kruskal–Wallis test, Sheirer–Ray–Hare test Chapter 5 Chi-squared test for heterogeneity

To help you choose which of these tests is appropriate you should check the details in the relevant chapters. We have also included examples of choosing statistical tests and how these inform your experimental design in Chapter 2 (2.2.8. and 2.3.5.), with further examples in the Online Resource Centre and a review in Appendix b.

Q1 **If you designed an experiment to find out if there is a difference in mineral content between organic and non-organic potatoes, which type of hypothesis would you be testing?**

A1 You do not start out with any expectation nor are you looking for an association but you do wish to compare two treatments (organic and non-organic). Therefore, you will be testing hypotheses iii.

Hypotheses may be either general or specific. For example, a general hypothesis might be 'there is no difference between the means of samples 1, 2, and 3'. Clearly any outcome from the hypothesis testing may tell you if there is probably a difference, but will not tell you anything more about that difference. Some hypothesis testing is more specific. For example, when using a chi-squared or G goodness of fit test (Chapter 5) you compare your observations to a specific ratio determined by biological principles, such as Mendelian laws of segregation or the Gaussian equation describing a normal distribution. The test of hypotheses here is more specific, asking if your observed values are significantly different from this *a priori* ratio. Similarly, in simple linear regression (Chapter 6) you may test the significance of an association between two variables by comparing the observed data with that predicted by a regression equation. In Chapters 7 and 8 there are statistical tests that allow you to make other specific comparisons. The Tukey's test allows you to compare pairs of means, in experiments with more than two samples, enabling you to gain a more detailed insight into the trend in your data. A similar test is considered in Chapter 8 for non-parametric data.

4.1.2. How to write hypotheses

A hypothesis is the formal phrasing of your objective (1.3.3.). You may have general or specific hypotheses. A hypothesis when written correctly should have a number of features, as the following checklist shows.

Checklist for writing hypotheses

a. Hypotheses come in pairs, the null hypothesis (H_0) and an alternate hypothesis (H_1).
The null hypothesis (H_0) is the negative hypothesis. The null hypothesis assumes that any variation in the data is due to sampling error. We explain what is meant by **sampling error** in 4.1.3. The alternate hypothesis (H_1) is worded in such a way that if H_0 is true then H_1 must be false. The alternate hypothesis indicates that the variation in the data is probably due to the factor(s) you are investigating in your experiment.

b. For a general hypothesis the phrasing of a hypothesis indicates the type of hypothesis being tested, i.e. 'expectation', 'association', or 'difference', and/or uses a term relating to the test itself, e.g. heterogeneity, correlation. A specific hypothesis includes details of the particular trend you believe is present in the data.

c. The hypothesis being tested refers to the statistical population under investigation rather than the sample (1.5, 1.6.).

d. Hypotheses usually indicate the statistic being examined. For example, most parametric statistical tests of hypotheses are testing the **means** of the data. When **non-parametric** tests of hypotheses are used, usually they are comparing the **medians**.

e. Hypotheses include details of the experiment, such as the numbers of samples or items, the variable(s), the species being studied, etc.

EXAMPLE 4.1. Germination of *Allium schoenoprasum* on three soil types

A researcher was interested in the effect of soil types on seed germination in the chive (*Allium schoenoprasum*). She sowed 100 seeds on each of three soil types and after 2 weeks recorded the number of seeds that had germinated. She expected that the number of seeds germinating on the three soils would be the same. In this instance the hypotheses would be written as:

> **H_0:** There is no difference in the median numbers of seed from *Allium schoenoprasum* germinating on the three soil types compared with the expected value.
>
> **H_1:** There is a difference in the median numbers of seed of *Allium schoenoprasum* germinating on the three soil types compared with the expected value.

We can compare these hypotheses with our checklist.

Checklist	Our example
a. Hypotheses come in pairs: the null hypothesis (H_0) and an alternate hypothesis (H_1).	Yes, there is a null and alternate hypothesis. The null hypothesis (H_0) is the negative hypothesis: 'There is *no* difference . . .'. The alternate hypothesis (H_1) is worded in such a way that if H_0 is true then H_1 must be false: 'There *is* a difference . . .'.
b. The phrasing of a hypothesis indicates the type of hypothesis being tested, i.e. 'expectation', 'association', or 'difference', and/or uses a term relating to the test itself, e.g. heterogeneity, correlation.	H_0: . . . 'compared with the *expected value*.'
c. The hypothesis being tested refers to the statistical population under investigation rather than the sample.	In this example the statistical population is seed produced by *Allium schoenoprasum*.
d. Hypotheses usually indicate the statistic being examined.	'. . . difference in *the median numbers* . . .'
e. Hypotheses include details of the experiment.	'. . . *seed* of *Allium schoenoprasum germinating on the three soil types*'.

Q2 **EXAMPLE 5.1. The distribution of holly leaf miners on *Ilex aquifolia***
As part of a student project the number of holly leaf miners was recorded at three heights on a holly tree. The student assumed that the holly leaf miners were distributed at random and so should be found in equal numbers in each part of the tree.

Write suitable hypotheses for this investigation.

A2 This is an example that we use later in the book. The hypotheses are given in BOX 5.1. Were you correct? If not, what did you miss out?

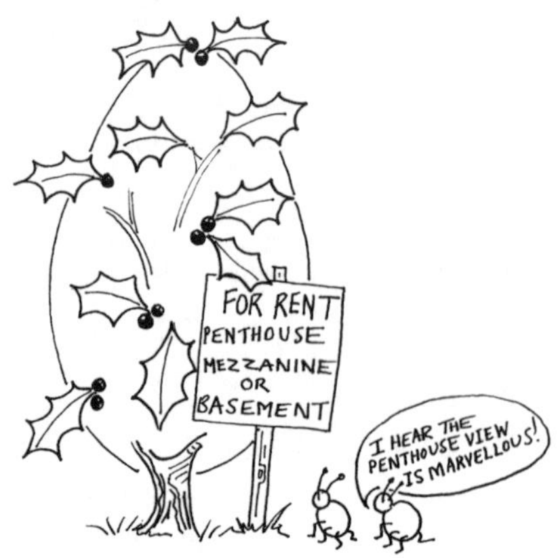

4.1.3. Why hypotheses come in pairs

To understand the process of hypothesis testing you need to understand what is meant by the term *sampling error*. When you record a series of observations these will vary. Sampling error is a measure of the variation in your data that is due either to chance and/or to **non-treatment variables** (1.9.).

i. Sampling error and chance

One cause of sampling error in your data is 'chance'. We can illustrate this best with an example.

EXAMPLE 4.2. Sampling error in a study of *Cepea nemoralis* in Jack's Wood

As part of a study of the ecological genetics of *Cepea nemoralis*, the shell patterns of the snails in a wood were recorded. First, every snail in the area (i.e. the whole population) was examined and their shell patterns recorded. Two days later three samples were collected at random and their shell patterns were also recorded (Table 4.1.).

Table 4.1. Percentage of *Cepea nemoralis* with particular shell patterns in Jack's Wood

	% *Cepea nemoralis* with a particular shell pattern and colour			
	Banded Yellow	**No bands Yellow**	**Banded Pink**	**No bands Pink**
Sample 1	90	0	5	5
Sample 2	20	5	30	45
Sample 3	0	50	50	0
Population	30	10	30	30

Clearly the samples differ from the population results and from each other. This variation is said to be due to sampling error. By chance, the samples collected did not reflect the actual population frequencies of shell pattern in this species. Two important effects of sampling error are first, that the samples may not reflect the statistical population values, and second, that the sampling error can lead to variation between samples by chance.

ii. Sampling error and experiments

Sampling error can also occur in your data as a result of factors that are not part of your experiment (non-treatment variables), but which also affect the characteristic you are measuring.

In Chapter 2 (2.2.) we designed an experiment to allow us to examine the maximum distance a seed was dispersed in a wind tunnel in relation to wind speed. We identified a number of non-treatment variables (2.2.6), including the effects of digging up the plants, the maturity of the seeds and inflorescence, the height of the inflorescence, the temperature and humidity of the wind tunnel, and the location of the plants during testing. All of these factors could influence how far the seed were dispersed. For example, when the plants were dug up: this may affect the plants turgor, which in turn might affect the adherence of the seed to the inflorescence; if the seeds were not mature their weight may vary, and the pappus might not be so effective in providing lift, etc. All these factors can influence the distance a seed might travel and therefore introduce a level of variation into the experiment that has nothing to do with wind speed. These all contribute to the sampling error, which is why you need to take steps when designing your investigation to minimize and/or equalize their impact on your observations (2.2.6).

iii. What does the term 'sampling error' mean?

The term 'sampling error' can mean one of three things:

- the variation between samples collected from a single population that has occurred by chance
- the variation between the population value and sample value that has arisen by chance
- some variation in the experimental observations that may be due to factors not being investigated by the researcher.

When you collect your data the variation between the observations will be due in part to one or more of these causes of sampling error. So in an experiment you have variation in your data due to the factor(s) you are investigating (treatments) and due to sampling error.

iv. Sampling error and hypotheses

This explains one aspect of hypotheses. They come in pairs: a null hypothesis and the alternate hypothesis. The null hypothesis describes the variation seen in the data in terms of sampling error. The alternate hypothesis describes the variation seen in your data in terms of the variable(s) you are investigating.

EXAMPLE 4.3. Attitudes to human cloning

An undergraduate investigated attitudes of students to different forms of human cloning. One question asked students taking a Biology degree and students taking a Health Studies degree if they agreed with somatic cell human cloning. Her hypotheses were:

H_0: There is no significant association between the median numbers of Biology and Health Studies students in their response to this question.

H_1: There is a significant association between the median numbers of Biology and Health Studies students in their response to this question.

In this example the factor under investigation is the education of the students. Are the students taking one course likely to give a different response compared with students on the other course? However, any variation in the data may have arisen due to sampling error. Perhaps the students on the Biology course all happen to be mature students and this influences their opinion. This is a non-treatment variable and therefore a contributor to the sampling error.

v. Hypothesis testing

Hypothesis testing allows us to examine our data and to use them to choose one of the two hypotheses: is the variation in our data due to sampling error (H_0) or is the variation in our data due to the factor(s) we are investigating (treatment) (H_1).

EXAMPLE 4.4. The effect of far-red light on the germination of tobacco seeds

An undergraduate exposed tobacco seed samples to a range of exposures of far-red light and recorded percentage germination after 2 weeks. His hypotheses were:

H_0: There is no significant difference between the median percentage germination of tobacco seeds in response to different times of exposure to far-red light.

H_1: There is a significant difference between the median percentage germination of tobacco seeds in response to different times of exposure to far-red light.

In this example the factor under investigation is the time that the seeds were exposed to far-red light. Variation in the percentage germination could be due to this treatment, but it could also be due to a number of other factors, such as inconsistent watering, variation in temperature during the experiment, age differences between the seeds in each sample, etc. In this experiment these factors are not under investigation and so their effects are all part of the sampling error.

By testing our hypotheses we will be able to indicate whether the null hypothesis (H_0) is most likely to be correct, i.e. we 'do not reject the null hypothesis', or we 'do reject the null hypothesis' in which case it is more likely that the alternate hypothesis (H_1) is correct and the difference between the samples is due to the treatment. This is done by calculating a test statistic and comparing it with a critical value. A rule for each statistical test then tells you whether to reject or not to reject the null hypothesis, depending on whether the calculated test statistic is larger than the critical value. We explain these stages in the rest of this chapter.

4.2. Working out the calculated value of the test statistic

4.2.1. What is a test statistic?

In hypothesis testing, your first step is to check your data against the criteria given for each statistical test. In doing this you are choosing a particular mathematical distribution to use in your hypothesis testing. Each distribution has certain characteristics; for example, a normal distribution can be described in terms of the mean and variance (3.3.– 3.5.). The mathematical characteristics of the distribution you choose to use in your hypothesis testing then determine the calculated and critical values of your test statistic.

The test statistic is usually given a letter or symbol that relates to the name of the statistical test you are using. If you were using a Mann–Whitney U test (Chapter 8) the test statistic is *U*. If you are using a chi-squared test then the test statistic is χ^2 (Chapter 5).

4.2.2. What is a calculated value?

A calculated value for a test statistic is a single number that provides a 'summary' of all your observations and is derived from the mathematical relationships of a particular chosen distribution.

In Example 4.1. the hypotheses were:

H_0: There is no significant difference in the median percentage germination of the *Allium schoenoprasum* seed on the three soil types.

H_1: There is a significant difference in the median percentage germination of the *Allium schoenoprasum* seed on the three soil types.

The null hypothesis states that there is 'no difference'. In other words, there should be little or no variation between the percentage germination of the seed sown on the three soil types other than that due to sampling error. If we work out how much the three samples differ from each other it may be very little, in which case we can probably agree with our null hypothesis. If the three samples are very different from each other then we can probably say that the null hypothesis should be rejected. But how much variation between our three samples do we need before we can say that we should probably reject our null hypothesis?

If you look at any statistical test in Chapters 5–8, you can see that each has a set of criteria. For example, if you wish to use a chi-squared goodness of fit test (5.1.1.) you:

1. Wish to compare your observed values with those predicted by an *a priori* expectation.
2. Have one treatment variable.
3. Have only one sample.
4. Have data that fall into more than two categories.
5. Have data that are counts or frequencies, not percentages or proportions.
6. Have observations that are **independent**.

If the data from your investigation satisfy the criteria, then the data may be examined in relation to a particular distribution. Distributions include *U*, *F*, chi-squared, etc. Having identified which distribution can be used to examine your data, you then use your observations in a series of mathematical steps to generate a *calculated* test statistic, e.g. chi-squared$_{calculated}$. This reflects the amount of variation within your observations in terms of a single value that in itself is drawn from a particular distribution.

There are many examples in subsequent chapters (Chapters 5–8) in which we show you how to work out a calculated value. It will help your understanding if you look briefly at several of these now (for example BOX 6.1. and BOX 7.1.) before you continue reading about hypothesis testing.

4.3. Finding the critical value of the test statistic

To find a critical value you first need to understand what a critical value is (4.3.1.). Most critical values are calculated for you, and there are many possibilities. If the values are not calculated you will need to interpolate your value from those provided (4.3.6.). If the value is provided, to find the correct critical value you need to select a p value (4.3.2.) and to know whether you are testing two tails or one tail of the distribution (4.3.3.). You may also need other values, such as the degrees of freedom (4.3.4.).

4.3.1. What is the critical value?

The critical value is also found from the mathematical relationships that are derived from the distribution you have chosen to use to test your hypotheses. It is the boundary mark against which you compare the calculated value. In most statistical tests, if your calculated value is greater than your critical value then you may reject the null hypothesis (e.g. F tests, t test, chi-squared test). In the Mann–Whitney U test (8.1.), if the calculated value is smaller than the critical value then you may reject the null hypothesis. Critical values have been calculated for you from the distribution you are using to test your hypotheses. Statistical tables with these critical values are included at the end of this book (Appendix d).

There is a plethora of critical values derived from any particular test distribution. You have to choose one of these critical values and to do this you will usually need a probability value (p), a measure of sample size, which is most often in terms of the degrees of freedom, and you will need to decide if you wish to use a one- or two-tailed test. We explain these three criteria in this section.

4.3.2. *p* values

Look at the critical values in the statistical table at the end of the book for the Spearman's rank test and at the excerpt on p. 96.

The first column is n, the number of pairs of observations. The remaining columns are titled 'probability' and there are several values. What is this probability all about?

Excerpt from the critical values for a Spearman rank test

	Critical values for a two-tailed* test for probability (p) values 0.10–0.01			
	0.10	0.05	0.02	0.01
n	Critical values for a one-tailed* test for probability (p) values 0.05–0.005			
	0.05	0.025	0.01	0.005
8	0.643	0.738	0.833	0.881
9	0.600	0.683	0.783	0.833
10 etc.	0.564	0.648	0.745	0.794

*The terms one-tailed and two-tailed are explained in 4.3.3.

i. Distributions and p *values*

As an example let us first consider one large sample of 1000 observations measured on a continuous scale with units from 0 to 30. The frequency distribution of this sample is shown in Fig. 4.1. These data have a normal distribution and can therefore be summarized using the mean and standard deviation (BOX 3.2.).

In this first sample you can see that most observations are central values. There are far fewer observations in the more extreme ends (e.g. 0–5 units and 25–30 units).

In fact since these data are normally distributed we would expect that about 68% of observations would have a value that falls between the mean ($\bar{x}$) + one standard deviation (s) to the mean ($\bar{x}$) − one standard deviation (s) (Table 4.2., rows a and b). We have looked at this in 3.8.

This relationship can be turned on its head and you could instead say that you would expect to find 32% of your observations being values that fall into one or both of the two, more extreme, ends of the distribution, i.e. the tails (Fig. 4.1.). In this example you would expect to find half of the 32% of observations in the tail at the left-hand end of the distribution and half of the 32% of observations in the right-hand tail. So each tail would have 16% of your observations and the central part would have 68% of your observations (Table 4.2., rows c and d).

You can rephrase this in yet another way. If you had 100 observations and you took one at random, then you would expect that the probability of this one observation having a value that falls in the range mean ± one standard deviation is 68/100 or $p = 0.68$; the probability that this one observation will have a value that falls outside the range mean ± one

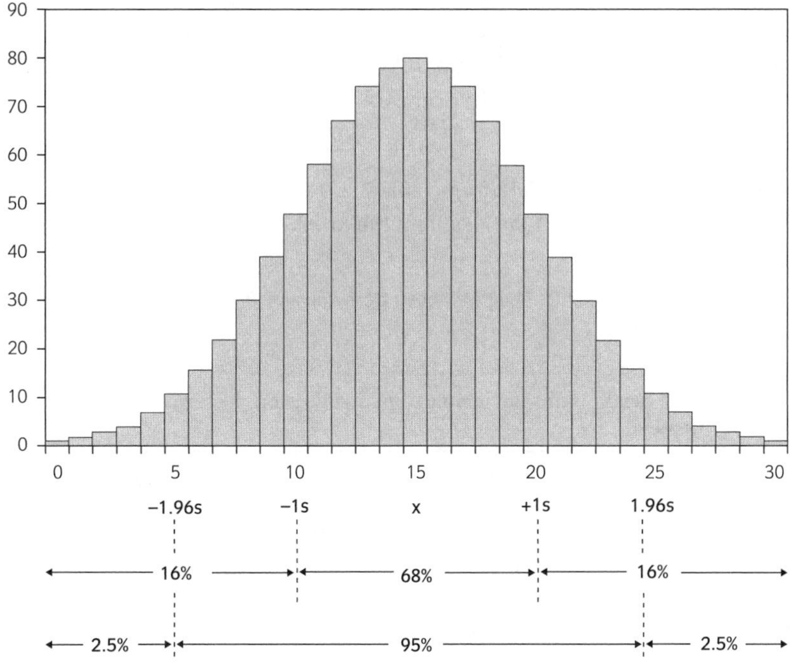

Fig. 4.1. Normal distribution, $n = 1000, \bar{x} = 15, s = 5$. The coloured dotted lines demarcate the number of observations that would be expected to fall into the centre of the distribution and two tails, when the range is defined by $\bar{x} \pm s$. The black dotted lines demarcate the number of observations that would be expected to fall into the centre of the distribution and two tails, when the range is defined by $\bar{x} \pm 1.96s$.

Table 4.2. Attributes of a normal distribution in relation to the sizes of the tails from $\bar{x} \pm 1.00s$ to $\bar{x} \pm 3.29s$

	Size of tails			
a. Size of the tails are determined by . . .	$\bar{x} \pm 1.00s$	$\bar{x} \pm 1.96s$	$\bar{x} \pm 2.58s$	$\bar{x} \pm 3.29s$
b. Number of observations in central range	68	95	99	99.9
c. Number of observations (%) in the two tails combined	32	5	1	0.1
d. Number of observations in each tail (%)	16	2.5	0.5	0.05
e. Probability of any observation falling within the central zone	$p = 0.68$	$p = 0.95$	$p = 0.99$	$p = 0.999$
f. Probability of any observation falling in either one of the two tails	$p = 0.32$	$p = 0.05$	$p = 0.01$	$p = 0.001$
f. Number of times on average where you might expect to be wrong	1/3	1/20	1/100	1/1000

standard deviation is 32/100 or $p = 0.32$ (Table 4.2., rows d–f). We are now thinking about probabilities. Hypothesis testing uses this idea to allow you to select the level of probability at which you choose whether to reject or not to reject the null hypothesis.

Examine some of the data from Table 7.2. The mean ($\overline{x}$) and standard deviation (s) of the periwinkles from the lower shore are $\overline{x}_1 = 5.89231$ and $s_1 = 1.4186$; $n = 13$.

Extract from **Table 7.2.** Shell height in periwinkles from the lower shore at Porthcawl, 2002

Shell height (mm) for sample of periwinkles on the lower shore ($n = 13$)			
4.0	5.0	6.0	8.4
4.8	5.0	6.2	
5.0	5.5	7.7	
5.0	5.6	8.4	

a. How many observations fall within the range $\overline{x} \pm s$, i.e. 4.5–7.3 mm?

b. How many observations fall outside this range?
Now divide the data so that 95% of your observations are included in the central block and only 5% of the observations are included in the tails.

c. How many observations would be included in the central block and how many would be in the tails?

d. What is the probability of the next periwinkle you measured also falling within these tails?

a. 9/13.

b. 4/13 (4.0, 7.7, 8.4, 8.4)

c. 12, 1 ($13 \times 95/100 = 12.35 \approx 12$, $13 \times 5/100 = 0.65 \approx 1$)

d. $p = 0.05$

ii. Hypotheses and p *values*

We have described the use of p values for a single set of data in 4.3.2.(i). (Fig. 4.1.). What then if we had a second sample and we wished to test the hypothesis that both samples came from the same population, i.e. there is no difference between the population means for the two samples? If we

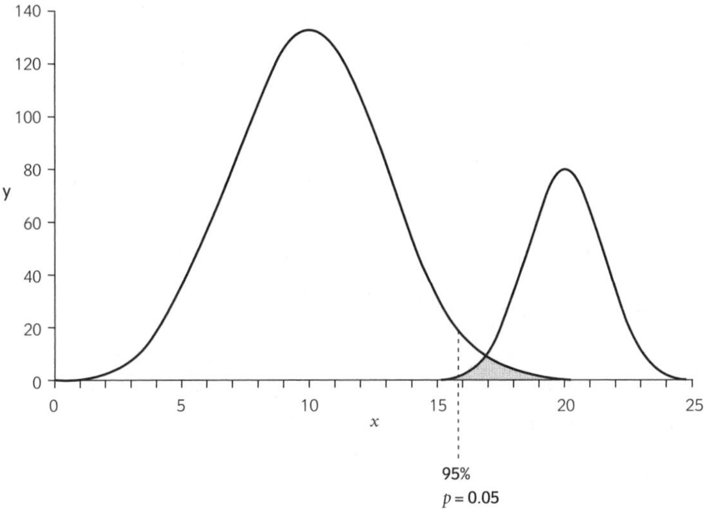

Fig. 4.2. Two samples with a normal distribution ($n = 1000$, $\bar{x} = 10$, $s = 3$ and $n = 300$, $\bar{x} = 20$, $s = 1.5$). The two distributions overlap in the upper tail of the larger distribution.

plotted the data from our second sample on top of the first sample (Fig. 4.2.) we would expect there to be an overlap if they came from the same population. We would need to make some allowance for sampling error, so we would not expect a perfect match but we would expect a reasonable match. If the samples were from different populations then the overlap between them would be less. If the two samples were very different then you would reach a point where the observations in sample 2 would only fall into one of the extreme ends of the first samples distribution. The point at which this lack of overlap is considered to be sufficiently significant as to indicate that the samples do not come from the same population is given in terms of *p* values.

iii. Which p *value?*

You have therefore to decide where the cut-off point will be. How small an overlap should there be between samples for you to reject the null hypothesis and to decide that the two samples probably come from different populations? Usually we use a *p* value of 0.05 or less. A probability of $p = 0.05$ is an arbitrary cut-off point which has been adopted in biological sciences. In other disciplines, such as medical science, a lower *p* value may be used. With a *p* value of 0.05 we will expect that 1/20 times the decision we come to will be incorrect (Table 4.2., row f). Making wrong decisions in hypothesis testing in this way is called a Type I error. We explain more about Type I and Type II errors in the glossary under 'Power of a test'.

If when testing your hypotheses you find that you do reject the null hypothesis at $p = 0.05$, you should see how small a *p* value you can adopt and still reject the null hypothesis. For example, if you can reject the null hypothesis at $p = 0.001$ then you can expect your decision to be probably incorrect only 1/1000 times (Table 4.2., row f). At this level of stringency you could carry out 1000 experiments and expect to probably only draw the wrong conclusion in one of them. Therefore, making a decision in hypothesis testing at a low p value increases the likelihood that the decision is correct (e.g. see BOX 5.1. and BOX 6.1). You may think then that a p value of 0 is the ideal. However, no hypothesis testing will set $p = 0.0$ as a cut-off point for rejecting or not rejecting the null hypothesis, because our experimental systems are never without some inaccuracies and you can never be absolutely sure that the decision you have made is correct. Where you do use a p value of less than 0.05 you may wish to indicate this either by giving the exact p value, such as $p = 0.01$, or using the 'less than' symbol $<$, i.e. $p < 0.05$. More details about such symbols are given in Appendix c.

iv. Critical values and p

The process of hypothesis testing starts then with you choosing a particular distribution e.g. chi-squared, F, t, z, etc. This distribution will have particular mathematical features that you use to summarize all your observations and determine a calculated value. The distribution you have chosen is then used to determine the critical value. This value is also dependent on the p value you have chosen for your decision-making. Therefore, there is a series of critical values for any one distribution for a range of p values. These are given in the statistical tables at the end of the book (Appendix d). You should start with the critical value for $p = 0.05$. We have examples of choosing critical values throughout Chapters 5–8 (e.g. BOX 5.1.)

4.3.3. **One-tailed and two-tailed tests**

In the table of critical values for the Spearman's rank test you will see the terms 'one-tailed' and 'two-tailed' tests. These terms are most easily explained in terms of a normal distribution, but apply to any statistics that assume any underlying distribution.

If you were carrying out hypothesis testing in which you were comparing two or more samples, the statistical test will provide an indication of the overlap between the samples and allow you to determine if there is sufficient overlap between the samples to indicate that they come from the same statistical population (Fig. 4.2).

You will remember that there is a mathematical relationship between the mean, standard deviation, and number of observations (4.3.1.) for a normal distribution so that 95% of your observations should fall within the range of the mean $+1.96s$ (Table 4.2). This means that 5% of your observations will fall outside this range. These extreme values may either fall evenly, 2.5% at each end (Fig. 4.3.), or all 5% in one or other of the tails only (Fig. 4.4.).

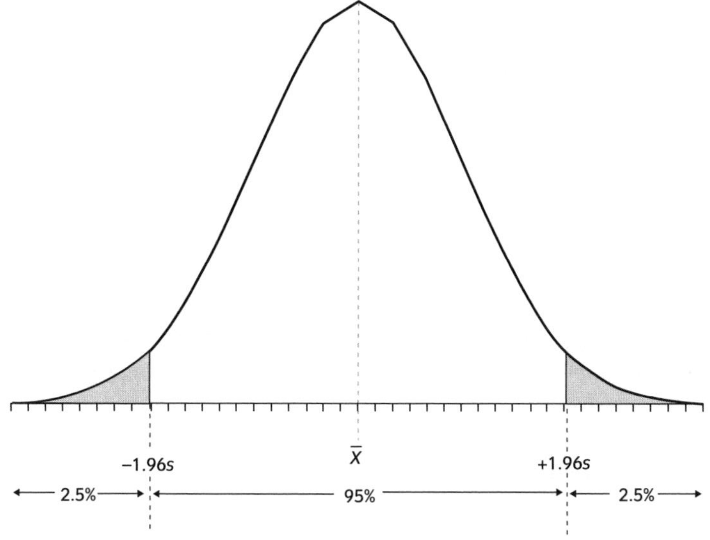

Fig. 4.3. Normal distribution showing the range $\overline{x} \pm 1.96s$, with 2.5% observations in each tail excluded from this range.

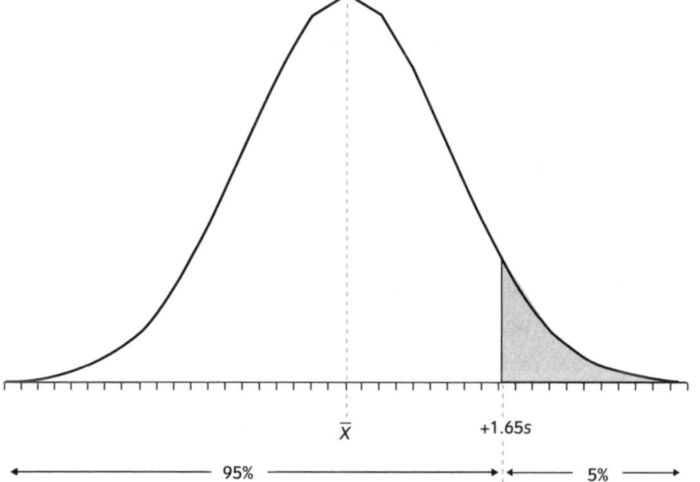

Fig. 4.4. Normal distribution showing the range $\overline{x} \pm 1.96s$, with 5% observations in the upper tail excluded from this range.

Imagine that you had two sets of data. In one set you had the weights of 250 adult female mice. In the other data set you had another 10 observations, but the gender and age of these mice were unknown. You wish to test the hypothesis that H_0: there is no significant difference between the mean weights (g) of the female mice and the unknown mice.

In the absence of any information about these 10 mice, the statistical test used would check to see if the data overlapped at the top or the bottom end of the distribution. This is a two-tailed test and is the most common form of statistical test.

However, you may know that if the mice do not belong to the adult female population then they will be adult males and belong to the adult male population, which has a different weight distribution with a higher mean. You could then use a one-tailed statistical test that checks for an overlap only against the top end of the distribution (Fig. 4.2.). Some statistical tests are by their nature one-tailed tests, e.g. when testing specific hypotheses using a non-parametric analysis of variance. However, most of the time we do not know enough about our investigative system to be able to predict which tail our second data set should fall into. So you should normally use a two-tailed test.

Q4 **Using the Spearman's rank critical values table at $p=0.05$, what are the critical values (r_s) at $n=10$ for a one-tailed and for a two-tailed test? Which value is the greater?**

A4 One-tailed test $n=10$, $p=0.05$, $r_s=0.564$
Two-tailed test $n=10$, $p=0.05$, $r_s=0.648$

The value for the two-tailed test is higher. This reflects the difference between the cut-off points. In a two-tailed test where $p=0.05$, the cut-off points are at 2.5% at each end of the distribution. In a one-tailed test where $p=0.05$, the cut-off point will be at 5% at one end only.

4.3.4. Degrees of freedom

Turn to the F_{max} table in Appendix d. There are three F tables, so you need to select the correct one. In the F_{max} table the critical value is found using a (the number of samples or treatments) and v (the degrees of freedom). What are these 'degrees of freedom'?

Example 7.1. has a frequently encountered experimental design where there are two treatments (periwinkles from the lower shore and periwinkles from the mid shore), and for each treatment there are several observations. We illustrate this design in Table 4.3. for $n=3$ observations.

Table 4.3. The impact of experimental design on the degrees of freedom based on Example 7.1. The evolution of *Littorina littoralis* at Aberystwyth, 2002

Observations	Periwinkles from lower shore	Periwinkles from mid shore
1		
2		
3		
Total	25	35

Table 4.4. The impact of experimental design on the degrees of freedom based on Example 5.5. Frequency of *Cepea nemoralis* and *Cepea hortensis* in a hedge and wood

	Species of snail		
	Cepea nemoralis	*Cepea hortensis*	Total
Hedgerow			148
Woodland			
Total	97		160

If you were to fill in this table for treatment 1 then you could choose any values for observation 1 and 2, but once these are chosen the third observation can only be one value. This final value is already determined and there is no 'freedom' for this final value. The same would be true for the observations for treatment 2. You only have freedom to choose $n - 1$ observations; the final observation is then fixed. It is this constraint within the experimental design that is recognised by the degrees of freedom. For this design the degrees of freedom for treatment 1 (periwinkles from the lower shore) are $v_1 = n_1 - 1 = 3 - 1 = 2$ and the same is true for treatment 2 (periwinkles from the mid shore) since n_2 also equals 3.

Examine Table 4.4. This represents one experimental design where there are two variables and for each variable there are two categories. (This design can be seen in Example 5.5 where the numbers of two snail species were recorded in a hedgerow and woodland.) You can see that once the grand total is fixed and you have put in one row total that the remaining row total can only be one number. (The missing woodland row total is 12.) The same is true of the columns. The table could be extended to three rows and two columns and again one row total is fixed once the grand total and two row totals are determined. For this experimental design this constraint is recognized by the degrees of freedom being calculated as (the number of rows – 1)(number of columns – 1).

Table 4.5. The impact on the degrees of freedom of the use of a Gaussian equation to determine expected values (5.1.3.)

	Variable				
	Treatment 1	Treatment 2	Treatment 3	Treatment 4	Total
Observation					106

Our final illustration of degrees of freedom is from 5.1.3. Here an additional constraint is recognized in that a mathematical formula has been used to determine values that were then used in the statistical analysis. In this experimental design there is one variable with several treatments (Table 4.5.) Usually for this experimental design the degrees of freedom would normally be the number of categories (a) – 1. However, because a mathematical formula in which the population parameter μ has been estimated using the sample statistic ($\overline{x}$), a further degree of freedom is lost and $v = a - 2$ (BOX 5.2.).

The degree of freedom, therefore, is a measure that reflects the number of observations in your calculation and the experimental design. You will be told for each statistical test how to calculate the degrees of freedom. It is not the same for all tests.

4.3.5. More than two criteria

There are a number of critical values that can only be found using three criteria. If you look now at the Mann–Whitney U table in Appendix d, you will see that rows are labelled n_1 and columns are labelled n_2. n_1 and n_2 are the number of observations for sample 1 and sample 2. The third criterion for choosing the critical value for the U distribution is p. When there are more than two criteria for choosing a critical value, as in this case, only one table is usually included for one p value. We have included the set of critical values where $p = 0.05$. So all the critical values in this table are at $p = 0.05$. If you wish to find the critical values for another p value for this test you will need to refer to other sources.

Q5 In a Mann–Whitney U table of critical values, if $n_1 = 7$ and $n_2 = 10$, what is the critical value of U at $p = 0.05$?

A5 $U = 14$ at $p = 0.05$, $n_1 = 7$, $n_2 = 10$.

Similarly, look at the F tables, either the F table that is used before a t or z test, or the F table that is used in a parametric analysis of variance test. Both these require a p value and two degrees of freedom values, so again each table of critical values is for a single p value.

Q6 If the degrees of freedom are $\nu_1 = 5$ and $\nu_2 = 10$, what are the critical F values if you were:

a. carrying out an F test for homogeneity of variances before a t test at $p = 0.05$.

b. carrying out an F test in a parametric analysis of variance, $p = 0.05$.

c. Are they the same?

d. What does this tell you?

a. $F_{critical} = 4.24$

b. $F_{critical} = 3.33$

c. No.

d. You don't want to get these two tables muddled up!

4.3.6. Interpolation

The final point we wish to make about finding critical values is relevant to several tables, including the F table for a parametric analysis of variance. A small excerpt is shown in Table 4.6.

Table 4.6 Excerpt from the table of critical values for an F test for a parametric analysis of variance

ν_2	ν_1		
	24	30	40
28	2.17	2.11	2.05
29	2.15	2.09	2.03
30	2.14	2.07	2.01

To locate a particular $F_{critical}$ value you use the two degrees of freedom, where the ν_1 column intersects the ν_2 row. However, it is not possible in this book to give all possible F values for all combinations of degrees of freedom. Clearly, if you needed to find the critical value at $\nu_1 = 29$, $\nu_2 = 29$ you will need to estimate it from the values provided. This process is called interpolation. It assumes that each missing degree of freedom increases the critical value in a regular way between the two known values.

For example, to find $F_{critical}$ at $\nu_1 = 29$, $\nu_2 = 29$, the nearest $F_{critical}$ values on the F table are:

ν_2	ν_1		
	24	30	40
29	2.15	2.09	

You therefore need to fill in the gaps between $v_1 = 24$ and $v_1 = 30$:

v_2	v_1						
	24	25	26	27	28	29	30
29	2.15						2.09

First take the difference between 2.15 and 2.09 = 0.06. The number of increments between 2.15 and 2.09 is 6 (for v_1, 24–25, 25–26, 26–27, 27–28, 28–29, 29–30).

Divide the difference between the known $F_{critical}$ numbers by the number of increments, 0.06 / 6 = 0.01. Keep adding this number to the smallest known $F_{critical}$ value until you have filled in the gaps. Therefore, the critical value for F when $v_1 = 29$, $v_2 = 29$ is 2.10.

v_2	v_1						
	24	25	26	27	28	29	30
29	2.15	2.14	2.13	2.12	2.11	**2.10**	2.09

4.4. The rule

This is the point at which the choice is made between the null and alternate hypotheses. The rule tells you about the relationship between your calculated test statistic and your critical test statistic. For example, if you were carrying out a chi-squared test for association the rule is 'If $\chi^2_{calculated}$ is *greater* than $\chi^2_{critical}$ you may reject the null hypothesis' (BOX 5.4.). Most tests have a rule that sounds similar to this. There are exceptions, however. For example, if you were carrying out a Mann–Whitney U test then the rule is 'If $U_{calculated}$ is *less* than $U_{critical}$ then you may reject the null hypothesis' (BOX 8.1.). The Mann–Whitney rule is the opposite to the majority of rules. We therefore include each rule in all our general and worked examples relating to all the statistical tests in this book.

Applying these rules allows you finally to choose between your two hypotheses and decide whether to reject or not reject the null hypothesis. You can then say given a certain level of probability whether there is or is not a significant difference, association, correlation, etc. between your data sets.

4.5. What does this mean in real terms?

The final step in hypothesis testing is the most important and yet most often forgotten. This is to refer back to your experiment. A common failing in hypothesis testing is for the first four steps to be completed and for a student to conclude that there is a significant difference. In what? What does this mean? Refer back to your hypotheses to conclude your hypothesis testing. We have included this step in all our examples of hypothesis testing to encourage you to not leave this out.

For example, if in the experiment to examine the effect of soil type on seed germination (Example 4.1.) a chi-squared goodness of fit test (5.1.) was used to analyse the data. The outcome was to reject the null hypothesis ($\chi^2_{\text{calculated}} = 9.00, p < 0.05$). But it is not enough to stop there. You should refer back to the alternate hypothesis and add details to this statement about the outcome. For this example we might conclude that there is a significant difference ($\chi^2_{\text{calculated}} = 9.00, p < 0.05$) in the median number of *Allium schoenoprasum* seed germinating on the three soil types compared with that expected by a random distribution of germination across the soil types.

Summary of Chapter 4

- You are introduced to the principles surrounding hypothesis testing and to the format used in Chapters 5–8 when a number of statistical tests are considered.
- There are three types of hypotheses: Do the data match an expected ratio; Is there an association between two or more variables; Do samples come from the same or different populations? (4.1.1. and developed in Chapters 5–8).
- Hypotheses may be either general or specific. The general hypotheses test questions such as, 'Is there a difference between the means of samples 1, 2, and 3?' The specific hypotheses may test questions such as, 'Is the mean of sample 1 significantly greater than the mean of sample 2 which is greater than the mean of sample 3?' (4.1.1. and developed in Chapters 5–8).
- Hypotheses have a number of features. Hypotheses come in pairs: the null and alternate hypotheses. They test the population parameter, indicate the type of data being analysed, indicate the type of hypothesis being tested, and give some details about the experimental design (4.1.2.).
- The null hypothesis describes the variation in the data in terms of sampling error. The alternate hypothesis describes the variation in terms of the treatment variable (4.1.3.).
- To enable you to decide if you may reject the null hypothesis you work out a calculated test statistic derived from a particular distribution (4.2.). This is compared with a critical value also derived from the same distribution (4.3.) and determined by the chosen level of stringency (usually $p = 0.05$) and other factors, such as the degrees of freedom (4.3.4. and developed in Chapters 5–8).
- For each statistical test there is a rule that determines the comparison you make between the calculated test statistic and the critical value of the test statistic. This

rule differs between statistical tests (4.4. and developed in Chapters 5–8.).

- When concluding your hypothesis testing you must refer back to your hypotheses and your original objective (4.5. and developed in Chapters 5–8 and Chapter 10).
- The Online Resource Centre includes interactive exercises that test your understanding of this chapter with other topics, particularly those considered in Chapters 5–8 and Chapter 10.

5

Hypothesis testing: Do my data fit an expected ratio?

This chapter tells you how you may analyse data from one or more **samples** where each observation can be assigned to one of two or more discrete categories and allows you to test either the **hypothesis** 'do my data fit an expected ratio' or the hypothesis 'is there an association between two or more variables'. The categories can be derived directly from the categories inherent in a **nominal** or **ordinal variable**, for example, flower colour categorized as purple and white (Example 5.2.), or imposed on an **interval** scale of measurement as classes, for example, the distribution of holly leaf miners on a holly tree (Example 5.1.). The data in these categories or classes are counts, i.e. the numbers of items in a category. The tests that are introduced here are the chi-squared test and the G test.

In all these tests, including those designed to test the hypothesis 'Is there an association?', the observed values are compared with calculated 'expected' values. The term 'expected' can be used in three ways:

a. For some experiments, you may have a reason for *expecting* certain outcomes from your investigation. This is often referred to as having an *a priori* expectation and arises when you carry out a test of the hypothesis 'there is an expectation'.
b. You can use your own (observed) data to generate '*expected*' values given certain rules relating to the statistical test and experimental design to allow you to test the hypothesis 'there is an association'.
c. You may have your own expectations for your experiment. We considered this in Chapter 4 (4.1.1.). We also return to this topic in 10.6., as a change in the approach to science in schools has led to an increase amongst undergraduates in putting forward such predictions. You **must never** confuse this, your own personal preference or 'expectation' which is usually for significant outcomes, with the idea of 'expected' values used in hypothesis testing.

The tests we are looking at in this chapter include those where you may have an *a priori* expectation (a) and those where you may not (b). (c) can lead to bias in carrying out an experiment and evaluating the results and should therefore be avoided.

Chi-squared test We look at three forms of this test; the goodness of fit test, a test for heterogeneity, and the test for association. With a chi-squared test you use an equation based on the chi-squared distribution to work out 'expected' values. These are then compared with the ones you recorded in your experiment. The three versions of the chi-squared test differ in how the 'expected' values are calculated. It is therefore important to be sure you have selected the right form of chi-squared test for your type of data and hypothesis.

G test The G test can be used in the same way as the chi-squared test. It has some advantages over the chi-squared test especially when numbers are large (5.5.). The chi-squared test tends to be most frequently used, so we have only included two forms of the G test: the goodness of fit test and an r × c test for association.

Worked examples are given for each test. If this is the first time you have used this test you should work through these examples and the questions and then check your answers before using the test on your own data. If your answer differs considerably from that given you should check your calculation by going to the Online Resource Centre, where we have included both the full calculation and more examples. If you work through the questions in this chapter it should take you about 3 hours to complete.

How to choose the correct test

Each version of each test has several requirements that must be met and these details are given at the start of each section. The following guide takes you to the most likely tests.

You have one variable and you have an *a priori* reason for expecting certain outcomes from your investigation. The variable has more than two categories.	Chi-squared goodness of fit test (5.1.) or G goodness of fit test (5.5.1.)
You have one variable and you have an *a priori* reason for expecting certain outcomes from your investigation. The variable has only two categories.	Chi-squared goodness of fit test with Yates's correction (5.4.1.) or G goodness of fit test (5.5.1.)
You have one variable with two or more categories. You have more than two samples in your data set and wish to know if the samples are similar or different from each other.	Chi-squared test for heterogeneity (5.2.)

You do not have an *a priori* expectation. You have two variables. At least one of these variables has more than two categories. You wish to test for an association between the variables.	Chi-squared test for association (5.3.) or G test for association (5.5.2.)
You do not have an *a priori* expectation. You have two variables. Both variables have only two categories. You wish to test for an association between the variables.	Chi-squared test for association with Yates's correction (5.4.2.) *or* G test for association (5.5.2.)

One of the criteria for using a chi-squared test is that the expected values are greater than 5. Having calculated your expected values, if you find any that are less than 5 you should refer to Section 5.6. The problem with small numbers.

5.1. **Chi-squared goodness of fit test**

There are three types of investigations where you may use a goodness of fit test:

1. You can reasonably argue that all samples should have the same value.

   This can arise when your sampling is very restricted, e.g. you may record the remains of shells from *Cepea nemoralis* and *C. hortensis* at the site where a thrush has been breaking them open (an anvil). You might hypothesize that there is no statistically significant difference between the numbers of each species, i.e. you would expect to see one *C. nemoralis* shell for every one *C. hortensis* shell at **this** thrush anvil. But beware. These samples may not be representative of thrush anvils as a whole. To examine a broader hypothesis you would need to collect more samples and use a different statistical test.

2. You wish to confirm that your data have a particular distribution, such as a normal or a Poisson distribution.

   In this instance your expected values are calculated using a particular formula that describes the distribution, such as the Gaussian equation (3.2.1.).

3. You have carried out a genetic cross or sampled in a population and have reason to expect a particular segregation ratio, sex ratio, allele frequency, etc.

In each of these examples you will be able to use known underlying principles, such as Mendelian genetics or the Hardy–Weinberg theorem, to determine the 'expected' values.

A chi-squared goodness of fit test is introduced in 5.1.1. in relation to an experimental design where the expected values are determined by a ratio resulting from 'random' events (investigation type 1). A second example is considered in 5.1.3. where this test is also used to confirm that data have a normal distribution. In 5.2. and 5.4.1. examples of testing an expectation arising from genetic crosses are shown. Other worked examples are included in the Online Resource Centre.

A goodness of fit test allows you to compare the data you have collected in an investigation with that predicted by an *a priori* expectation and to calculate how probable the match is. As in all chi-squared tests there are several common steps to be followed. The first of these is to arrange your data in a contingency table.

EXAMPLE 5.1. The distribution of holly leaf miners on *Ilex aquifolia*

As part of a student project the number of holly leaf miners was recorded on a holly tree. The heights of the holly leaf miners within the tree were recorded (Table 5.1.). If the distribution was random you would expect equal numbers at each height.

Table 5.1. is a contingency table for the data from Example 5.1. A contingency table can be the same as a frequency table, as in this example. At this stage you will only have 'observed' values. The observed data have been organized into classes: 0.00–1.99 m, 2.00–3.99 m and 4.00–6.99 m. There is no ambiguity about the categories: they are discrete. Categories are also known as classes.

Table 5.1. Contingency table for Example 5.1. The distribution of holly leaf miners on a single *Ilex aquifolia* tree

	Height on holly tree (m)			Total number of holly leaf miners
	0.00–1.99	2.00–3.99	4.00–5.99	
Observed number of holly leaf miners	131	38	2	171

5.1.1. To use this test you:

1. Wish to compare your observed values to those predicted by an *a priori* expectation.
2. Have one treatment variable.
3. Have only one sample.
4. Have data that fall into more than two categories.
5. Have data that are counts or frequencies and are not percentages or proportions.
6. Have observations that are **independent**.
7. Have expected values that are more than 5.

The data from Example 5.1. meet these criteria. The *a priori* expectation is that the holly leaf miners should by chance be present in equal numbers at the different tree heights. There is only one treatment variable (height on tree) and one sample. The observations fall into three discrete categories (0.00–1.99 m, 2.00–3.99 m, and 4.00–6.99 m) and the scale of measurement is an interval scale (m). Each holly leaf miner was recorded only once, therefore each observation is independent of all other observations in the sample. The expected values are only worked out as part of the calculation, so cannot be easily checked in advance. If you look at Table 5.2 on the 'expected' row, these values are greater than 5. So this criterion is also met. If the criterion is not met then you should refer to 5.7.

5.1.2. The general calculation

Having organized your observed values in a contingency table and checked that the criteria for using this test are met, you can now proceed with the test. In BOX 5.1. we show you the general calculation and a specific example of a chi-squared goodness of fit test. In addition, you will need to refer to the contingency calculation table (Table 5.2.). Where steps have been abbreviated the full calculation is included in the Online Resource Centre.

BOX 5.1. How to calculate a chi-squared goodness of fit test

GENERAL DETAILS	THIS EXAMPLE
1. Hypotheses to be tested **H_0:** There is no difference between the expected and observed values. **H_1:** There is a difference between the expected and observed values.	**1. Hypotheses to be tested** **H_0:** There is no difference between the numbers of holly leaf miners found at various heights (m) on the tree compared with those expected. **H_1:** There is a difference between the numbers of holly leaf miners found at various heights (m) on the tree compared with those expected.
2. How to work out expected values Expected values are calculated using the numerical formula you are expecting your data to conform to.	**2. How to work out expected values** We are expecting a random distribution of the holly leaf miners throughout the tree. This means that we should find equal numbers in each height category i.e. 171 / 3 = 57 (Table 5.2.).
3. How to work out chi-squared $(\chi^2)_{calculated}$ $\chi^2_{calculated} = \sum \frac{(observed - expected)^2}{expected}$	**3. How to work out chi-squared $(\chi^2)_{calculated}$** $\chi^2_{calculated} = \frac{(131-57)^2}{57} + \frac{(38-57)^2}{57} + \frac{(2-57)^2}{57}$ $= 96.070175 + 6.3333333 + 53.070175$ $= 155.47368$
4. How to find chi-squared$_{critical}$ To find the critical value of χ^2 you need to know the degrees of freedom (v). In this goodness of fit test the degrees of freedom are the number of categories (a) – 1. Use a χ^2 table of critical values to locate the value at $p = 0.05$.	**4. How to find chi-squared$_{critical}$** The categories in this example are 0.00–1.99 m, 2.00–3.99 m and 4.00–6.99 m and therefore $\nu = 3 - 1 = 2$. When $p = 0.05$, $\nu = 2$, then $\chi^2_{critical}$ is 5.99.
5. The rule If $\chi^2_{calculated}$ is greater than $\chi^2_{critical}$ you may reject the null hypothesis.	**5. The rule** $\chi^2_{calculated}$ (155.47) is greater than $\chi^2_{critical}$ (5.99) at $p = 0.05$ and therefore we reject the null hypothesis. In fact at $p = 0.001$ the $\chi^2_{critical}$ is 13.82. Therefore we can reject the null hypothesis at this higher level of significance.
6. What does this mean in real terms?	**6. What does this mean in real terms?** There is a significant difference ($\chi^2_{calculated} = 155.47$, $p < 0.001$) between the numbers of holly leaf miners found at the various levels on the tree compared with those expected such that the holly leaf miners are not found in equal numbers at all heights.

Table 5.2. Expected values for a goodness of fit chi-squared test using the data from Example 5.1. The distribution of holly leaf miners on a single *Ilex aquifolia* tree

Number of holly leaf miners	Height on holly tree (m)			Total number of holly leaf miners
	0.00–1.99	2.00–3.99	4.00–6.99	
Observed	131	38	2	171
Expected	57	57	57	

Q1 **In an investigation into the visual responses of beetles, a number of coloured pitfall traps were placed at random in grassland. After 24 hours the pitfall traps were collected and the number of beetles recorded. The results for each trap were red 20, yellow 34, white 10, and black 40. What is your *a priori* expectation? Calculate the expected values.**

A1 The *a priori* expectation is that there would be equal numbers of beetles in the pitfall traps. The total number of beetles observed was $20+34+10+40=104$. We would expect $104/4=26$ beetles in each trap.

5.1.3. How to check if your data have a normal distribution using a chi-squared goodness of fit test

In Chapter 3 we discussed distributions and in particular the normal distribution. Data with a normal distribution can be analysed using **parametric** statistics and it is therefore important to be able to check if your data are normally distributed. Section 3.8. and BOX 3.2. explained a number of ways to check your data to see if they are normally distributed. The best support for deciding whether your data are normally distributed is to calculate expected values using the Gaussian equation and then use a chi-squared goodness of fit test to check that your observed data are not statistically significant from that predicted by the equation. To do this requires some extra steps in addition to those described in BOX 5.1. and a change to how you work out the degrees of freedom, so we have included a worked example here. The observed values we will use come from Example 3.2. Length (mm) of two-spot ladybirds (*Adalia bipunctata*).

Although we only include an example relating to the normal distribution, the principles are the same for checking against any known distribution, such as the Poisson distribution. All you need to know is the equation that describes the distribution you are interested in. You then follow the same steps using this other equation.

Table 3.9. Frequency table of length of two-spot *Adalia bipunctata* (ladybirds)

	Size classes for length (mm) of *Adalia bipunctata* (ladybird)								
	1.0–1.9	2.0–2.9	3.0–3.9	4.0–4.9	5.0–5.9	6.0–6.9	7.0–7.9	8.0–8.9	9.0–9.9
Mid-point of class (m)	1.45	2.45	3.45	4.45	5.45	6.45	7.45	8.45	9.45
Frequency (f)	2	3	5	8	12	9	5	4	2

i. Arrange your data into a contingency table

The contingency data for this example are the same as the frequency table (Table 3.9.). We reproduce this again here in an amended form so that the whole of this process is represented here for convenience.

ii. Do the data meet all the criteria for using a chi-squared goodness of fit test?

For this example the criteria are met. The *a priori* expectation is determined by the Gaussian equation. There is one treatment variable (length), one sample, and the variable falls into discrete categories. The data are measured on an interval scale (mm) and are not percentages or proportions. Each ladybird was only measured once. We explain how the last criterion is met in BOX 5.2.

iii. To calculate the expected values

Like the previous example for the chi-squared goodness of fit test you use your *a priori* expectation to determine the expected values. In this case we use the Gaussian equation, which is the mathematical equation that describes the normal distribution, where:

$$y = \frac{1}{\sqrt{(2\pi s^2)}} e^{-h}$$

and

$$h = \frac{(x - \bar{x})^2}{2s^2}$$

The terms in this equation are explained in 3.2.1. and Appendix c. The symbols e and π have values of 2.72 and 3.14 respectively. The $\bar{x}$ (mean) and s^2 (variance) were calculated in BOX 3.2. and are $\bar{x} = 5.08$ mm, $s^2 = 3.74857$ mm^2. x is any one observation in your sample. So all these terms have known values apart from y. These y values can be calculated for each x in your sample as follows.

Choose an x value that fits the lowest category of your data. As we have classes we use the mid-point of that class, so the first $x = 1.45$. By including this x value and all the other known values you can now work out y. Don't be put off, even though it looks complicated. Break the calculation down into smaller steps.

The first part of the equation is:

$$\frac{1}{\sqrt{(2\pi s^2)}} = \frac{1}{\sqrt{(2 \times 3.14 \times 3.74857)}}$$

$$= \frac{1}{\sqrt{23.5529}} = \frac{1}{4.8531} = 0.20605$$

The second part of the equation involves the exponential term e. First work out the value for h:

$$h = \frac{(x - \bar{x})^2}{2s^2} = \frac{(1.45 - 5.08)^2}{2 \times 3.74857} = \frac{13.1769}{7.49714} = 1.75759$$

Use your calculator function buttons to find $e^{-1.75759} = 0.17246$
So $y = 0.20605 \times 0.17246 = 0.03554$

This y value is worked out as a proportion and our final step is to calculate expected numbers from these proportions. Since the sample size for our ladybird example is 50 then the expected number for y when $x = 1.45$ is $50 \times 0.03554 = 1.77678$

Repeat this calculation for all mid-point values from 1.45 to 9.45. These are your expected values (Table 5.3.) and you can now carry on with the chi-squared goodness of fit test (BOX 5.2.).

Table 5.3. Contingency table for length of two-spot *Adalia bipunctata* (ladybirds) with expected numbers calculated from the Gaussian equation

	Size classes for length (mm) of *Adalia bipunctata* (ladybird)								
	1.0–1.9	2.0–2.9	3.0–3.9	4.0–4.9	5.0–5.9	6.0–6.9	7.0–7.9	8.0–8.9	9.0–9.9
Mid-point of class (m)	1.45	2.45	3.45	4.45	5.45	6.45	7.45	8.45	9.45
Observed number of ladybirds	2	3	5	8	12	9	5	4	2
Expected number of ladybirds	1.7768	4.0951	7.2283	9.7714	10.1162	8.0209	4.8705	2.2649	0.8066

BOX 5.2. To check if your data are normally distributed using a chi-squared goodness of fit test

GENERAL DETAILS	THIS EXAMPLE
1. Hypotheses to be tested **H_0:** There is no difference between the expected and observed values. **H_1:** There is a difference between the expected and observed values.	**1. Hypotheses to be tested** **H_0:** There is no difference between the observed lengths of ladybirds compared with that expected if the data are normally distributed. **H_1:** There is a difference between the observed lengths of ladybirds compared with that expected if the data are normally distributed.
2. How to work out expected values Expected values are calculated using the numerical formula you are expecting your data to conform to.	**2. How to work out expected values** See 5.1.3.iii. and Table 5.3. It is clear that 5/9 expected values are less than 5. To overcome this we will add together the expected values for the lower two classes $(1.7768 + 4.0951 = 5.8719)$ and the expected values in the upper two classes $(2.2649 + 0.8066 = 3.0717)$. The reasons for this are explained at the end of section 5.6. These expected values will be compared with observed values that have been combined in the same way.
3. How to work out chi-squared$_{\text{calculated}}$ $\chi^2_{\text{calculated}} = \sum\left[\frac{(observed - expected)^2}{expected}\right]$	**3. How to work out chi-squared$_{\text{calculated}}$** $\chi^2_{\text{calculated}} = \frac{(5-5.8719)^2}{5.8719} + \frac{(5-7.2283)^2}{7.2283} + \ldots + \frac{(6-3.0717)^2}{3.0717} = 4.40299$
4. How to find chi-squared$_{\text{critical}}$ To find the critical value of χ^2 at $p = 0.05$ you need to know the degrees of freedom (v). Usually for a chi-squared goodness of fit the degrees of freedom would be the number of categories (a) − 1. However, because a mathematical formula in which the population value of μ has been estimated using the sample mean ($\bar{x}$), a further degree of freedom is lost. Therefore for this chi-squared goodness of fit test $v = a - 2$.	**4. How to find chi-squared$_{\text{critical}}$** Since we combined the lower two classes and the upper two classes there were only 7 classes in this calculation, so $\nu = 7 - 2 = 5$. When $p = 0.05$, $\nu = 7$, then χ^2_{critical} is 14.07.
5. The rule If $\chi^2_{\text{calculated}}$ is greater than χ^2_{critical} you may reject the null hypothesis.	**5. The rule** $\chi^2_{\text{calculated}}$ (4.40) is less than χ^2_{critical} (14.07) at $p = 0.05$ and therefore we do not reject the null hypothesis.

BOX 5.2. Continued

GENERAL DETAILS	THIS EXAMPLE
6. What does this mean in real terms?	**6. What does this mean in real terms?** There is no significant difference ($\chi^2_{\text{calculated}} = 4.40$, $p = 0.05$) between the observed lengths of ladybirds compared with that expected if the data are normally distributed. The data can be said to be normally distributed.

5.2. Heterogeneity in a goodness of fit test

There are some circumstances where you have several samples and wish to know if they can be pooled. For example, you may have carried out an investigation into the genetic inheritance of flower colour where you had several pairs of parent plants. Crossing within these pairs would produce F1 offspring. If you kept the seed from each cross separate from the others, grew these F1 plants up, and then crossed these, you could collect the results from several different lineages. These separate lineages are known as accessions.

EXAMPLE 5.2. The genetics of flower colour in *Allium schoenoprasum*

Crosses were carried out in three pairs of plants of *Allium schoenoprasum*. In each cross one parent had purple flowers and the other was white flowered. Seeds from each cross were collected and grown on. The offspring from each cross were kept as separate accessions. These F1 plants were all purple flowering and thought to be heterozygous for a single gene that controlled the pigmentation in the flowers, and purple was believed to be the dominant allele. These F1 plants were then crossed amongst themselves within an accession and the seed from these crosses grown up and the flower colours recorded (Table 5.4.). If the genetic theory is correct then the F2 plants should occur in the ratio of three purple flowering plants : one white flowering plant.

In examining Table 5.4. it appears that the total values (487 purple and 175 white) are reasonably close to a 3:1 ratio. When a goodness of fit chi-squared test is carried out on just these total values there is no significant difference ($\chi^2 = 0.73$, $p < 0.05$) between the observed values and that expected for a three purple : one white ratio (Table 5.5). But should you combine the data from your different samples in this way? You can get some idea by carrying out a goodness of fit chi-squared test on each accession separately (Table 5.5.). These analyses confirm that the results from accessions 1, 2, and 4 also conform to a 3:1 ratio at $p = 0.05$.

Table 5.4. Flower colour exhibited by F2 *Allium schoenoprasum* in four accessions

	Flower colour in the F2 generation		Total number of plants flowering in the F2
	Purple	White	
Accession 1	127	41	168
Accession 2	123	39	162
Accession 3	107	53	160
Accession 4	130	42	172
Total	487	175	662

Accession 3, however, is different and the test there indicates that the observed values depart significantly from the predicted 3:1 ratio. Does this difference between accession 3 and accession 1, 2, and 4 mean that the data should not be pooled? In these circumstances you may use the chi-squared test to see if the data are statistically heterogeneous (different).

5.2.1. To use this test you:

1. Wish to test for heterogeneity between samples.
2. Have one treatment variable.
3. Have two or more samples.
4. Have data that fall into categories.
5. Have data that are counts or frequencies and are not percentages or proportions.
6. Have observations that are independent of each other.
7. Do not have expected values smaller than 5.

We have checked the data from Example 5.2. against the criteria for using this test and all the criteria are met. We do wish to test for heterogeneity between samples. There is one treatment variable (flower colour) and four samples (accessions). The observations fall into two discrete categories (purple and white) and the number of plants has been recorded. Each plant has only been counted once. All the expected values are greater than 5 (Table 5.5.). If this were not the case you should refer to 5.6.

5.2.2. The calculation

At this point you have constructed a contingency table, checked the criteria for using this test, and wish to compare your samples to see if they are heterogeneous or can be pooled. You now proceed with the chi-squared test for heterogeneity. Where steps have been abbreviated the full calculation is included in the Online Resource Centre.

Table 5.5. Contingency table with expected values for a chi-squared test for heterogeneity for Example 5.2. Flower colour exhibited by F2 *Allium schoenoprasum* in four accessions

		Flower colour in the F2 generation		Total number of plants flowering in the F2	χ^2 from the goodness of fit test for each accession and for the total values
		Purple	White		
Accession 1	Observed	127	41	168	0.03175 (NS)
	Expected	126	42		
Accession 2	Observed	123	39	162	0.07407 (NS)
	Expected	121.5	40.5		
Accession 3	Observed	107	53	160	5.63333 $0.05 > p > 0.01$
	Expected	120	40		
Accession 4	Observed	130	42	172	0.03101 (NS)
	Expected	129	43		
Total	Observed	487	175	662	0.72709 (NS)
	Expected	496.5	165.5		

NS, not significant

BOX 5.3. How to calculate a chi-squared test for heterogeneity

GENERAL DETAILS	THIS EXAMPLE
1. Hypotheses to be tested **H_0:** There is no heterogeneity between the samples. **H_1:** There is heterogeneity between the samples.	**1. Hypotheses to be tested** **H_0:** There is no heterogeneity between the F2 accessions of *Allium schoenoprasum*. **H_1:** There is heterogeneity between the F2 accessions of *Allium schoenoprasum*.
2. How to work out expected values These are calculated for each accession and for the total values. The expected values are calculated using the numerical formula you are expecting your data to conform to.	**2. How to work out expected values** We are expecting three purple flowering plants : one white flowering plant. Since the grand total number of plants examined was 662 then 3/4 should be purple flowering, i.e. 496.5, and ¼ plants should be white flowering i.e. 165.5. Using the same ratio, calculate expected values for each observed value as well as the total (Table 5.5.).

BOX 5.3. Continued

GENERAL DETAILS	THIS EXAMPLE
3. How to work out chi-squared$_{calculated}$ i. First calculate chi-squared values for each separate sample and for the total. Where: $\chi^2_{calculated} = \sum\left[\frac{(observed - expected)^2}{expected}\right]$ It is at this point that the process differs from that described for the goodness of fit test for one sample and a few further calculations are required. ii. 'Summed' chi-squared First sum all the chi-squared values calculated for each accession. iii. 'Deviation' chi-squared This is the chi-squared value found when examining the total values. iv. 'Heterogeneity' chi-squared or chi-squared calculated is found by subtracting the 'deviation' value from the 'summed' value. This is $\chi^2_{calculated}$.	**3. How to work out chi-squared$_{calculated}$** i. e.g. Accession 1 $\chi^2 = \frac{(127-126)^2}{126} + \frac{(41-42)^2}{42}$ $= 0.0079365 + 0.0238095$ $= 0.03175$ Total $\chi^2 = \frac{(487-496.5)^2}{496.5} + \frac{(175-165.5)^2}{165.5}$ $= 0.18177 + 0.54532$ $= 0.72709$ The other values are shown in Table 5.5. ii. 'Summed' chi-squared In our current example this 'summed' value is: $0.03175 + 0.07407 + 5.63333 + 0.03101$ $= 5.77016$ iii. 'Deviation' chi-squared $= 0.72709$ iv. $\chi^2_{calculated} = 5.77016 - 0.72709 = 5.04307$
4. How to find chi-squared$_{critical}$ First calculate the degrees of freedom (ν). Again there are several steps. i. ν for the 'deviation' value is the number of categories – 1. ii. ν for the 'summed' chi-squared value is the number of rows of observed values excluding the total. iii. ν for the 'heterogeneity' value is the value from (i) – the value from (ii). The critical value is found at $p = 0.05$ and the heterogeneity degrees of freedom from iii.	**4. How to find chi-squared$_{critical}$** i. There are two categories (purple and white) so $\nu = 2 - 1 = 1$ ii. There are four rows of observed values, the accessions, therefore $\nu = 4$. iii. For this example $\nu = 4 - 1 = 3$. The critical value to test for heterogeneity in the data is found in the chi-squared table at $\nu = 3$, $p = 0.05$ and is $\chi^2_{critical} = 7.81$.
5. The rule If $\chi^2_{calculated}$ is greater than $\chi^2_{critical}$ you may reject the null hypothesis. What do you do if your data are heterogeneous? Go to Chapter 8.	**5. The rule** $\chi^2_{calculated}$ (5.04) is less than $\chi^2_{critical}$ (7.81) at $p = 0.05$ and therefore we do not reject the null hypothesis.

BOX 5.3. Continued

GENERAL DETAILS	THIS EXAMPLE
6. What does this mean in real terms?	**6. What does this mean in real terms?** There is no statistically significant heterogeneity ($\chi^2_{\text{calculated}}$ 5.04, $p = 0.05$) between the accessions of *Allium schoenoprasum* and it is therefore reasonable to sum the data across all accessions and use a goodness of fit chi-squared test on the totals.

5.3. Chi-squared test for association

A chi-squared test for association is used when you have categorical data for two variables and you wish to examine the possibility of an association between these two variables. The term **'association'** has a very specific meaning. Before proceeding with this test you should read the first paragraphs of Chapter 6 to ensure you understand how this term is being used.

As in all chi-squared tests the data are arranged in a contingency table. If your data have more than two discrete categories, either for variable 1 (columns) and/or for variable 2 (rows), you should use the method in BOX 5.4. This method is often referred to as a generalized or r × c test. If your contingency table has only two categories for variable 1 and two categories for variable 2 you should refer to 5.4.

EXAMPLE 5.3. Shell colour in *Cepea nemoralis* in coastal and hedgerow habitats

An investigation was carried out into the frequency of banding and colour patterns in snail shells and habitat. Four patterns were observed in the two populations studied: pink with bands, pink with no bands, yellow with bands and yellow with no bands. These results are shown in Table 5.6. The association that is under investigation is therefore between the variables 'shell pattern' and 'habitat'.

Table 5.6. Shell colour in *Cepea nemoralis* in coastal and hedgerow habitats

	Shell pattern and colour in *C. nemoralis*				
Habitat	Banded yellow	No bands yellow	Banded pink	No bands pink	Total
Coastal Observed	10	19	5	16	50
Hedgerow Observed	17	8	19	11	55
Total	27	27	24	27	105

5.3.1. To use this test you:

1. Wish to test for an association between two treatment variables.
2. Have data that are organized into more than two categories for at least one of the variables and into two or more categories for the second variable. (If you have two categories for each variable go to 5.4.)
3. Have data that are counts or frequencies and are not percentages or proportions.
4. Have observations that are independent of each other.
5. Have expected values that are more than 5.

For the data from Example 5.3. we do wish to test for an association between two treatment variables (shell pattern and habitat). Shell pattern has four categories (banded yellow, no bands yellow, banded pink, no bands pink) and the habitat has two categories (coastal and hedgerow). The data are numbers of snails in each category. When the expected values are calculated (Table 5.7.) they are greater than 5, so all the criteria for using this test are met.

Table 5.7. Contingency table with expected values for a chi-squared test for association using data from Example 5.3. Shell colour in *Cepea nemoralis* in coastal and hedgerow habitats

	Shell pattern and colour in *C. nemoralis*				
Habitat	**Banded yellow**	**No bands yellow**	**Banded pink**	**No bands pink**	**Total**
Coastal Observed	10	19	5	16	50
Coastal Expected	$27/105 \times 50 = 12.85714$	$27/105 \times 50 = 12.85714$	$24/105 \times 50 = 11.42857$	$27/105 \times 50 = 12.85714$	
Hedgerow Observed	17	8	19	11	55
Hedgerow Expected	$27/105 \times 55 = 14.14206$	$27/105 \times 55 = 14.14206$	$24/105 \times 55 = 12.57143$	$27/105 \times 55 = 14.14206$	
Total	27	27	24	27	105

5.3.2. The calculation

This is the calculation for an r × c chi-squared test. At this point you will have arranged your data in a contingency table and have checked the criteria for using this test. You can now proceed to test for an association between the two variables (BOX 5.4. and Table 5.7.). Where steps have been abbreviated the full calculation is included in the Online Resource Centre.

BOX 5.4. How to calculate an r × c chi-squared test for association

GENERAL DETAILS	THIS EXAMPLE
1. Hypotheses to be tested **H_0:** There is no association between the two variables. **H_0:** There is an association between the two variables.	**1. Hypotheses to be tested** **H_0:** There is no association between the distribution of shell patterns observed and the habitat (coastal and hedgerow) of *Cepea nemoralis*. **H_1:** There is an association between the distribution of shell patterns and the habitat (coastal and hedgerow) of *Cepea nemoralis*.
2. How to work out expected values In chi-squared tests for association you have no *a priori* expectation against which to compare your observed data. Instead, the expected values are calculated from the totals of each column and row.	**2. How to work out expected values** Look first at the expected value in row 1, column 1 in Table 5.7. To calculate this expected value take the column total for banded yellow (27) ÷ grand total (105) × row total for coastal snails (50) = 12.85714 This calculation is repeated for each row × column combination.
3. How to work out chi-squared$_{calculated}$ The rest of the procedure is the same as that described for the goodness of fit chi-squared test. Where: $\chi^2_{calculated} = \sum\left[\frac{(observed - expected)^2}{expected}\right]$	**3. How to work out chi-squared$_{calculated}$** $\chi^2 = \frac{(10 - 12.85714)^2}{12.857143} + \frac{(19 - 12.85714)^2}{12.85714} + \frac{(24 - 11.42857)^2}{11.42857} + \frac{(16 - 12.85714)^2}{12.85714} + \frac{(17 - 14.14206)^2}{14.14206} + \frac{(8 - 14.14206)^2}{14.14206} + \frac{(19 - 12.57143)^2}{12.571433} + \frac{(11 - 14.14206)^2}{14.14206}$ $\chi^2_{calculated} = 15.18472$
4. How to find chi-squared$_{critical}$ As before, you now need to calculate the degrees of freedom before looking up the critical value. In this case v is the (number of rows − 1) × (the number of columns − 1). The critical value is found in the statistical table at $p = 0.05$ and the degrees of freedom just calculated.	**4. How to find chi-squared$_{critical}$** In our example there are two rows (coastal and hedgerow). Do not include your 'expected' rows as these are part of your calculation. There are four columns (banded yellow, no bands yellow, banded pink, no bands pink). Therefore: $v = (2 - 1) \times (4 - 1) = 1 \times 3 = 3$ The critical value to test for an association between the snail shell patterns and habitat is found in the chi-squared table where $\nu = 3$, $p = 0.05$, and $\chi^2_{critical} = 7.81$.
5. The rule If $\chi^2_{calculated}$ is greater than $\chi^2_{critical}$ you may reject the null hypothesis.	**5. The rule** $\chi^2_{calculated}$ (15.18) is greater than $\chi^2_{critical}$ (7.81) at $p = 0.05$ and therefore we reject the null hypothesis.

BOX 5.4. Continued

GENERAL DETAILS	THIS EXAMPLE
6. What does this mean in real terms?	**6. What does this mean in real terms?** There is a significant association ($p = 0.05$) between the distribution of shell patterns and habitat (coastal and hedgerow) of *Cepea nemoralis*. In fact at $p = 0.01$, $\chi^2_{\text{critical}} = 11.34$ and at $p = 0.001$, $\chi^2_{\text{critical}} = 16.27$. Therefore, you may reject the null hypothesis at $p = 0.01$ but not at $p = 0.001$. This can be written as: there is a highly significant association ($\chi^2_{\text{calculated}}$ 15.18, $0.01 > p > 0.001$) between the distribution of shell patterns and habitat (coastal and hedgerow) of *Cepea nemoralis*.

Q2 Two groups of students were asked to rate on a scale of 1 (not at all) to 10 (outstanding) how helpful they had found this book (Table 5.8.). The investigators wished to test whether there was an association between the answers and the subject studied. Calculate appropriate expected values.

Table 5.8. The numbers of Microbiology and Forensic Science students giving particular responses to the question: 'How helpful have you found this book?'

	Answers on a scale 1 (not at all) to 10 (outstanding)									
	1	2	3	4	5	6	7	8	9	10
Number of Microbiology students	1	0	1	2	2	3	3	4	3	2
Number of Forensic Science students	0	0	0	0	7	5	4	3	1	0

A2 This is an example we included in Chapter 2 (Table 2.7.). Did your expected values agree with those given? If not why not?

5.4. Chi-squared test with one degree of freedom

In any chi-squared test when there is only one degree of freedom the $\chi^2_{\text{calculated}}$ value is too high if calculated in the ways described earlier in this chapter. These calculations therefore have to be modified using a Yates's correction. The Yates's correction reduces the $\chi^2_{\text{calculated}}$ value by subtracting 0.5 from the numerator as shown:

$$\chi^2_{\text{calculated}} = \sum \left[\frac{(|observed - expected| - 0.5)^2}{expected} \right]$$

The symbols | | indicate that the **absolute** value is used. So when you take the expected value from the observed value it does not matter if the outcome is negative. You can ignore this sign. For example, if the observed value was 10 and the expected value was 15 then $|observed - expected| = |10 - 15| = 5$, not -5. The negative sign is ignored. For further examples of using absolute values see Appendix c.

There are two occasions when you may come across a chi-squared test with one degree of freedom: in a goodness of fit test where the variable has only two categories or in an r × c test for association where both r and c = 2. The methods for applying a Yates's correction in these two cases are explained in this section. Further details about goodness of fit tests and tests for association are included earlier in this chapter (5.1. and 5.3.).

5.4.1. Chi-squared goodness of fit test when there is one degree of freedom

In the earlier examples relating to a goodness of fit test we considered the distribution of holly leaf miners on a single holly tree (BOX 5.1.) and we checked to see if the distribution of a set of data is normal (BOX 5.2.). Another application of a goodness of fit test is when you have a genetic theory that is tested by carrying out a controlled cross. The observed

EXAMPLE 5.4. The genetics of flower colour in *Allium schoenoprasum*

A cross was carried out between two plants of *Allium schoenoprasum*. As in Example 5.2. the flower colours were purple and white respectively. The genetic model proposed for the inheritance of the flower colour means that the F2 plants should occur in the ratio three purple flowering plants : one white flowering plant.

Table 5.9. Flower colour in the F2 of a single *Allium schoenoprasum* cross

	Flower colour in the F2		Total number of plants flowering in the F2
	Purple	White	
Number of plants	131	37	168

results are then compared with those predicted by your theory. We used an example like this in 5.2. where there were several accessions. Here we only have a single accession and one degree of freedom.

A contingency table for the data relating to this example is shown in Table 5.9. If you analysed this data using a chi-squared goodness of fit test then the degrees of freedom will be the number of categories $-1 = 2 - 1 = 1$; therefore, you should use the Yates's correction as show in BOX 5.5. The criteria for using this modified version of the chi-squared goodness of fit test are the same as described in 5.1.1. apart from the number of categories. For this modified test you will have one variable with only two categories.

Q3 **Are all the criteria for using this modified chi-squared goodness of fit test met in this example?**

A3 Yes, we do have an *a priori* expectation that the plants will be present in a three purple : one white ratio. There is one treatment variable (flower colour), one sample, and the observations only fall into two discrete categories (purple and white). The number of plants has been recorded and each plant has only been recorded once, the data are independent. The final criterion relating to expected numbers can only be confirmed when you begin the calculation. If you look at Table 5.10. on the 'expected' row, these values are greater than 5. So this criterion is also met.

Having organized your observed values in a contingency table and checked that the criteria for using this test are met you can now proceed with the test (BOX 5.5. and Table 5.10.).

Table 5.10. Contingency table with expected values calculated for a goodness of fit test using a Yates's correction for the Example for 5.4. Flower colour in the F2 *Allium schoenoprasum* cross

	Flower colour in the F2		Total number of plants flowering in the F2
	Purple	White	
Observed	131	37	168
Expected (BOX 5.5)	126	42	168

BOX 5.5. How to calculate a chi-squared goodness of fit test when there is one degree of freedom

GENERAL DETAILS	THIS EXAMPLE
1. Hypotheses to be tested **H_0:** There is no difference between the expected and observed values. **H_1:** There is a difference between the expected and observed values.	**1. Hypotheses to be tested** **H_0:** There is no difference between the observed numbers of plants with purple or white flowers in the F2 of *Allium schoenoprasum* plants and that expected from the *a priori* prediction of three purple flowering plants : one white flowering plant. **H_1:** There is a difference between the observed numbers of plants with purple or white flowers in the F2 of *Allium schoenoprasum* plants and that expected from the genetic prediction of three purple flowering plants : one white flowering plant.
2. How to work out expected values Expected values are calculated using the numerical formula you are expecting your data to conform to.	**2. How to work out expected values** We are expecting three purple flowering plants : one white flowering plant. Since the total number of plants examined was 168 then 3/4 should be purple flowering, i.e. 126, and ¼ plants should be white flowering, i.e. 42 (Table 5.10.).
3. How to work out chi-squared$_{\text{calculated}}$ $\chi^2_{\text{calculated}} = \sum\left[\frac{(\lvert observed - expected\rvert - 0.5)^2}{expected}\right]$	**3. How to work out chi-squared$_{\text{calculated}}$** $\chi^2_{\text{calculated}} = \frac{(\lvert 131 - 126\rvert - 0.5)^2}{126} + \frac{(\lvert 37 - 42\rvert - 0.5)^2}{42}$ $= 0.64286$
4. How to find chi-squared$_{\text{critical}}$ To find the critical value of χ^2 at $p = 0.05$ you need to know the degrees of freedom (v). In this goodness of fit test the degrees of freedom are the number of categories $- 1$.	**4. How to find chi-squared$_{\text{critical}}$** The categories in this example are 'purple' and 'white' and therefore: $\nu = 2 - 1 = 1$. When p=0.05, $\nu = 1$, then χ^2_{critical} is 3.84.
5. The rule If $\chi^2_{\text{calculated}}$ is greater than χ^2_{critical} you may reject the null hypothesis.	**5. The rule** $\chi^2_{\text{calculated}}$ (0.64) is less than χ^2_{critical} (3.84) at $p = 0.05$ and therefore we do not reject the null hypothesis.
6. What does this mean in real terms?	**6. What does this mean in real terms?** The flower colours in the F2 generation of *Allium schoenoprasum* plants do not differ significantly ($\chi^2_{\text{calculated}}$ 0.64, $p = 0.05$) from the predicted ratio of three purple : one white. This indicates that the genetic model is probably correct.

5.4.2. Chi-squared test for association when there is only one degree of freedom

Some experiments are designed to test for an association between two variables, both of which have discrete categories. In our first example (Example 5.3.) we carried out a test for association between the numbers of snails with particular shell patterns and their habitat. But there are many occasions when our two variables will each only have two categories. This is often referred to as a 2 × 2 chi-squared test for association.

EXAMPLE 5.5. Frequency of *Cepea nemoralis* and *Cepea hortensis* in a woodland and a hedgerow

A survey of two species of snail was carried out at two habitats: a woodland and a hedgerow. The numbers of snails at each location was recorded (Table 5.11.)

Table 5.11. The distribution of *Cepea nemoralis* and *Cepea hortensis* in a woodland and a hedgerow

	Species of snail		Total
	C. nemoralis	*C. hortensis*	
Hedgerow Observed	89	59	148
Woodland Observed	16	8	24
Total	105	67	172

In an r × c chi-squared test for association the degrees of freedom are (r – 1)(c – 1) (BOX 5.4.). For this example, the degrees of freedom will therefore be (2 – 1)(2 – 1) = 1, so a Yates's correction needs to be applied. The criteria for using this modified version of the chi-squared test for association are the same as described in 5.3.1. apart from the number of categories. For this modified test you will have two variables each with only two categories.

Q4 Are all the criteria for using this modified chi-squared test for association met in this example?

A4 Yes, we do wish to test for an association between two treatment variables (species and habitat) and each have only two categories. The number of snails has been recorded and each snail has only been recorded once, so the data are independent. If you look at Table 5.12. on the 'expected' rows all of these values are greater than 5. So this criterion is met.

Table 5.12. Contingency table with expected values calculated for a 2 × 2 chi-squared test for association using Yates's correction for the Example 5.5. The distribution of *Cepea nemoralis* and *Cepea hortensis* in two habitats

	Species of snail		Total
	C. nemoralis	*C. hortensis*	
Hedgerow Observed	89	59	148
Hedgerow Expected	90.34884	57.65116	
Woodland Observed	16	8	24
Woodland Expected	14.65116	9.34884	
Total	105	67	172

Having organized your observed values in a contingency table and checked that the criteria for using this test are met you can now proceed with the modified test (BOX 5.6., Table 5.12.).

BOX 5.6. How to calculate a 2 × 2 chi-squared test for association

GENERAL DETAILS	THIS EXAMPLE
1. Hypotheses to be tested **H_0:** There is no association between the two variables. **H_1:** There is an association between the two variables.	**1. Hypotheses to be tested** **H_0:** There is no association between the distribution of the two snail species (*Cepea nemoralis* and *C. hortensis*) and habitat (hedgerow and woodland). **H_1:** There is an association between the distribution of the two snail species (*C. nemoralis* and *C. hortensis*) and habitat (hedgerow and woodland).
2. How to work out expected values In chi-squared tests for association you have no *a priori* expectation against which to compare your observed data. Instead, the expected values are calculated from the totals of each column. The totals for rows and columns are first calculated and the expected values are then determined.	**2. How to work out expected values** Look first at the expected value in row 1, column 1 in Table 5.12. To calculate this expected value take the column total for *C. nemoralis* (105) ÷ grand total (172) × row total for hedgerow snails (148) = 90.34884 This calculation is repeated for each row × column combination as shown in Table 5.12.
3. How to work out chi-squared$_{calculated}$ The rest of the procedure is the same as that described for the modified goodness of fit chi-squared test. Where: $\chi^2_{calculated} = \sum\left[\frac{(\lvert observed - expected\rvert - 0.5)^2}{expected}\right]$	**3. How to work out chi-squared$_{calculated}$** $\chi^2_{calculated} = \frac{(\lvert 89-90.34884\rvert -0.5)^2}{90.34884} + \frac{(\lvert 59-57.6511\rvert -0.5)^2}{57.65116} + \frac{(\lvert 16-14.65116\rvert -0.5)^2}{14.65116} + \frac{(\lvert 8-9.34884\rvert -0.5)^2}{9.34884} = 0.14672$

BOX 5.6. Continued

GENERAL DETAILS	THIS EXAMPLE
4. How to find chi-squared$_{critical}$	**4. How to find chi-squared$_{critical}$**
As before you now need to calculate the degrees of freedom (v) before looking up the critical value. In this case v is the (number of rows − 1) × (the number of columns − 1).	In our snails example there are two rows (hedgerow and woodland). **Do not** include your 'expected' rows these are part of your calculation. There are two columns (*C. nemoralis* and *C. hortensis*) Therefore $\nu = (2-1) \times (2-1) = 1 \times 1 = 1$.
The critical value can be found in a chi-squared table at $p = 0.05$ and the appropriate degrees of freedom.	The critical value to test for an association between the snail species and habitat is found in the chi-squared table where $\nu = 1$, $p = 0.05$, and $\chi^2_{critical} = 3.84$.
5. The rule	**5. The rule**
If $\chi^2_{calculated}$ is greater than $\chi^2_{critical}$ you may reject the null hypothesis.	$\chi^2_{calculated}$ (0.15) is less than $\chi^2_{critical}$ (3.84) at $p = 0.05$ and therefore we do not reject the null hypothesis.
6. What does this mean in real terms?	**6. What does this mean in real terms?**
	There is no significant association ($\chi^2_{calculated} = 0.15$, $p = 0.05$) between the distribution of snail species and the two habitats.

Q5 In a small review of resources two groups of students were asked to select one answer in response to the question 'Have you found this book helpful?' They could select 'Yes' or 'No' (Table 5.13.). The investigators wish to test if there is an association between the course that the students are taking and their answer. What are the expected values and what is $\chi^2_{calculated}$?

Table 5.13. Closed answers given by Microbiology and Forensic Science students in their response to the question: 'Have you found this book helpful?'

	Answers to the question 'Have you found this book helpful?'		
	YES	NO	Total
Number of Microbiology students	15	5	20
Number of Forensic Science students	10	10	20
Total observed	25	15	40

A5 There are two variables and each has only two categories; therefore, there is only one degree of freedom. The investigators wish to test for an association and there is no *a priori* expectation; therefore, a chi-squared test for association with a Yates's correction should be used. We included this example in Chapter 2 and the expected values are given in Table 2.5.

$$\chi^2_{\text{calculated}} = \frac{(|15 - 12.5| - 0.5)^2}{12.5} + \frac{(|5 - 7.5| - 0.5)^2}{7.5} + \frac{(|10 - 12.5| - 0.5)^2}{12.5} + \frac{(|10 - 7.5| - 0.5)^2}{7.5}$$

$$= 1.70667$$

5.5. G tests

An alternative test to the chi-squared tests described above is the G test. The advantages of this statistic are that in larger data sets it is easier to work out using a calculator and mathematicians believe that extensions of this test have advantages over the chi-squared test (Sokal & Rohlf, 1981). $G_{\text{calculated}}$ tends to be slightly larger than $\chi^2_{\text{calculated}}$, so if the calculated value is close to the critical value the G tests will tend to reject the H_0 more often.

There are several versions of the G test and we include two of these: a goodness of fit test and a test for association. The criteria that need to be satisfied are the same as those for the equivalent version of the chi-squared test. These G tests can therefore be used to examine frequency distributions of nominal, ordinal, and interval data that are organized into discrete categories.

When observed numbers are low (between 1 and 4) *G* rejects H_0 too often and is likely to generate **Type I errors**. Therefore, a correction (Williams, 1976) similar in principle to the Yates's correction can be used. When the total sample size is very small (less than 25) see Sokal & Rohlf (1981).

Since the chi-squared tests are still the tests of choice we have only included two applications of the G test. However, *G* can be used for tests for heterogeneity and 2 × 2 tests for association. If you wish to use these types of tests you should refer to other books, such as Sokal & Rohlf (1981).

5.5.1. G goodness of fit test

G tests are similar to chi-squared tests in many respects; therefore, only the points at which they differ are shown below. Keep an eye on BOX 5.1. to see where the similarities and differences lie. The data being evaluated are found in Table 5.1. The criteria for using this test are the same as those given in 5.1.1. Unlike chi-squared tests it is common practice to apply the Williams's correction routinely, although it has little effect when sample sizes are large.

BOX 5.7. How to calculate a G goodness of fit test

GENERAL DETAILS	THIS EXAMPLE
1. Hypotheses to be tested	**1. Hypotheses to be tested**
H_0: There is no difference between the expected and observed values. **H_1:** There is a difference between the expected and observed values.	BOX 5.1.
2. How to work out expected values	**2. How to work out expected values**
Expected values are calculated using the numerical formula you are expecting your data to conform to.	BOX 5.1.
3. How to work out $G_{calculated}$	**3. How to work out $G_{calculated}$**
i. For each pair of observed (O) and expected (*E*) values calculate O ln (*O/E*). The term 'ln' is the natural logarithm. Calculators usually have a function key for converting values to this scale.	i. First calculate *O/E*, then take the natural log (ln). Multiply the natural log value by *O*. $131/57 = 2.29825$ $\ln 2.29825 = 0.83215$ $131 \times 0.83215 = 109.01113$ $38/57 = 0.66666$ $\ln 0.66666 = -0.40547$ $37 \times -0.40547 = -15.00221$ $2/57 = 0.03509$ $\ln 0.03509 = -3.34990$ $2 \times -3.34990 = -6.69981$
ii. Add these values together $\Sigma[O \times \ln[O/E]]$ and multiply by $2 = G$	ii. $G = 2 \times [109.01113 + (-15.00221) + (-6.69981)]$ $= 2 \times 87.30911 = 174.61823$

BOX 5.7. Continued

GENERAL DETAILS	THIS EXAMPLE
4. Williams's correction	**4. Williams's correction**
The Williams's correction requires two steps. First calculate W and then use this term to revise the G value. We call this corrected G value $G_{\text{calculated}}$. $W = 1 + (a^2 - 1)/6nv$ $G_{\text{calculated}} = \frac{G}{W}$ where a is the number of categories, n is the total number of observations in the sample, and v the degrees of freedom $= a - 1$.	For this example there are three categories so $a = 3$. The total number of observations in the sample is 171 and the degrees of freedom are $a - 1 = 3 - 1 = 2$. Using these values we can now work out W. $W = 1 + (3^2 - 1) / (6 \times 171 \times 2) = 1 + 8/2052$ $= 1 + 0.00390 = 1.00390$ $G_{\text{calculated}} = \frac{174.61823}{1.00390} = 173.94011$
5. How to find G_{critical}	**5. How to find G_{critical}**
To find the critical value of G at $p = 0.05$ you need to know the degrees of freedom (v). In this goodness of fit test the degrees of freedom are the number of categories (a) – 1. Use a chi-squared table to find the critical value.	When $p = 0.05$, $\nu = 2$, then G_{critical} is 5.99
6. The rule	**6. The rule**
If $G_{\text{calculated}}$ is greater than G_{critical} you may reject the null hypothesis.	$G_{\text{calculated}}$ (173.94) is more than G_{critical} (5.99) at $p = 0.05$ and therefore we reject the null hypothesis.
7. What does this mean in real terms?	**7. What does this mean in real terms?**
	BOX 5.1.

In this example, $G_{\text{calculated}}$ (173.94) is greater than the $\chi^2_{\text{calculated}}$ (155.47) (BOX 5.1) showing that this method for calculating a goodness of fit produces a higher test statistic, but since the critical value ($\chi^2_{\text{critical}}/G_{\text{critical}}$) is much smaller than either calculated values the outcome for both tests is the same.

5.5.2. An $r \times c$ *G* test for association

Here we are using the data from Example 5.3. where the association between habitat and distribution of snail shell patterns is examined. The observed data are found in Table 5.6. The criteria for using this test are given in 5.3.1. The process of calculating the $G_{\text{calculated}}$ statistic differs critically from the chi-squared. Again the Williams's correction has been used. Where steps are similar to those in BOX 5.4. this is indicated.

BOX 5.8. How to calculate an $r \times c$ G test for association

GENERAL DETAILS	THIS EXAMPLE
1. Hypotheses to be tested **H_0:** There is no association between the two variables. **H_1:** There is an association between the two variables.	**1. Hypotheses to be tested** BOX 5.4.
2. How to work out expected values Not needed	**2. How to work out expected values**
3. How to work out $G_{calculated}$ i. First calculate the totals for each row and column and the grand total. ii. Calculate $\Sigma(O \times \ln O)$ for all categories and sum. (The term ln refers to the natural logarithm. This function button can be found on most statistical calculators.) iii. Calculate the same for the grand total. i.e. $N \times \ln N$ iv. For each row total and column total calculate the same and add together. v. Add the value from (ii) to the value from (iii) and take away the value from (iv). vi. $G = 2 \times$ value from (v.).	**3. How to work out $G_{calculated}$** i. Table 5.6. ii. First take the natural log of an observed value. Then multiply this by the same observed value. e.g. $10 \times \ln 10 = 10 \times 2.30258 = 23.02585$ Repeat this for each observed value and add all these values together = 278.50015 iii. The grand total is 105. So this step is: $105 \times \ln 105 = 105 \times 4.65396 = 488.66584$ iv. The first row total is 50, so for this row: $50 \times \ln 50 = 195.60115$ The second row total is 55, so for this row: $55 \times \ln 55 = 220.40333$ Continue for the four column totals (27, 27, 24 and 27) and add all the resulting values together The total for rows and columns in this example = 759.24055. v. In this example this step will be: $278.50015 + 488.66584 - 759.24055 = 7.92544$ vi. $G = 2 \times 7.92544 = 15.85088$
4. Williams's correction i. First work out 1/each row total and add these values together. Multiply by the grand total. Subtract 1.	**4. Williams's correction** i. $1/50 + 1/55 = 0.03818$ $0.03818 \times 105 = 4.00909$ $4.00909 - 1 = 3.00909$

BOX 5.8. Continued

GENERAL DETAILS	THIS EXAMPLE
ii. Work out 1/each column total and add these together. Multiply by the grand total. Subtract 1.	ii. $1/27 + 1/27 + 1/24 + 1/27 = 0.15277$ $0.15277 \times 105 = 16.04167$ $16.04167 - 1 = 15.04167$
iii. Calculate (i) × (ii)	iii. $(i) \times (ii) = 3.00909 \times 15.04167 = 45.26173$
iv. Next calculate $6n(\text{rows} - 1)(\text{columns} - 1)$	iv. $6 \times 105 \times (2-1) \times (4-1) = 1890$
v. $W = 1 + \frac{(iii)}{(iv)}$ $G_{calculated} = \frac{G}{W}$	v. $W = 1 + \frac{45.26173}{1890} = 1.02395$ $G_{calculated} = \frac{15.85088}{1.02395} = 15.48016$
5. How to find $G_{critical}$ v is the (number of rows − 1) × (the number of columns − 1). Look up $G_{critical}$ in the chi-squared table at $p = 0.05$.	**5. How to find $G_{critical}$** When $\nu = 3$ and $p = 0.05$, $G_{critical} = 7.81$.
6. The rule If $G_{calculated}$ is greater than $G_{critical}$ you may reject the null hypothesis.	**6. The rule** $G_{calculated}$ (15.48) is greater than $G_{critical}$ (7.81) at $p = 0.05$ and therefore we reject the null hypothesis.
7. What does this mean in real terms?	**7. What does this mean in real terms?** BOX 5.4.

If you compare the results from using the chi-squared test for association (BOX 5.4.) and the G test for association (BOX 5.8.) it is again clear that the two calculated values differ. For the same data set the $\chi^2_{calculated}$ (15.19) is less than $G_{calculated}$ (15.48). However, since the critical value is much smaller (7.81) the decision that is made is the same in both cases.

5.6. The problem with small numbers

One of the criteria for using the tests described in this chapter is a requirement for all expected values to be greater than 5. There is some debate about this requirement and it is generally accepted that as long as no more than 20% of all expected numbers are less than 5 and none is less than 1 then you may use these tests.

Clearly, therefore, you need to bear this in mind when designing your investigations if you are planning to use any of the tests described in this chapter. Ideally you need to carry out a preliminary investigation to get a sense of the relative numbers you would expect in each category. If you look at Example 5.5. you can see that not many *C. hortensis* are sampled from the woodland. The sampling strategy used must therefore ensure that adequate numbers of *C. hortensis* are included in the sample for analysis. Another example that we frequently come across is when students wish to use a questionnaire in their honours year research project. For example, a student planned to ask 21 people a closed question and this closed question had 7 possible answers. If these were chosen at random you would expect only 3 respondents to pick each answer. A sample size of 21 is therefore not adequate. This can be clearly seen in Q2/A2 earlier in this chapter.

Even with the best planning you may still find yourself with expected values less than 5. In these circumstances you may, if it is sensible, combine categories. For example, in Table 5.14. the results from a survey of oak trees (*Quercus petraea*) at the Wyre Forest and Mortimer Forest, Shropshire, are shown. Initially the investigator placed the trees in one of four categories: seedling, sapling, immature tree, and mature tree.

You can see that two of the expected values are less than 5. This is more than 20% of the expected values. In addition, one of the values is about 1. To enable you to analyse these data with a chi-squared or G test it is therefore necessary to combine some of these categories and for these data it would appear both biologically acceptable and mathematically useful to combine the data from the immature and mature trees. The revised contingency table and new expected values are shown in Table 5.15.

Table 5.14. Maturity of *Quercus petraea* at the Wyre Forest and Mortimer Forest: the problem with small numbers in a chi-squared or G test

Number of trees	Maturity of *Quercus petraea*				
	Seedling	Sapling	Immature tree	Mature tree	Total
Wyre Forest Observed	51	21	29	4	105
Wyre Forest Expected	51.3	23.3	26.4	**3.8**	
Mortimer Forest Observed	15	9	5	1	30
Mortimer Forest Expected	14.6	6.6	7.5	**1.1**	
Total	66	30	34	5	135

Table 5.15. Revised contingency table with combined classes for the maturity of *Quercus petraea* at the Wyre Forest and Mortimer Forest

Number of trees	Maturity of *Quercus petraea*			
	Seedling	Sapling	Mature and Immature trees	Total
Wyre Forest Observed	51	21	33	105
Wyre Forest Expected	51.3	23.3	30.3	
Mortimer Forest Observed	15	9	6	30
Mortimer Forest Expected	14.6	6.6	8.6	
Total	66	30	39	135

The expected values are now all greater than 5. This does mean, however, that you can only comment on the immature and mature trees in the two woodlands as you will have no statistics in which these two groups are examined.

The example demonstrates in Table 5.14. data that are measured on a nominal scale and the test is a test for association. In this case the combining of classes can be done before the revised expected values are calculated. The same is true for a test for association for data measured on an ordinal or interval scale and where you have an *a priori* expectation and ordinal or nominal data. When you have an *a priori* expectation and interval data the mid-point is used when calculating the expected values from the predicted ratio (5.1.3.iii.). The relative relationship between the mid-point and predicted ratio become disturbed when classes are combined and incorrect expected values will be calculated. In these circumstances you should work out the expected values for the original classes and then add the expected values together in the classes you are combining (BOX 5.2.).

5.7. Experimental design, and chi-squared and G tests

Chi-squared tests in particular are often used inappropriately because they are mathematically simple to carry out and may be used for this reason and no other. Chi-squared and G goodness of fit tests can be used to compare the observed values with those predicted by an *a priori* expectation. The test for heterogeneity allows you to compare several samples and to determine if they are significantly different. The third group of tests will test for an association.

These tests can be used when you have count data organized into two or more categories; for example, the number of trees in a forest at different stages of maturity (Table 5.14.). The categories can be either derived directly from the variable being observed (Example 5.3.) or imposed as 'classes' (Example 5.1.). These categories could include a control, although the information gained about the differences between the effect of the treatment and the control would be limited.

The goodness of fit tests require a single sample with one treatment variable where there are two or more categories (5.1., 5.4.1., and 5.5.1.). In Table 5.1., for example, there is one row of observations in three categories. This experimental design has the generic structure shown in Table 5.16.

Table 5.16. Experimental design that may be suitable for analysis by a chi-squared or G goodness of fit test if all other criteria are met (5.1.1.)

	Category 1	Category 2 etc.
Sample		

This simple design can be extended in that there may be more than one sample as seen in 5.2. Heterogeneity in a goodness of fit test (Table 5.17.). These samples within the chi-squared or G tests for heterogeneity are replicates within the experimental design.

Table 5.17. Experimental design that may be suitable for analysis for heterogeneity in a goodness of fit chi-squared test if all other criteria are met (5.2.1.)

	Category 1	Category 2 etc.
Sample 1		
Sample 2 etc.		

The final type of design that may be suitable for testing with a chi-squared or G test for association (5.3., 5.4.2., and 5.5.2.) is one with two variables each having two or more categories (Table 5.18.).

Table 5.18. Experimental design that may be suitable for analysis by a chi-squared or G test for association if all other criteria are met (5.3.1.)

	Treatment variable 1	
Treatment variable 2	Category 1	Category 2 etc.
Category 1		
Category 2 etc.		

From these you can see that the applications of chi-squared and G tests are limited to only a few experimental designs and you may need to investigate other more flexible tests to accommodate your design.

Q6 Which of the following designs are suitable for analysis by a chi-squared or G test, and which of these tests would be most appropriate?

a. An investigator recorded percentage germination in three varieties of seed and wished to examine whether they were significantly different.

b. Science students on three different courses were surveyed and asked if they agreed with the government reinstating the student grant system. They could answer either 'Yes' or 'No'. The investigator wished to compare the three groups of students.

c. In an investigation of visual responses in insects, bowls of different colours were placed in a grassy area and the numbers of insects attracted to each bowl were recorded. In total three different types of insects (bugs, beetles, flies) were attracted to the bowls. The numbers of each type of insect in each bowl were recorded and the researcher wished to compare the groups of insects in their responses to the particular colours.

A6 a. The first point to note is that the data are a derived variable (percentage) and therefore should not be analysed using a chi-squared or G test. If the researcher had the original data, where the number of seeds sown were recorded and of these the number germinated, then a table could be drawn up (Table 5.19.).

Table 5.19. Generic outline of the experimental design from Q3a but using count data not percentages

	Number of seeds that germinated	Number of seeds that did not germinate
Variety 1		
Variety 2		
Variety 3		

This design is in accord with Table 5.18., and if all other criteria are met (5.3.1.) then these data may be analysed using a chi-squared or G test for association.

b. The data here reflect two variables, the 'answer to the question' and the 'three courses', which might suggest that a chi-squared or G test for association is appropriate. The data could be organized in two ways (Tables 5.20. and 5.21.).

Table 5.20. First arrangement of data. Comparing the courses for the numbers of students giving a yes or no answer

	Number of students on the course		
Answers	**Course 1**	**Course 2**	**Course 3**
Yes			
No			

Table 5.21. Second arrangement of data. Comparing the answers given to a particular question by students on particular courses

	Number of students giving this response	
	Yes	**No**
Course 1		
Course 2		
Course 3		

In the chi-squared and G tests it is the rows that are compared, the distribution of values in one row being contrasted with the distribution of values in the other row(s). For Table 5.20. this does not make sense. Row 1 cannot help but be the direct opposite to row 2 since a student could only answer yes or no. In the second version (Table 5.21.) the three rows can sensibly be compared: what is the difference in relative responses to the question by the students on the three courses? We have come across many chi-squared analyses where inappropriate comparisons of the sort indicated in Table 5.20. are made. Check that you organize your data so that it really does make sense to compare the rows. Therefore, for this example the data should be organized in the form shown in Table 5.21. and then a chi-squared or G test for association may be carried out if all other criteria are met (5.3.1.).

c. There are two variables here: the colour of the bowls and the number of insects of each type (Table 5.22.).

Here we have organized the data so that we are comparing the numbers of insects found in the bowls for each given insect type. As we explained in A6b the chi-squared or G test for association will test whether the relative numbers of insects within each coloured bowl is statistically similar for all groups of insects. As in all chi-squared or G tests for association the comparison is between the rows and you learn little about the difference between the columns. If you were interested in this then you could use a

Table 5.22. Experimental design for a comparison between the numbers of different groups of insects attracted to different coloured bowls

	Number of insects attracted to each coloured bowl			
Insect type	**Red**	**White**	**Green**	**Blue**
Bugs				
Beetles				
Flies				

chi-squared goodness of fit test for each group of insect (e.g. bugs, then beetles, and then flies as three separate calculations) or we suggest you reconsider your statistical test and have a look at the two-way non-parametric ANOVA or the Scheirer–Ray–Hare test (Chapter 8).

Summary of Chapter 5

- The two groups of statistical tests considered in this chapter are the more commonly used chi-squared test (5.1.–5.4.) and the G test (5.5.). The latter is more useful when you have a large data set.
- The chi-squared and G tests may be used to test two hypotheses: 'Do the data match an expected ratio?' (5.1., 5.4.1., and 5.5.1.) or 'Is there an association between two or more variables?' (5.3., 5.4.2., and 5.5.2.). These tests may also be used to check if your samples are heterogeneous (5.2.).
- The chi-squared and G tests may be used if you have count data and one or two treatment variables organized into categories (5.7.). If there are only two categories for the variables then a Yates's correction should be used for the chi-squared test. In the G test the Williams's correction is always used (5.4. and 5.5.).
- Both the chi-squared test and G test require you to work out expected values which are then compared with your observations. In the goodness of fit tests these expected values are derived from your *a priori* expectation (5.1.) and in the test for association the expected values are determined by totals from the complete data set (5.3.).
- There are only three experimental designs suitable for chi-squared and G tests (5.7.).
- Common faults in the use of these tests are the use of percentage data and organizing the data incorrectly in the contingency table so that the wrong comparisons are made (5.7., Q6).
- The Online Resource Centre includes interactive exercises that test your understanding of this chapter with other topics, particularly those considered in Chapters 2–8.

6

Hypothesis testing: Associations and relationships

In your research you often investigate the effect of one **treatment variable**, for example, a comparison of snail shell patterns found in coastal and woodland populations of *Cepea nemoralis* (Table 5.6.). The treatment variable here is habitat. But there are some investigations where you will have more than one variable, such as when comparing leg length and arm length in humans (Example 6.3.). In this chapter, therefore, we look at testing **hypotheses** where you wish to evaluate the possibility of an association between these two variables. This is the second of the three types of hypotheses we discussed in Chapter 4 (4.1.2.). We cover two groups of tests: correlations and regressions. If you work through this chapter and the questions it should take about 2 hours. We have included further examples in the Online Resource Centre.

Associations and relationships

It is important to be clear about what we mean by the terms 'association' and 'relationship'. In some textbooks these terms are used interchangeably, but we believe it is important to make a distinction.

Association: Where one variable is found to change in a similar manner to another variable (e.g. Fig. 6.14 and Fig. 6.15.).

Relationship: Where one variable is ***directly*** responsible for causing the change in another variable.

Most samples of humans in which arm and leg length are recorded show a significant association between these two variables: if you have relatively long arms you will also have relatively long legs (Example 6.3.). But you would not argue that your arm length directly causes your leg length. This is not sensible. Instead, this association reflects the actions of other phenomena that have not been observed. Therefore, although you may have a significant association, this does not indicate a relationship. You can contrast this example with a significant association observed between the hardness of eggshells and the amount of feed supplement eaten by pullets (Example 6.1.). In this case you could argue that nutritional factors in the feed supplement directly influence eggshell hardness. This is an association for which you can argue that there is a relationship.

A correlation or regression analysis tests for the presence of a statistically significant linear association. As such the change in one **variable** (y) may well be caused by the change in the second variable (x), or the change in the second variable may be caused by the change in the first variable. These are both relationships. In addition to this a significant association may indicate that the changes in the first and second variables are due to a change in a third, unrecorded variable, as described in our arms and legs example. Alternatively, the statistical association between two variables may be a coincidence. You will need to bear these points in mind when you are interpreting the results from any significant test for association.

How to choose the correct test

In this section we first discuss when you might consider using a chi-squared or G test for association (Chapter 5) and when a correlation or regression (Chapter 6). We then consider why you might choose a correlation and/or a regression to analyse your data. Finally we identify the key points to enable you to choose between the correlations and regressions included in this chapter.

Should I use a chi-squared test for association or a correlation/regression analyses?

In Chapter 5 (5.3., 5.4.2., and 5.5.2.) you were introduced to the chi-squared and G tests for association. For example, we examined the frequency of *Cepea nemoralis* and *C. hortensis* in a hedge and a wood (Example 5.5.). This was an example of an investigation where we wished to test for an association in the data. This was done by comparing the relative distributions in the two samples to an average 'expected' distribution. These tests are particularly useful if either of your variables is measured on a **nominal** scale or if the apparent association is not linear. In this chapter we look at another group of tests where you test for an association. For example, is there an association between the concentration of zinc extracted in soil samples and the distance from an electricity pylon (Example 6.2.). These variables are not measured on a nominal scale and the apparent association is linear (Fig. 6.12.) and therefore a correlation or regression is appropriate. However, if you are choosing a test and your variables are measured on an interval or ordinal scale you should check both the chi-squared and G tests for association and the tests introduced in this chapter before making your choice.

Should I use a correlation and/or a regression analysis?

When making this decision you need to consider your reasons for choosing your analysis. These tests of association can be used for three things: modelling, prediction, and establishing if there is a significant association. The line that may be used to describe the trend in the data is a model of this association. For example, in Fig. 6.15. the line that best fits the data from Table 6.1. is shown and indicates a positive association between the amount of feed supplement eaten by pullets and eggshell hardness. This line is described by the equation $y = -2.4 + 0.5x$ (BOX 6.6.). This line can also be used to predict a y value for any given x value within the range of your data. For example, if the pullets were given 10 g of feed supplement (x) then the hardness of their eggshells would be expected to be about 2.6 units. If these eggshells were hard enough for the eggs to be sold then the feed supplement given to the pullets could be more economically controlled. For both of these applications (modelling and prediction) you should choose a regression analysis. However, if you wish to establish whether there is a significant association between the two variables then a correlation analysis is more likely to be appropriate, but you should read on before making your final choice.

Which correlation or regression?

Each test introduced in this chapter has several requirements that must be met and these details are given at the start of each section. The following guide takes you to the most likely tests for your data. The regression analyses fall into two groups. The first (Model I) may be appropriate when you, the investigator, determine one of the variables. For example, you may set up an experiment to examine the number of pollen grains falling on agar plates set at specific distances from a field of oilseed rape. In this investigation the location of the agar plates will be under your control and is not therefore subject to **sampling error**. If neither variable is under your control then you should consider using a Model II regression. However, there is some debate as to which is the most appropriate test to use for a given set of data. We have endeavoured to summarize this debate in the following table, but for a fuller discussion you should refer to Legendre & Legendre (1998).

You have two treatment variables. The data are **non-parametric**.* You wish to test for an association.	Spearman's rank correlation (6.2.)
You have two treatment variables. The data are **parametric**. You wish to test for an association.	Pearson's product moment correlation (6.3.)

You have two treatment variables, one of which is under the control of the investigator. You wish to test for an association or you wish to model the association and/or predict *y* values for given *x* values within the range of your observations.	Simple linear regression (6.5.)
You have two treatment variables and neither is under the control of the investigator. The two variables are measured on the same scale, e.g. mm. You wish to model the association and/or predict *y* values for given *x* values within the range of your observations.	Principal axis regression (6.6.)
You have two treatment variables and neither is under the control of the investigator. The two variables are measured on different scales, e.g. grams and arbitrary units. You wish to model the association and/or predict *y* values for given *x* values within the range of your observations.	Ranged principal axis regression (6.7.)
None of these tests seems to be right for your data. For example, you have three or more treatment variables; you have more than one *y* value for each *x* value; you do not have a linear distribution.	Grafen & Hails, 2002; Legendre & Legendre, 1998; Sokal & Rohlf, 1981.

* Parametric tests are more **powerful**; therefore, if you have parametric data you should always use the parametric test. If you have non-parametric data you should consider **transforming** them (3.9.).

6.1. Correlations

If you have data with two treatment variables and you plot this on a scatter plot, it may suggest that there is a linear association between the two variables. A correlation analysis examines the data mathematically to see how probable this is. The two important values for this analysis are the correlation coefficient (r) and the probability value (p). The correlation coefficient (r) may be either positive in sign or negative. A positive sign indicates that there is a positive association between the two variables: as one increases so does the other (Fig. 6.1.). A negative sign indicates a negative association, as one variable increases the other decreases (Fig. 6.2.).

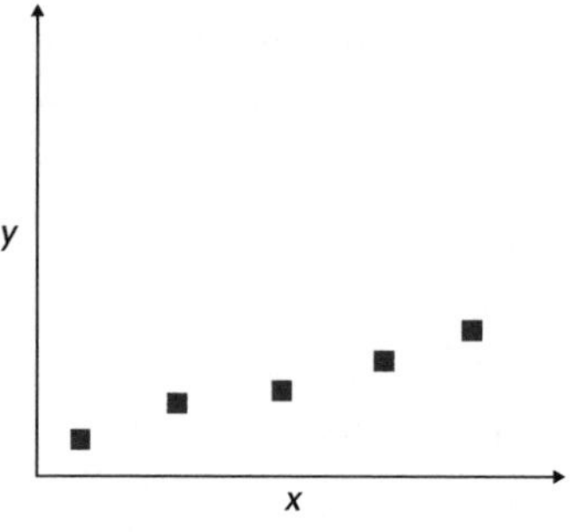

Fig. 6.1. A positive linear association. As x increases, y increases.

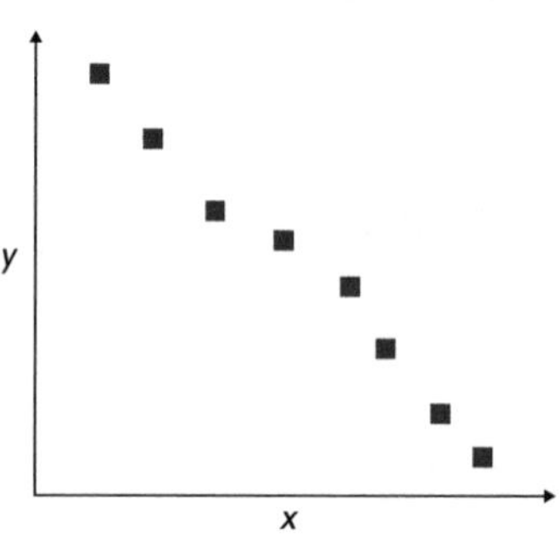

Fig. 6.2. A negative linear association. As x increases, y decreases.

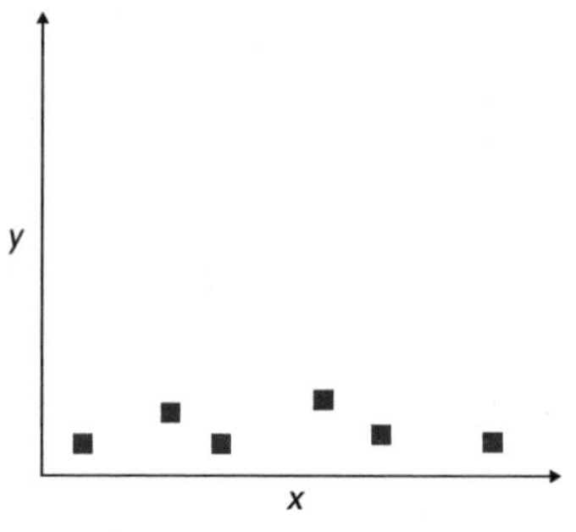

Fig. 6.3. Is there an association between x and y?

Q1 **Is there an apparent linear association between the two variables illustrated in Fig. 6.3.?**

A1 No. As the x values increase the value of y stays the same. Therefore, there is no association between these two variables.

A correlation coefficient will lie somewhere between -1 and $+1$. If you have a value greater than $+1$ or less than -1 then you have made a mistake somewhere in your calculation. A value of $r=0$ means that there is zero correlation, i.e. no association between the two variables, whereas r = +1 or r = −1 indicates a 'perfect' association between the two variables.

A correlation coefficient in itself does not tell you whether this is a statistically significant association: a comparison between the calculated value and a critical value for a given level of probability must be made. For example, $r = 0.2$ indicates a weak association and $r = 0.8$ indicates a stronger association. A strong association is likely to be statistically significant even with a small sample size. However, a small correlation such as $r = 0.2$ may be statistically significant but only if the sample size is large enough. You need to bear this in mind when interpreting findings from your own or other people's research.

6.2. Spearman's rank correlation

This is a ranking test. There are many ranking tests which can be used to test hypotheses. These tests are commonly used for the analysis of non-parametric data where the data does not have to fit a particular type of distribution (3.8.). These tests are collectively known as ranking tests since the first step, which they all have in common, requires you to assign ranks to your observations. If you are not familiar with this process you should first read BOX 3.3.

6.2.1. To use this test you:

1. Wish to test for an association between the treatment variables.
2. Need two treatment variables recorded for each item.
3. Have non-parametric data.
4. Have an apparently linear distribution where the points are reasonably scattered (Q2).
5. Do not need to have an independent and a dependent variable.
6. Have variables that are measured on rankable scales.
7. Have 7 or more pairs of observations. Having more than 30 pairs of observations does not add to the accuracy of this test.
8. Should have few tied values, i.e. if many of the numbers in one of the data sets are the same. This may look like an association, but it is not (Q1).

Q2 **Examine Figs 6.1., 6.2., and 6.4a–d. Which of these fulfil criterion 4? For those distributions that do not meet criterion 4, what can be done about it?**

A2 Only Figs 6.1. and 6.2. meet criterion 4.

Fig. 6.1. illustrates a positive linear distribution, whereas Fig. 6.2. illustrates a negative linear distribution.

Fig. 6.4a. has a gap in the middle of the distribution. To resolve this you would need to repeat the experiment ensuring that observations are recorded for the middle part of the distribution.

In Fig. 6.4b. the data are bunched and again you would gain a better understanding of your topic of investigation by repeating the experiment and expanding the range of measurements taken.

In Fig. 6.4c. there is one **outlying** observation. You cannot ignore this datum point. If you have data like this you should first carry out your data analyses with this datum point included. You can then repeat the analysis with this datum point excluded. You should report on the outcomes of both analyses.

Fig. 6.4d. shows an apparent curvilinear association. Curvilinear data can be analysed using correlation and/or regression analyses if one or both sets of observations are transformed (3.9.) so that a linear distribution results, or using methods not described in this book.

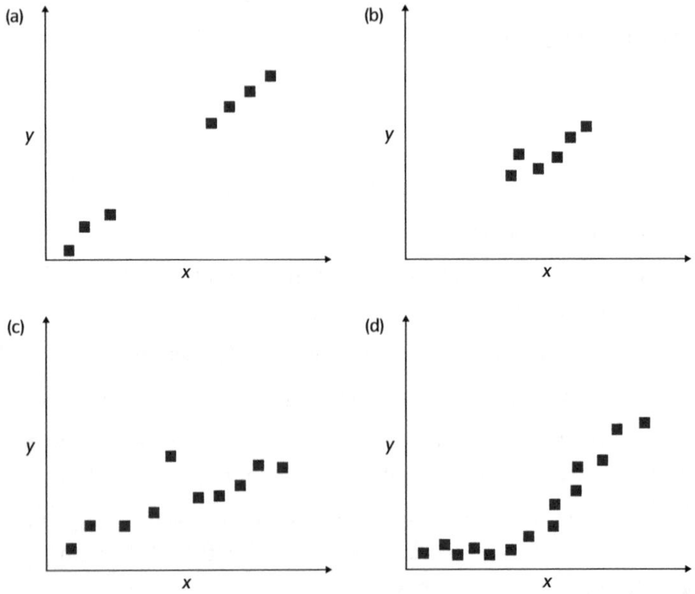

(a) y x

(b) y x

(c) y x

(d) y x

Fig. 6.4. a–d. Do these distributions meet the criteria for using a correlation analysis?

EXAMPLE 6.1. The hardness of eggshells in pullets

A student believes that there is a relationship between the hardness of shells in eggs laid by Maren pullets and the amount of a particular layer's pellet eaten as part of their mixed diet. She selected 13 of her pullets at random and on one day recorded the amount of layer's pellets consumed and the hardness of the eggs laid, on a (scale of 0.0 (soft) to 10.0 (hardest) (Table 6.1.).

Table 6.1. The hardness of eggshells produced by 13 Maren pullets and their consumption of a food supplement

Pullet	Amount of food supplement (g)	Hardness of shells
1	19.5	7.1
2	11.2	3.4
3	14.0	4.5
4	15.1	5.1
5	9.5	2.1
6	7.0	1.2
7	9.8	2.1
8	11.6	3.4
9	17.5	6.1
10	11.2	3.0
11	8.2	1.7
12	12.4	3.4
13	14.2	4.2

Do the results in Example 6.1. meet all the criteria for using a Spearman rank correlation? Yes. For each pullet there are two treatment variables: the hardness of shells and the amount of food supplement eaten. Both these scales of measurement can be ranked. There are 13 pairs of observations. Both sets of data show reasonable amounts of variation and when plotted on a scatter plot there appears to be a linear association and the points are reasonably scattered (Fig. 6.5.).

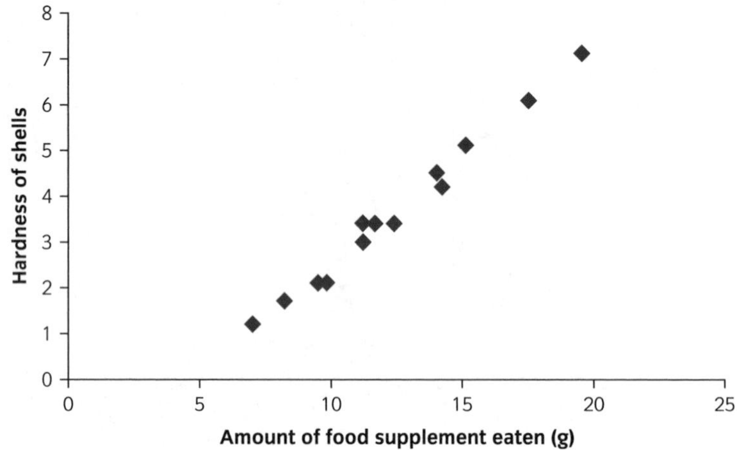

Fig. 6.5. Amount of food supplement eaten (g) and hardness of eggshells in Maren pullets.

6.2.2. The calculation

We show you the general calculation and a specific example (Example 6.1.) for a Spearman's rank correlation in BOX 6.1. In addition, you will need to refer to the calculation table (6.2.). The full details of all steps in these calculations are given in the Online Resource Centre.

BOX 6.1. How to carry out a Spearman's rank correlation

GENERAL DETAILS	THIS EXAMPLE
1. Hypotheses to be tested **H_0**: There is no correlation between variable 1 and variable 2 in the **population** from which the **samples** are taken. **H_1**: There is a correlation between variable 1 and variable 2 in the population from which the samples are taken.	**1. Hypotheses to be tested** **H_0**: There is no correlation between the hardness of shells and the amount of food supplement (g). **H_1**: There is a correlation between the hardness of shells and the amount of food supplement (g).
2. How to work out $r_{s\ \text{calculated}}$ i. For this test it does not matter if you have a dependent and an independent variable therefore let the variable on the x-axis be variable 1 and the variable on the y-axis be variable 2. ii. Arrange the data for variable 1 in numerical order from the smallest to the largest. iii. Assign each observation a rank. Where there is a tie assign the middle (average) rank. iv. Repeat steps i. and ii. for variable 2. v. Return to your original table of data and note down the ranks for each observation. Make sure you do not reorganize the data. The two observations for each item must remain in the same row as each other. vi. Calculate the difference between the ranks for each pair of observations (d). vii. Square each d value and then sum all the d^2 to work out the Σd^2. viii. Calculate r_s: $$r_s = 1 - \frac{6\Sigma d^2}{n(n^2 - 1)}$$ where n is the number of pairs of observations.	**2. How to work out $r_{s\ \text{calculated}}$** i. Let the amount of food supplement eaten by the pullets be variable 1 and shell hardness be variable 2 (Table 6.1). ii., iii., iv., v. If you are working this calculation out by hand it is simplest to use a calculation table. The outcome from steps ii., iii., iv., v., and vi. are shown in Table 6.2. using the data from Example 6.1. A common error is to reorganize the data in this table to reflect the ranks. This can lead to the two observations from one item no longer being paired on the same row. If you are not sure, compare Table 6.1. with Table 6.2. The observations remain in the same order in relation to each other. vi., vii. The difference between the ranks and sum of ranks has been added to the calculation table (Table 6.2.). viii. In this example $n = 13$ pairs of observations. So: $$r_s = 1 - \frac{6 \times 6.0}{13(13^2 - 1)} = 0.98352$$
3. How to find $r_{s\ \text{critical}}$ Find $r_{s\ \text{critical}}$ in a Spearman's rank r table for a two-tailed test at $p = 0.05$ and where $n =$ the number of pairs of observations.	**3. How to find $r_{s\ \text{critical}}$** $r_{s\ \text{critical}}$ at $p = 0.05$, $n = 13$ is 0.566

BOX 6.1. Continued

GENERAL DETAILS	THIS EXAMPLE
4. The rule If the absolute, calculated value of r_s is greater than the critical r_s we reject H_0.	**4. The rule** In our example r_s was positive so the fact that we take the absolute value will not change r_s in this example. $r_{s\ calculated}$ (0.984) is greater than $r_{s\ critical}$ (0.566) at $p = 0.05$. In fact at $p = 0.01$ the $r_{s\ critical}$ is 0.745. Therefore we can reject the null hypothesis at this higher level of significance.
5. What does this mean in real terms?	**5. What does this mean in real terms?** There is a strong and highly statistically significant positive correlation ($r_s = 0.989$, $p = 0.01$) between the hardness of shells of eggs and the amount of food supplement (g) eaten by Maren pullets.

Table 6.2. Calculating a Spearman's rank correlation between the hardness of shells and consumption of food supplement in Maren pullets

Amount of food supplement (g)	Rank	Hardness of shells	Rank	Difference (d)	Difference^2 (d^2)
19.5	13	7.1	13	0.0	0.0
11.2	5.5	3.4	7	−1.5	2.25
14.0	9	4.5	10	−1.0	1.0
15.1	11	5.1	11	0.0	0.0
9.5	3	2.1	3.5	−0.5	0.25
7.0	1	1.2	1	0.0	0.0
9.8	4	2.1	3.5	0.5	0.25
11.6	7	3.4	7	0.0	0.0
17.5	12	6.1	12	0.0	0.0
11.2	5.5	3.0	5	0.5	0.25
8.2	2	1.7	2	0.0	0.0
12.4	8	3.4	7	1.0	1.0
14.2	10	4.2	9	1.0	1.0
$n = 13$					$\Sigma d^2 = 6.0$

6.3. Pearson's product moment correlation

This is a parametric test and unlike the Spearman's rank test you do not rank the data; however, you do assume that your data are parametric (3.8.). Like the Spearman's rank correlation the coefficient (r) may indicate the strength and sign of the association that is apparent in your data (6.1.). However, to test whether this correlation arose by chance or is statistically significant a modified t test is used (BOX 6.2.).

6.3.1. To use this test you:

1. Wish to test for an association between treatment variables.
2. Have two treatment variables measured for each item.
3. Have parametric data. Both variables must be approximately normally distributed (BOX 3.2.).
4. Have a distribution that appears to be linear when plotted, with points that are reasonably scattered (e.g. Fig. 6.1., Fig. 6.2., Q2).
5. There is no requirement for there to be an independent or a dependent variable.

Although the Maren pullet data was used in 6.2. to illustrate the Spearman rank correlation, these data also meet the more stringent criteria for using a Pearson's correlation and so are used here. If you have data that meet the criteria for a parametric test this is the one you should choose since these tests are more powerful (3.8.1.). The data from Example 6.1. meet the criteria for the Pearson's correlation. There are two treatment variables (amount of food supplement eaten and hardness of shells); we wish to test for an association between these two variables; both columns of data meet the criteria for normally distributed data and are therefore parametric; and as we have already seen in Fig. 6.5. the distribution is apparently linear and the points are reasonably scattered.

6.3.2. The calculation

We show you the general calculation and a specific example (Example 6.1.) for a Pearson's correlation in BOX 6.2. and the calculation table (6.3.). Where steps have been abbreviated the full calculation is included in the Online Resource Centre.

BOX 6.2. How to carry out a Pearson's product moment correlation

GENERAL DETAILS	THIS EXAMPLE
1. Hypotheses to be tested $\mathbf{H_0}$: There is no correlation between variable 1 and variable 2 in the population from which the samples are taken. $\mathbf{H_0}$: There is a correlation between variable 1 and variable 2 in the population from which the samples are taken.	**1. Hypotheses to be tested** $\mathbf{H_0}$: There is no correlation between the hardness of shells and the amount of food supplement. $\mathbf{H_1}$: There is a correlation between the hardness of shells and the amount of food supplement.
2. How to work out *r* and $t_{calculated}$ i. Again there is no requirement for there to be a dependent or an independent variable, so you may choose which variable should be *x* and which *y*.	**2. How to work out *r* and $t_{calculated}$** i. Let the amount of food supplement eaten be the *x* variable and hardness of shells be the *y* variable.
ii. Calculate the sums of squares of *x* (SS (*x*)): $SS(x) = \Sigma x^2 - \frac{(\Sigma x)^2}{n}$ where *n* is the number of pairs of observations. This term should be familiar to you as it has figured in the calculation for standard deviation and variances (BOX 3.1.). Alternatively, many calculators will work out these terms for you. If you are using a calculator or computer to work out the whole correlation, make sure you use the sample statistics (1.7.).	ii. These are important values which you will come across frequently in this book so we have shown you again in this example how to work these out long-hand using a calculation table (Table 6.3.). $SS(x) = 2153.88 - \frac{25985.44}{13}$ $= 155.0$
iii. Calculate the sums of squares *y* (SS(*y*)). For the SS(*y*) substitute the '*y*' variable data into the same formula as above, i.e. $SS(y) = \Sigma y^2 - \frac{(\Sigma y)^2}{n}$	iii. For details see Table 6.3 $SS(y) = 208.35 - \frac{2237.29}{13}$ $= 36.25077$
iv. Calculate the sums of products (SP(*xy*)). For your data, multiply each *x* with its related *y* to produce a column of values (*xy*). Add these together = Σxy. Using this and values from ii and iii. calculate SP(*xy*) as follows: $SP(xy) = \Sigma xy - \frac{(\Sigma x)(\Sigma y)}{n}$	iv. For details see Table 6.3. $SP(xy) = 661.0 - \frac{161.2 \times 47.3}{13}$ $= 74.48$

BOX 6.2. Continued

GENERAL DETAILS	THIS EXAMPLE
v. Now you can calculate r. $r = \frac{SP(xy)}{\sqrt{[SS(x)\ SS(y)]}}$ Remember r is the correlation coefficient and should fall in the range ± 1. The sign indicates whether this is a positive or negative association. r does not indicate the level of significance. For this a modified t test is used.	v. $r = \frac{74.48}{\sqrt{(155.0 \times 36.25077)}} = 0.99361$ This indicates a strong positive correlation. But is it statistically significant?
vi. To work out t calculated $t_{\text{calculated}} = \frac{r\sqrt{(n-2)}}{\sqrt{(1-r^2)}}$	vi. $t_{\text{calculated}} = \frac{0.99361\sqrt{(11)}}{\sqrt{\left[1-(0.99361)^2\right]}} = 29.19724$
3. How to find t_{critical} Look up the value in a t table, for a two-tailed test at $p = 0.05$ where the degrees of freedom (ν) are $n-2$.	**3. How to find t_{critical}** When $\nu = 13 - 2 = 11$, at $p = 0.05$ then $t_{\text{critical}} = 2.201$.
4. The rule If the absolute value of $t_{\text{calculated}}$ is greater than the critical value of t, then you may reject H_0.	**4. The rule** The calculated value of t (29.197) is greater than the critical value of t (2.201), so you may reject the null hypothesis. In fact t_{critical} at $p = 0.001$ is 4.437 and you may therefore reject the null hypothesis at this highly significant level.
5. What does this mean in real terms?	**5. What does this mean in real terms?** There is a strong ($r = 0.99$) and highly significant positive correlation ($p < 0.001$) between the amount of food supplement (g) given to the Maren pullets and the hardness of the eggshells that they produce.

6.3.3. Coefficient of determination

$r^2 \times 100$ is a useful measure called the coefficient of determination (%). It is a measure of the proportion of variability in one treatment variable that is accounted for by variation in the other treatment variable and therefore indicates to what extent other factors are influencing x and y. The use of

Table 6.3. Calculating a Pearson's product moment correlation between the hardness of shells and food consumption in Maren pullets

Variable x Amount of food supplement (g)	x^2	Variable y Hardness of shells	y^2	$x \times y$
7.0	49.0	1.2	1.44	8.4
8.2	67.24	1.7	2.89	13.94
9.5	90.25	2.1	4.41	19.95
9.8	96.04	2.1	4.41	20.58
11.2	125.44	3.0	9.0	33.6
11.2	125.44	3.4	11.56	38.08
11.6	134.56	3.4	11.56	39.44
12.4	153.76	3.4	11.56	42.16
14.0	196.0	4.5	20.25	63.0
14.2	201.64	4.2	17.64	59.64
15.1	228.01	5.1	26.01	77.01
17.5	306.25	6.1	37.21	106.75
19.5	380.25	7.1	50.41	138.46
Σx 161.2	$\Sigma(x^2)$ 2153.88	Σy 47.3	$\Sigma(y^2)$ 208.35	Σxy 661.0
$(\Sigma x)^2$ 25985.44		$(\Sigma y)^2$ 2237.29		
$\bar{x}$ 12.4		$\bar{y}$ 3.63846		
n 13		n 13		

the coefficient of determination is dependent on whether your data satisfy the criteria for a Pearson's product moment correlation and an r value can be calculated.

From BOX 6.2. we know that $r = 0.99361$. Therefore $r^2 \times 100 =$ 98.7%. This indicates that in Example 6.1. nearly all the variation observed in the hardness of the eggshells is due to the variation in the amount of food supplement eaten.

6.4. Regressions

A regression analysis differs from a correlation analysis in that it calculates the mathematical equation that describes a straight line for your data. This model may be used as a description of any significant association between the two variables and/or used to predict values. We describe two methods for modelling a regression line. In the first (simple linear regression) the line is determined by mathematically minimizing the distance between the observations and the line (Fig. 6.6.). Clearly, in this process some distances will be negative and some positive so these values are squared. This regression is also therefore known as least squares regression. The two important values that are calculated in the simple linear regression are the slope or gradient of the linear association (b) and the point at which the line cuts the y-axis (a). The general mathematical formula which describes this line is $y = a + bx$ (Fig. 6.7.). This is a Model I regression. In the principal axis regression and the ranged version of the principal axis regression (Model II) the line is determined by 'drawing' an ellipse around the data. The flatter the ellipse, the more likely it is that the association will be significant (Fig. 6.8.). The line which passes along the longest axis of the ellipse is called the principal axis, and this can also be found by calculating the slope b_1. In these regressions the general mathematical formula for the line (the principal axis) is $y = \bar{y} + b_1(x - \bar{x})$. As in correlation analyses the slope b or b_1 can be

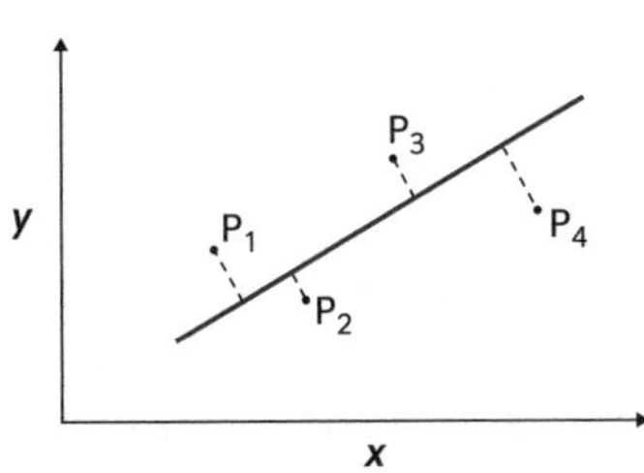

Fig. 6.6. Minimization of differences from regression line. Differences from P_1 and P_3 to the line are positive; differences from P_2 and P_4 are negative.

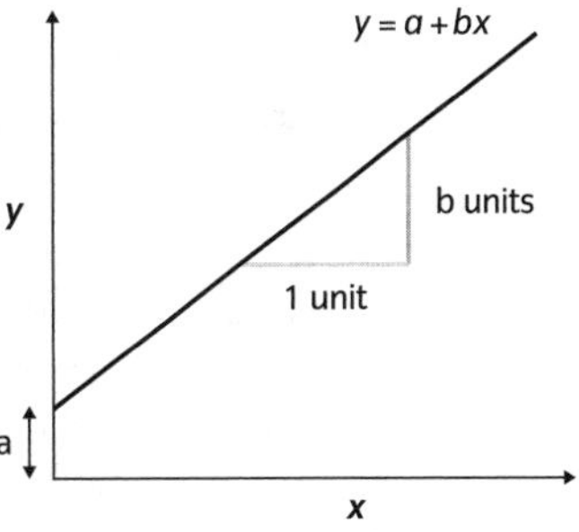

Fig. 6.7. The general line described by a Model I simple linear regression.

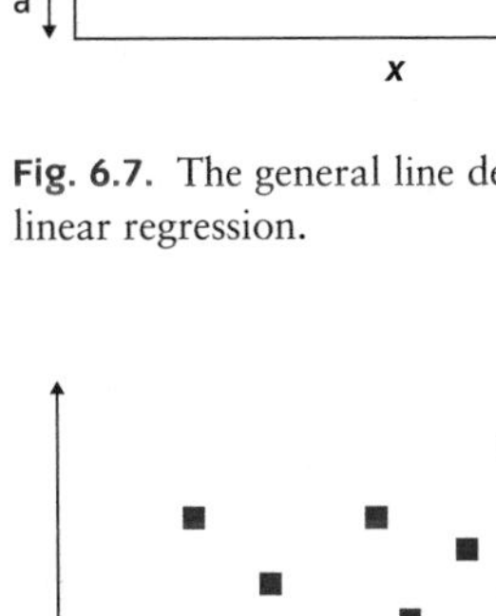

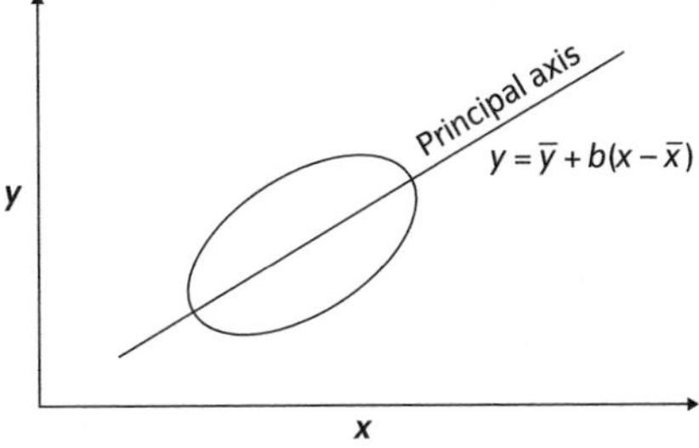

Fig. 6.8. The general line described by the Model II principal axis and ranged principal axis regressions.

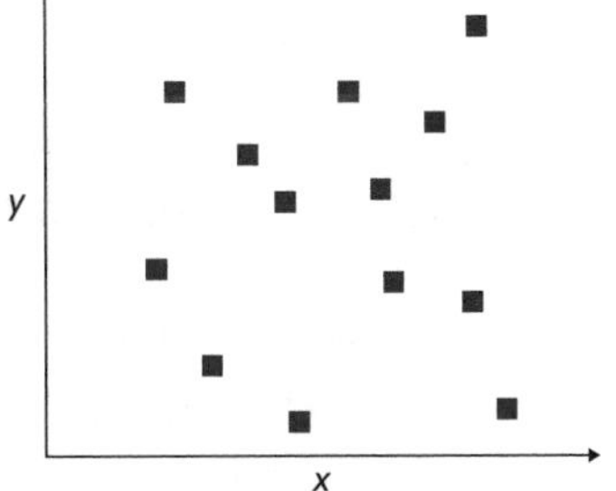

Fig. 6.9. Distribution of data with no obvious linear association.

either positive or negative. These b values are sometimes referred to as regression coefficients.

You could fit a line through any data on a scatter plot. For example, you could draw a line on Fig. 6.9., although clearly this would be meaningless. Just drawing a line is not enough to indicate an association. It is therefore necessary to show whether this line represents a statistically significant trend in the data. For a simple linear regression a modified t test is used (BOX 6.4.). However, for the principal axis regression and ranged principal axis regression a correlation should be carried out to test the significance of the association before making use of any regression analysis.

In the following sections we describe the methods for drawing and testing a regression line when you have two variables with no replicates. However, regressions can be modified in a number of ways, for example, when you have replicates of the y variable or if you have three treatment variables. An explanation of these and other uses of regression analyses are given in, for example, Sokal & Rohlf (1981).

6.5. Model I: Simple linear regression

A simple linear regression analysis is the most commonly used type of regression. In this analysis first the coefficients a and b are calculated that allow you to mathematically describe and to draw the regression line. To test the significance of this line a modified t value is calculated. One of the criteria that must be met to use a simple linear regression is that one of the variables is taken without sampling error. This is always denoted the x variable, and since regressing y on x is not the same as regressing x on y you must take care to identify which is the variable taken without sampling error (x) and which is not (y). Although a change in x may not directly cause a change in y (6.1.), y is often referred to as the dependent variable and x as the independent variable.

6.5.1. To use this test you:

1. Wish to test for an association between two treatment variables.
2. Have an association that appears to be linear, with the points reasonably scattered (Q2).
3. Assume that one variable (x) has been taken without sampling error and as such is usually under the investigator's control.

4. Should know that each value of y varies normally with each value of x and they should have a similar variance. (This sounds complicated but you can tell if your data probably meet this criterion if the points on your scatter plot are fairly evenly distributed along the whole line (Fig. 6.10. compared with Fig. 6.11.).)
5. Have three or more pairs of measurements.
6. Should note that if you are using the modified t test to test the significance of the association then the data should be parametric.

Example 6.2. meets the criteria for using a simple linear regression analysis. For each item (sample of soil) there are two treatment variables (distance from pylon and concentration of zinc). Fig. 6.12. shows that the association appears to be linear and the points are reasonably scattered. The x variable (distance from pylon) was under the investigator's control and there is no sampling error when determining the distance from the pylon. The concentrations of zinc are the y or dependent variables. There are five pairs of measurements and no replicates.

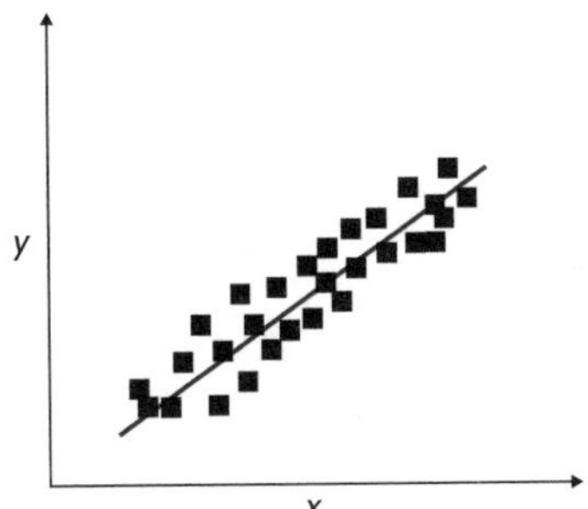

Fig. 6.10. Points evenly distributed about the line, indicating that y varies normally with x.

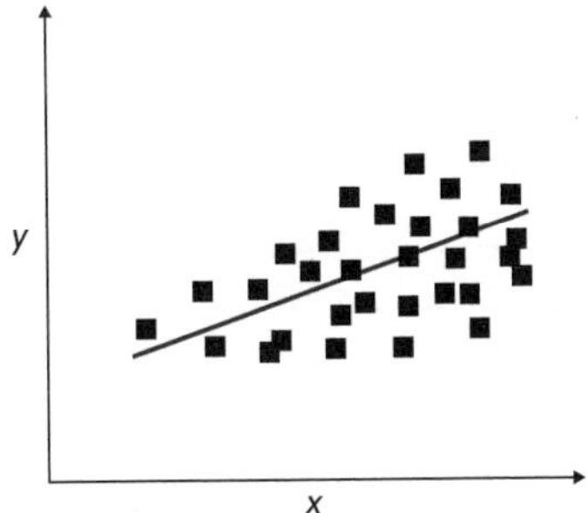

Fig. 6.11. Points becoming more spread out as x increases, indicating that y does not vary normally with x.

EXAMPLE 6.2. Heavy metal contamination of soil under electricity pylons

An undergraduate investigated heavy metal tolerance in plants growing under electricity pylons. As part of her study she recorded the concentration of zinc in soil samples taken at regular intervals moving away from the pylons (Table 6.4). The aim of her investigation was to see if there was an association between the distance from the pylon and the concentration of zinc in the soil.

Table 6.4. Zinc concentrations in soil at specific distances from an electricity pylon

Distance from pylon (m)	Zinc concentration (μg Zn/g soil)
1.0	648
1.5	610
2.0	534
2.5	500
3.0	472

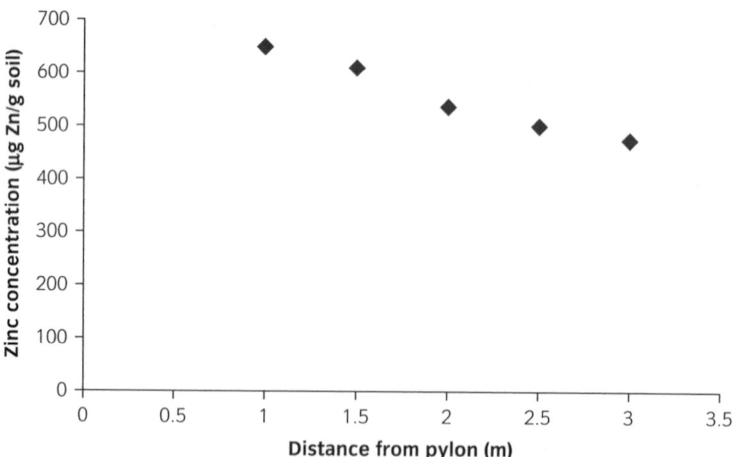

Fig. 6.12. Zinc concentration in soil at specific distances from an electricity pylon.

6.5.2. The calculation

There are two parts to this regression analysis. In the first part you calculate the coefficients that allow you to draw the regression line through

your data. BOX 6.3. explains how to calculate the regression line for a simple linear regression. Full details of the calculation are included in the Online Resource Centre.

The second step of the simple linear regression is to see if there is a statistically significant association that is described by this line. BOX 6.4. shows you this part of the calculation.

BOX 6.3. How to carry out a Model I. Simple linear regression: drawing a regression line

GENERAL DETAILS	THIS EXAMPLE
1. How to work out the regression coefficient *b*	**1. How to work out the regression coefficient *b***
i. The independent variable is that plotted on the x-axis. The dependent variable is plotted on the y-axis.	i. Variable x, the distance from the pylon is under the investigator's control and is the independent variable. Variable y is the concentration of zinc in the soil.
ii. Calculate the following: Σx Add all x values together $(\Sigma x)^2$ Square the Σx value Σx^2 Square all the x values and sum Σy Add all the y values together Σxy Multiply each x by its corresponding y. Add all these xy values together.	ii. Table 6.5. summarizes these calculations.
iii. $b = \dfrac{n\Sigma xy - \Sigma x\Sigma y}{n\Sigma x^2 - (\Sigma x)^2}$	iii. In this example $b = \dfrac{(5 \times 5297) - (10 \times 2764)}{(5 \times 22.5) - 100} = -92.4$
2. How to work out *a*	**2. How to work out *a***
i. Examine your data and record the means for each variable.	i. The means for the data are shown in Table 6.5.
ii. a can be calculated using the general formula for this straight line. So that: $a = \bar{y} - b\bar{x}$	ii. Therefore: $a = 552.8 - (-92.4 \times 2) = 552.8 + 184.8 = 737.6$
3. What is the regression equation?	**3. What is the regression equation?**
For a Model I regression the general equation of the line is: $y = a + bx$	$y = a + bx$ $y = 737.6 + (-92.4)x$ $= 737.6 - 92.4x$

BOX 6.3. Continued

GENERAL DETAILS	THIS EXAMPLE
4. How to draw the line The regression line indicates the apparent trend in your data. If you want to draw this line on your scatter plot you will need to calculate three pairs of data points as follows: Take any x value from your observed data and place this and the values for a and b into the general equation $y = a + bx$. The only unknown value will be y. Work out the y value and plot the x you selected and this y value on your graph. Repeat for at least two more points and draw a line through them within the range of your data set. This is the regression line and should be labelled with the regression equation. GO TO BOX 6.4.	**4. How to draw the line** When $x = 1$ m $y = 737.6 - (92.4 \times 1) = 645.2$ µg Zn/g soil $x = 2$ m $y = 737.6 - (92.4 \times 2)$ $= 552.8$ µg Zn/g soil $x = 3$ m $y = 737.6 - (92.4 \times 3)$ $= 460.4$ µg Zn/g soil Fig. 6.13. illustrates the scatter plot with the regression line added. But you need to confirm that this this is a statistically significant association: see BOX 6.4.

Table 6.5. Calculating a Model I: Simple linear regression using the data from Example 6.2. Heavy metal contamination of soil under electricity pylons

Distance from pylon (m) (x)	Zinc concentration (µmol Zn/g soil) (y)	$x \times y$
1.0	648	648
1.5	610	915
2.0	534	1068
2.5	500	1250
3.0	472	1416
n 5	n 5	n 5
Σx 10	Σy 2764	Σxy 5297
$(\Sigma x)^2$ 100	$(\Sigma y)^2$ 7639696	
Σx^2 22.5	Σy^2 1549944	
$\bar{x}$ 2.0	$\bar{y}$ 552.8 7	

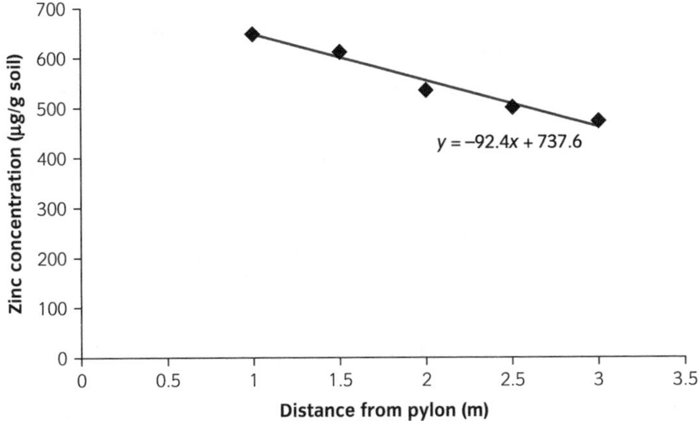

Fig. 6.13. Highly significant association ($0.01 > p > 0.001$) between the concentration of zinc in soil and the distance from the pylon so that $y = -92.4x + 737.6$.

BOX 6.4. How to carry out a Model I. Simple linear regression: testing the significance of the association

GENERAL DETAILS	THIS EXAMPLE
1. Hypotheses to be tested **H$_0$**: there is no linear association between x and y. **H$_1$**: there is a linear association between x and y, described by $y = a + bx$.	**1. Hypotheses to be tested** **H$_0$**: there is no linear association between the distance from the pylons and concentrations of zinc in the soil. **H$_1$**: there is a linear association between the distance from the pylons and concentrations of zinc in the soil.
2. How to work out $t_{calculated}$ i. First calculate basic terms SS(x), SS(y) and SP(xy). $SS(x) = \Sigma x^2 - \frac{(\Sigma x)^2}{n}$ $SS(y) = \Sigma y^2 - \frac{(\Sigma y)^2}{n}$ $SP(xy) = \Sigma xy - \frac{(\Sigma x)(\Sigma y)}{n}$ More details about how to do this are included in BOX 6.2. ii. Calculate the residual variance (s_r^2) $s_r^2 = \frac{1}{n-2}\left[SS(y) - \frac{(SP(xy))^2}{SS(x)}\right]$	**2. How to work out $t_{calculated}$** i. See Table 6.5. $SS(x) = 22.5 - \frac{100}{5} = 2.5$ $SS(y) = 1549944 - \frac{7639696}{5}$ $= 22004.8$ $SP(xy) = 5297 - \frac{(10 \times 2764)}{5} = -231.0$ ii. The residual variance is: $s_r^2 = \frac{1}{5-2}\left[22004.8 - \frac{(-231.0)^2}{2.5}\right]$ $s_r^2 = 220.13333$

BOX 6.4. Continued

GENERAL DETAILS	THIS EXAMPLE
iii. Calculate the standard error of b (SE(b).). $$\mathrm{SE}(b)=\sqrt{\frac{s_r^2}{\mathrm{ss}(x)}}$$ $$t_{\text{calculated}}=\frac{b}{\mathrm{SE}(b)}$$	iii. $$\mathrm{SE}(b)=\sqrt{\frac{220.13333}{2.5}}=9.38367$$ $$t_{\text{calculated}}=\frac{-92.4}{9.38367}=-9.84689$$
3. How to find $t_{critical}$	**3. How to find $t_{critical}$**
Use a t table for a **two-tailed test** where $p=0.05$. The **degrees of freedom** (v) are $n-2$.	When $\nu=5-2=3$ and at $p=0.05$, $t_{\text{critical}}=3.182$
4. The rule	**4. The rule**
If the absolute value of $t_{\text{calculated}}$ is greater than the critical value of t then you may reject H_0.	Since $t_{\text{calculated}}$ (9.847) is more than t_{critical} (3.182) at $p=0.05$ you may reject the null hypothesis. In fact at $p=0.01$ $t_{\text{critical}}=5.841$ and at $p=0.001$ $t_{\text{critical}}=12.941$ so you may reject the null hypothesis at $p=0.01$ but not at $p=0.001$.
5. What does this mean in real terms?	**5. What does this mean in real terms?**
	There is a highly significant ($0.01>p>0.001$) negative linear association between the distance from the pylons (m) and concentrations of zinc in the soil (μg Zn/g soil) described by $y=737.6-92.4x$.

6.6. Model II: Principal axis regression

Unlike the simple linear regression model, the principal axis regression model can be used when both variables are subject to sampling error. However, the principal axis regression cannot be used to test for an association and in this instance a correlation should be used. The principal axis regression and ranged axis regression (6.7.) are part of a group of tests called principal component analyses. In these Model II regressions the line is determined by placing an ellipse around the data and drawing a line through the longest axis – the principal axis. The general formula for this principal axis is $y=\bar{y}+b_1(x-\bar{x})$. As in simple linear regression there is a regression coefficient (b), but to denote the difference in the method used to calculate b compared with the Model I regression we call this regression coefficient b_1.

6.6.1. To use this test you:

1. Need two treatment variables recorded for each item.
2. Have an association that appears to be linear and the points reasonably scattered.
3. Have confirmed using a correlation analysis that the association is significant.
4. Have neither treatment variable under the investigator's control. Therefore, both treatment variables are subject to sampling error.
5. Have two treatment variables that are measured on the same scale.
6. Have three or more pairs of measurements.

EXAMPLE 6.3. The lower arm (cm) and lower leg (cm) length of a small cohort of female undergraduates

In an investigation into human morphology the lower arm and lower leg length of 12 female undergraduates was recorded (Table 6.6.) on one particular day.

Table 6.6. Lower arm (cm) and lower leg (cm) length in a small cohort of female undergraduates

Student	Lower arm length (cm)	Lower leg length (cm)	Student	Lower arm length (cm)	Lower leg length (cm)
1	24	39	7	26	41
2	25	40	8	25.5	40
3	23	36	9	24.2	36.7
4	26	43	10	25	38
5	24	38	11	24	41
6	23	35	12	24.5	40

The data from this example meet the criteria for a principal axis regression as there are two treatment variables (arm length and leg length). The association appears to be linear and the points are reasonably scattered (Fig. 6.14.). Neither variable was under the control of the investigator. There are 12 pairs of observations and both are measured on the same scale (cm). A Spearman's rank correlation (BOX 6.5.) confirms that this is a significant association ($r_s = 0.725$, $p = 0.05$).

6.6.2. The calculation

We show you the general process and a specific example of a Model II principal axis regression in BOX 6.5. In addition, you will need to refer to the calculation table 6.7. Full details of the calculation are included in the Online Resource Centre.

BOX 6.5. How to carry out a Model II. Principal axis regression: drawing a regression line

GENERAL DETAILS	THIS EXAMPLE
1. How to work out b_1	**1. How to work out b_1**
i. First decide which of the variables to call x and which to call y.	i. Let the measurements of arm length (cm) be x and the measurements of leg length (cm) be y (Fig. 6.14.).
ii. Calculate basic terms.	ii. Calculate basic terms.
To carry out this regression analysis you need to first calculate the following for the two samples: sum of x values (Σx) sum of y values (Σy) means ($\bar{x}$ and $\bar{y}$) the variances (s_x^2 and s_y^2) the number of pairs of observations n The method for calculating these is given in BOX 3.1.	The general methods for calculating such terms as Σx and s_x^2 are described in Chapter 3 (BOX 3.1.) and therefore are not included here. The basic terms are: $\bar{x} = 24.51666$ $\bar{y} = 38.975$ $\Sigma x = 294.2$ $\Sigma y = 467.7$ $s_x^2 = 1.03061$ $s_y^2 = 5.38932$ $n = 12$

BOX 6.5. Continued

GENERAL DETAILS	THIS EXAMPLE
iii. Calculate the variance for products xy, (s_{xy}^2) Where $s_{xy}^2 = \frac{SP(xy)}{n-1}$ The term SP(xy) has figured already in this chapter (BOX 6.2) and is $SP(xy) = \Sigma xy - \frac{(\Sigma x)(\Sigma y)}{n}$	iii. Variance for the products xy, $\left(s_{xy}^2\right)$. First note or calculate each of the terms you need. Σx, Σy and n are above (ii). To calculate Σxy see Table 6.7. $\Sigma xy = 11487.14$ $sp(xy) = \Sigma xy - \frac{(\Sigma x)(\Sigma y)}{n}$ $= 11487.14 - \frac{(294.2)(467.7)}{12}$ $= 20.695$ $s_{xy}^2 = \frac{SP(xy)}{n-1} = \frac{20.695}{11}$ $= 1.88136$
iv. The slope of the principal axis (b_1) is calculated as: $b_1 = \frac{s_y^2 - s_x^2 + \sqrt{\left[\left(s_y^2 - s_x^2\right)^2 + 4(s_{xy}^2)^2\right]}}{2s_{xy}^2}$	iv. The slope of the principal axis b_1. First calculate the components of this equation. $s_y^2 - s_x^2 = 5.38932 - 1.03061$ $= 4.35871$ $(s_y^2 - s_x^2)^2 = (4.35871)^2$ $= 18.99837$ $4(s_{xy}^2)^2 = 4 \times (1.88136)^2$ $= 14.15812$ $2s_{xy}^2 = 2 \times 1.88136$ $= 3.76273$ Therefore: $b_1 = \frac{4.35871 + \sqrt{(18.99837 + 14.15812)}}{3.76227}$ $= 2.68871$

BOX 6.5. Continued

GENERAL DETAILS	THIS EXAMPLE
2. What is the regression equation? The equation for the principal axis is: $y = \bar{y} + b_1(x - \bar{x})$ Substitute the known numerical values where possible.	**2. What is the regression equation?** Using the values you have already calculated: $y = \bar{y} + b_1(x - \bar{x})$ $= 38.975 + 2.68871(x - 24.51667)$ $= 38.975 + 2.68871x - 65.91822$ $= -26.94319 + 2.68871x$
3. How to draw the line In the same way as other regressions you need to calculate the x and y values for three points using x values that lie within your data set, and your regression equation. Your line then passes through these points within the range of your data set. This is the regression line and if you add it to your figure it should be labelled with the regression equation.	**3. How to draw the line** The full regression equation for this example is: $y = -26.94319 + 2.68871x$ Therefore if: $x = 23$ cm, $y = -26.94319 + (2.68871 \times 23)$ $= 34.9$ cm and if $x = 24.5$ cm then $y = 38.9$ cm, and if $x = 26$ cm then $y = 43$ cm. The data from Table 6.6. and the regression line are shown on Fig. 6.14.
4. Testing the significance of the line The principal axis regression allows you to identify a regression line that reflects the linear nature of the association between your two variables. However, unlike the simple linear regression there is no simple associated process for testing a hypothesis relating to this association. Therefore, you should also use a correlation analysis to confirm the strength and significance of the association between the two variables.	**4. Testing the significance of the line** An examination of the data indicates that they are non-parametric; therefore, a two-tailed Spearman's rank correlation was carried out. From this it was concluded that there is a moderate ($r_s = 0.725$) statistically significant ($p = 0.05$) positive correlation between the lower arm and lower leg length in a cohort of female students which can be described by $y = -26.94319 + 2.68871x$ (Fig. 6.14.).

Table 6.7. Calculating a Model II: Principal axis regression using the data from Example 6.3. Arm and leg length (cm) in a small cohort of female students

Length of arm (cm) (x)	Length of leg (cm) (y)	$x \times y$
24	39	936
25	40	1000
23	36	828
26	43	1118
24	38	912
23	35	805
26	41	1066
25.5	40	1020
24.2	36.7	888.14
25	38	950
24	41	984
24.5	40	980
n 12	n 12	n 12
Σx 294.2	Σy 467.7	Σxy 11487.14
$\bar{x}$ 24.51667	$\bar{y}$ 38.975	

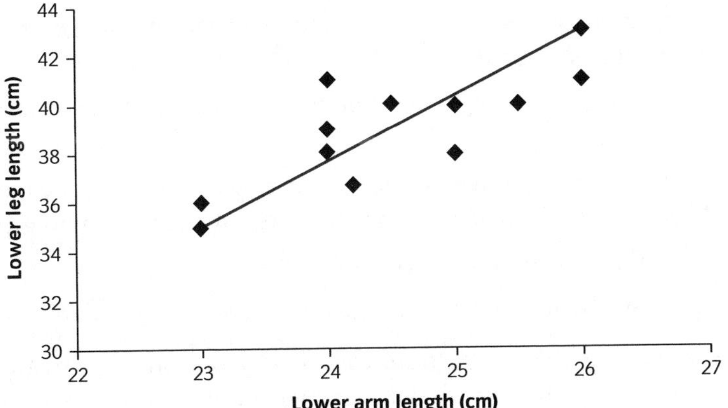

Fig. 6.14. Significant association ($r_s = 0.725$, $p = 0.05$) between the lower arm length (cm) and lower leg length (cm) of female undergraduates, such that $y = -26.94 + 2.69x$.

6.7. Model II: Ranged principal axis regression

Many investigators wish to determine if there is an association between two variables, neither of which has been manipulated by the experimenter. In Example 6.1. neither hardness of eggshell nor the amount of food supplement eaten was under the investigator's control; therefore, a Model II regression is appropriate. Where the units of measurement for the two variables differ, as in this example, a Model II ranged principle axis regression may be used (Legendre & Legendre, 1998). Unlike the principal axis regression (6.6.) the first step in the calculation is to transform the data so that each scale of measurement is revised to one that only extends from 0 to 1. These ranged data then satisfy the criteria for using a principal axis regression. The principal axis regression is carried out on this ranged data and the slope b' is then transformed back to the original units of measurement before the regression line is calculated. Like the previous principal axis regression, the line is the one that passes longitudinally through an ellipse which encompasses the data. This ranged principal axis regression cannot easily be used as the basis for a test for association so a correlation should be carried out to test the strength and significance of any association before using this test.

6.7.1. To use this test you:

1. Need two treatment variables recorded for each item.
2. Have an association that appears to be linear and the points reasonably scattered.
3. Have confirmed using a correlation analysis that the association is significant.
4. Have neither treatment variable under the investigator's control. Therefore, both treatment variables are subject to sampling error.
5. Have three or more pairs of measurements.
6. Have two treatment variables that are measured on different scales.

The example we use to illustrate this Model II method of regression analysis is Example 6.1. These data satisfy the criteria for using a ranged principal axis regression because there are two treatment variables (hardness of eggshells and amount of food supplement eaten). The association appears to be linear and the points reasonably scattered (Fig. 6.5.). Neither the hardness of eggshell nor the amount of food additive eaten was manipulated by the investigator. There are 13 pairs of observations. The hardness of eggshell is an arbitrary unit of measure and the amount of food additive eaten is measured in grams. Therefore, the two units of measure for the two treatment variables are not the same. Using the Pearson's product moment correlation (BOX 6.2.) we have confirmed that there is a strong ($r = 0.99$) and highly significant positive correlation ($p < 0.001$).

6.7.2. The calculation

As discussed earlier, there are more steps in this ranged principal axis regression than in the principal axis regression. However, the regression itself is the same. As in the principal axis regression it is necessary to carry out a correlation analysis to test the significance of the association. The ranged principal axis regression is shown in general and with a specific example (Example 6.1.) in BOX 6.6. with some calculations included in Table 6.8. Where steps have been abbreviated the full calculation is included in the Online Resource Centre.

BOX 6.6. How to carry out a Model II. Ranged principal axis regression: drawing a regression line

GENERAL DETAILS	THIS EXAMPLE
1. How to work out b′	**1. How to work out b′**
i. First decide which of the variables to call x and which to call y.	i. Example 6.1. has been used to illustrate the Pearson's and the Spearman's rank correlation (BOX 6.1. and BOX 6.2.). Therefore, we will let the amount of food supplement eaten (g) be the x variable and hardness of shells be the y variable.
ii. Transform each y and x value into x' and y' as follows: $$x' = \frac{x - \text{lowest value of } x}{\text{highest value of } x - \text{lowest value of } x}$$ $$y' = \frac{y - \text{lowest value of } y}{\text{highest value of } y - \text{lowest value of } y}$$	ii. The transformed data are given in Table 6.8. The first original x observation was 19.5 g. The lowest x value was 7.0 g and the highest value was 19.50 g. Therefore: $$x' = \frac{19.5 - 7.0}{19.5 - 7.0} = 1.0$$ The second original x observation was 11.20 g. Therefore: $$x' = \frac{11.2 - 7.0}{19.5 - 7.0} = 0.336$$ The first original y observation was 7.1 units, the lowest y value was 1.2 units and the highest was 7.1 units. Therefore: $$y' = \frac{7.1 - 1.2}{7.1 - 1.2} = 1.0$$ etc. (Table 6.8.).

BOX 6.6. Continued

GENERAL DETAILS	THIS EXAMPLE
iii. Calculate basic terms To carry out this regression analysis you need to first calculate the following for the two samples: sum of x' values ($\Sigma x'$) sum of y' values ($\Sigma y'$) means ($\bar{x}'$ and $\bar{y}'$) the variances ($s^2_{x'}$ and $s^2_{y'}$) number of pairs of observations n The method for calculating these is given in BOX 3.1.	iii. Calculate basic terms Since this method has been demonstrated elsewhere (BOX 3.1., BOX 6.2.) we will not repeat it here. The basic terms are: $\bar{x}' = 0.432, \bar{y}' = 0.41320$ $\Sigma x' = 5.616, \Sigma y' = 5.37288$ $s^2_{x'} = 0.08266, s^2_{y'} = 0.08678$ $n = 13$
iv. Calculate the variance ($s^2_{x'y'}$) for the products $x'y'$ Where $s^2_{x'y'} = \dfrac{\text{SP}(x'y')}{n-1}$ The term SP($x'y'$) has figured already in this chapter and is: $\text{SP}(x'y') = \Sigma x'y' - \dfrac{(\Sigma x')(\Sigma y')}{n}$	iv. Variance ($s^2_{x'y'}$) for the products $x'y'$. First note or calculate each of the terms you need. $\Sigma x'$, $\Sigma y'$, and n are above (iii). To calculate $\Sigma x'y'$ see Table 6.8. $\Sigma x'y' = 3.33098$ $\text{sp}(x'y') = \Sigma x'y' - \dfrac{(\Sigma x')(\Sigma y')}{n}$ $= 3.33098 - \dfrac{(5.616)(5.37288)}{13}$ $= 1.009896$ $s^2_{x'y'} = \dfrac{\text{SP}(x'y')}{n-1} = \dfrac{1.009896}{13-1}$ $= 0.08416$
v. The slope of the principal axis (b') for the ranged data is $\text{b}' = \dfrac{s^2_{y'} - s^2_{x'} + \sqrt{[(s^2_{y'} - s^2_{x'})^2 + 4(s_{x'y'})^2]}}{2s^2_{x'y'}}$	v. To find the slope of the principal axis (b') for the ranged data. First calculate the components of this equation. $s^2_{y'} - s^2_{x'} = 0.08678 - 0.08266 = 0.00412$ $(s^2_{y'} - s^2_{x'})^2 = (0.00412)^2$ $= 1.69398 \times 10^{-5}$

BOX 6.6. Continued

GENERAL DETAILS	THIS EXAMPLE
	$4(s^2_{x'y'})^2 = 4 \times (0.08416)^2$ $= 0.02833$ $\sqrt{[(s^2_{y'} - s^2_{x'})^2 + 4(s^2_{x'y'})^2]}$ $= \sqrt{(1.69398 \times 10^{-5} + 0.02833)}$ $= 0.16837$ $2s^2_{x'y'} = 2 \times 0.08416$ $= 0.16832$ Therefore: $b' = \dfrac{0.00412 + 0.16837}{0.16837}$ $= 1.02475$
vi. Transform b' to original units to obtain b_1. $b_1 = b' \times \dfrac{\text{highest } y \text{ value} - \text{lowest } y \text{ value}}{\text{highest } x \text{ value} - \text{lowest } x \text{ value}}$	vi. Transform b' to original units to obtain b_1. $b_1 = 1.02475 \times \dfrac{7.1 - 1.2}{19.5 - 7.0}$ $= 0.48368$
2. What is the regression equation?	**2. What is the regression equation?**
The equation for the principal axis is: $y = \bar{y} + b_1(x - \bar{x})$ You need first to calculate the means for the original x and y values then substitute these and b_1 in the equation.	Calculate the means for the original observations (Table 6.3). Then: $y = \bar{y} + b_1(x - \bar{x})$ $= 3.63846 + 0.48368(x - 12.4)$ $= 3.63846 + 0.48368x - 5.99763$ $= -2.35920 + 0.48368x$
3. How to draw the line	**3. How to draw the line**
In the same way as other regressions, you need to calculate the x and y values for three points using x values that lie within your data set and your regression equation. Your line then passes through these points within the range of your data set. This is the regression line and should be labelled with the regression equation.	The regression equation for this example is $y = -2.35977 + 0.48368x$ Therefore if: $x = 8.2$ g $y = -2.35977 + 0.48368 \times 8.2$ $= 1.6$ units and when $x = 19.50$ g, $y = 7.1$ units and if $x = 15.10$ g, $y = 4.9$ units.

BOX 6.6. Continued

GENERAL DETAILS	THIS EXAMPLE
4. Testing the significance of the line	**4. Testing the significance of the line**
The ranged principal axis regression allows you to identify a regression line that reflects the linear nature of the association between your two variables. However, unlike the simple linear regression there is no simple associated process for testing a hypothesis relating to this association. Therefore, you should also use a correlation analysis to confirm the strength and significance of the association between the two variables.	This data were used to illustrate both the Spearman rank correlation (BOX 6.1.) and the Pearson correlation (BOX 6.2.). Using the more powerful parametric Pearson's correlation it can be seen that there is a strong ($r = 0.99$) and highly significant positive correlation ($p < 0.001$) between the amount of food supplement (g) given to the Maren pullets and the hardness of the eggshells that they produce. This linear association can be described by the regression equation $y = -2.35977 + 0.48368x$ (Fig. 6.15.).

Table 6.8. Calculation of ranged values for a ranged principal axis regression using data from Example 6.1. The association between the amount of food supplement eaten and eggshell hardness in Maren pullets

Pullet	Amount of food supplement (g) (x)	Ranged x' values	Hardness of shells (y)	Ranged y' values	$x' \times y'$
1	19.5	1.0	7.1	1.0	1.0
2	11.2	0.366	3.4	0.37288	0.12529
3	14.0	0.56	4.5	0.55932	0.31322
4	15.1	0.648	5.1	0.66102	0.42834
5	9.5	0.2	2.1	0.15254	0.03051
6	7.0	0.0	1.2	0.0	0.0
7	9.8	0.224	2.1	0.15254	0.03417
8	11.6	0.368	3.4	0.37288	0.13722
9	17.5	0.84	6.1	0.83051	0.69763
10	11.2	0.336	3.0	0.30508	0.10251
11	8.2	0.096	1.7	0.08475	0.00814
12	12.4	0.432	3.4	0.37288	0.16108
13	14.2	0.576	4.2	0.50847	0.29288
					$\Sigma xy = 3.33098$

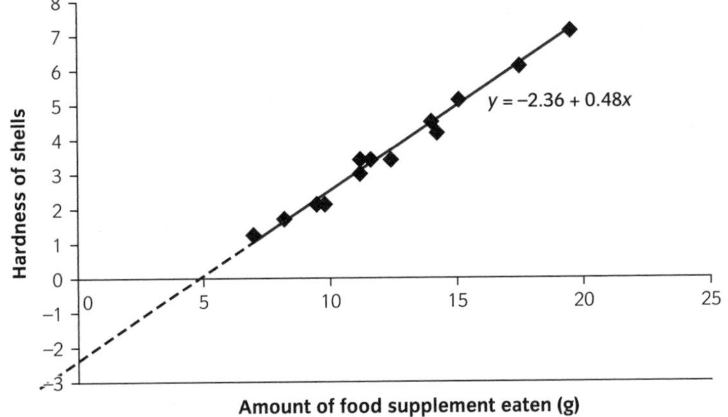

Fig. 6.15. Significant association ($r = 0.99$; $p < 0.001$) between the amount of food supplement eaten and the hardness of eggshells in Maren pullets, described by $y = -2.36 + 0.48x$. The figure also illustrates why you should not extrapolate beyond your known data (Q3).

6.8. About regression lines

Reporting correlations and regressions is usually achieved using a scatter plot. For a correlation this is sufficient if the results from the correlation, including the correlation coefficient and p value, are included in the figure legend. If a regression analysis has been used then it is common practice to add the regression line on the scatter plot and label this with the regression equation. We consider the topic of drawing lines on figures in more detail in Chapter 10, but whilst we are on the subject of regressions we would like to make two points:

- A trend may not reflect an association or a relationship. Lines (we refer to them as trend lines) are often drawn on figures, especially by the common software packages. These lines indicate the apparent trend but have no statistical meaning. There are many common errors associated with these trend lines (10.8.2.). One way to overcome these errors, if you have appropriate data, is to carry out a regression analysis to model a significant association. A significant regression line is a reasonable mathematical description of the trend in your data. You should show how important the line is by adding the regression equation (e.g. Fig. 6.15. compared with Fig. 6.5.). A regression line that is shown not to be significant should not be used.
- Do not extend the regression line beyond your data. A common error is to extend the significant regression line beyond the data set. Q3 illustrates why this is not appropriate.

Q3 **The regression equation that we calculated for Example 6.1. in BOX 6.6. between the amount of food supplement eaten by Maren pullets and the hardness of the eggshells is $y = -2.35977 + 0.48373x$ (Fig. 6.15.). If the farmer wishes to economize and only gives the pullets 1 g of food supplement, how hard would you predict the eggshells will be?**

A3 In our original investigation the minimum feed supplement eaten was 7.0 g but in theory you could use the regression equation to calculate this unknown y value. In this example if $x = 1.0$ g then $y = -2.35977 + (0.48373 \times 1.0) = -1.87604$ units. But what would an egg with a negative shell hardness look like? This illustrates the folly of working outside the boundaries of your known data set. For all we know the pullets may die unless they eat a minimum of food supplement, or they may do brilliantly without the supplement and produce really hard eggshells. You must therefore never extrapolate outside the range of your existing observations.

6.9. Experimental design, correlation, and regression

As we explained at the beginning of this chapter, regression and correlation analyses are designed to test for an association between two variables and/or model this association so allowing you to describe the association and make predictions. To use such tests you will design an investigation so that for each item you will have a pair of observations, one observation for each treatment variable. This differs from the chi-squared and G tests for association where you may also be examining two variables, but here the data recorded are the number of items in particular categories.

The experimental designs for correlation and regression are often quite simple (Table 6.9.), the number of items being determined by the type of correlation or regression. For example, to use a Spearman's rank correlation you would design an experiment that had between 7 and 30 pairs of observations (6.2.1.). For a regression analysis the minimum number of items is 3, although you will only detect strong associations with a small sample size. To detect a weak association will require a large sample size. There is no simple way to determine sample size in this instance without carrying out a relatively comprehensive pilot study from which r may be ascertained.

When designing your experiment one of the variables may be measured without sampling error; usually this means that it is under the investigator's control. In this case you may consider using the Model I simple linear

Table 6.9. Experimental design for correlation and regression if all other criteria are met

Item	Observations for variable 1	Observations for variable 2
1		
2		
3		
4 etc.		

regression. If this is not so, then you may use a Model II regression. The scale of measurement is also important when determining which test you may use. For example, if you expect to have non-parametric data you should plan your investigation around the Spearman's rank correlation rather than the Pearson's product moment correlation. Normalizing non-parametric data may be possible (3.9.), although interpreting the outcome of the analysis of two treatment variables using transformed data is complex and you should refer to other texts.

Correlations and regression are used to examine whether there is a linear association between two or more variables, only one of which may be under the control of the investigator. You may have a 'null state' within your treatment but it cannot be called a control as the analysis makes no comparison between the various observations within one variable (i.e. it does not compare a control value with a treatment value). You may have replication and can carry out a multiple regression analysis but we do not cover this here (see, for example, Sokal & Rohlf, 1981).

Summary of Chapter 6

- In this chapter we consider correlation and regression analyses which may be used to test for a linear association between two variables. The term association indicates a statistical and not necessarily a biological relationship between the two variables (Introduction).
- These tests differ from the chi-squared and G tests in that for each item two observations are recorded, one for each treatment variable. In the chi-squared and G tests the number of items in each category is determined. The chi-squared and G tests may be used for data that are measured on a nominal scale. In the correlations and regressions neither variable can be measured on a nominal scale (Introduction, 5.7., and 6.9.).
- A correlation analysis will generate a coefficient which indicates the strength of association. This coefficient may then be used to test the significance of the association. The coefficient is affected by sample size; therefore, a weak correlation may not be significant if the sample size is small but may be significant if the sample size is large. We consider two correlation

analyses: the Spearman's rank test for non-parametric data and the Pearson's product moment test for parametric data (6.1., 6.2., 6.3.).

- Regression analyses are primarily used to determine the mathematical equation for the line that may be drawn through the data. The significance of this line as a line of best fit can be determined only in the simple linear regression using a modified *t* test. In the principal axis and ranged principal axis regression the significance of the association can be determined using a correlation analysis (6.1., 6.5., 6.6., and 6.7.).
- The regression analyses fall into two groups: Model I and Model II. In the Model I regression (simple linear regression) one variable must be measured without sampling error and is therefore usually under the control of the investigator. In the Model II regressions (principal axis and ranged principal axis regressions) both variables may be measured with some sampling error, either using the same scale (principal axis) or different scales (ranged principal axis) (6.5., 6.6., and 6.7.).
- Correlation and regression analysis may be extended to testing or describing the association in non-linear distributions, or where you have more than one y for each x. We are not able to include these tests here and refer you to other texts, such as Grafen & Hails (2002) and Sokal & Rohlf (1981).
- The Online Resource Centre includes interactive exercises that test your understanding of this chapter with other topics, particularly those considered in Chapters 2–8 and Chapter 10.

Hypothesis testing: Do my samples come from the same population? Parametric data

This chapter tells you how you may analyse **parametric** data to test the **hypothesis** that two or more **samples** come from the same **statistical population.** All these tests use the underlying mathematical relationships that come from the data having a normal distribution (3.2.). Therefore, an essential step is to calculate the sums of squares and the variance. If you are not familiar with these terms you should refer to BOX 3.1. In many of these tests a variance ratio or *F* test is calculated. For example, an *F* test is used to check one of the criteria for using a *z* test. Similarly, an *F* test is used before an ANOVA to check that the data are suitable for using this test. An *F* test is also used at the end of the ANOVA enabling you to test your hypotheses. These *F* tests differ in the tables of critical values that are used to compare the calculated *F* values to. You must therefore always check to make sure you are using the correct table for the particular *F* test.

Ten tests are covered in this chapter and worked examples are given for each. If this is the first time you have used this test you should work through these examples and check your answers before using the test on your own data. If your answer differs considerably from that given, you should check your calculation by going to the Online Resource Centre. If you work through these examples we would expect it to take you 8 hours. If you prefer to work on parts of this chapter we recommend you look at the *t* and *z* tests first, then the one- and two-way ANOVAs, and finally the three-way ANOVA and two-way nested ANOVA. To study these we would expect you to take 2 hours, 3 hours, and 3 hours respectively. As always there are worked examples in the Online Resource Centre and an explanation as to how you may use statistical software to carry out the calculations.

How to choose the correct test

Each test has several requirements that must be met and details of these are given at the start of each section. The following guide takes you to the most likely tests. It is assumed that you have parametric data.

You have one treatment **variable.** You are going to compare two samples. The data are **unmatched.**	*t* or *z* test for unmatched data (7.1. or 7.2.)
You have one treatment variable. You are going to compare two samples. The data are **matched.**	*t* or *z* test for matched data (7.3.)
You have one treatment variable. You are going to compare more than two samples. You wish to test general and specific hypotheses.	One-way ANOVA and Tukey's test (7.5. and 7.6.)
You have two treatment variables. Each variable has at least two categories or classes and all categories from one variable are combined with all categories from the second variable. You wish to test general and specific hypotheses.	Two-way ANOVA and Tukey's test (7.7. and 7.8.)
You have two treatment variables. Each variable has at least two categories. One variable is randomized or nested with regard to the second variable. You wish to test general hypotheses.	Two-way nested ANOVA (7.9.)
You have three treatment variables. Each variable has at least two categories and all categories from each variable are combined with all other categories from the other variables. You wish to test general and specific hypotheses.	Three-way ANOVA (7.10.)
None of the above.	Chapter 8 and Sokal & Rohlf (1981)

7.1. *z* test with unmatched data

The z test relies on the known relationship between the **mean** and **standard deviation** in normally distributed data and between the mean and standard error of the mean (3.2.–3.7.). If samples are from the same population then you would expect the difference between the sample means, to be close to zero. The greater the difference between the sample means, the more unlikely it is that the means are from the same population. The z test estimates, in terms of standard errors, how far apart the means are. This then indicates the probability that the means are from the same population.

There are two forms of this z test: one for unmatched data and one for matched data. These two tests are quite different from each other. When samples sizes are small a t test is recommended. The t and z tests are similar and the important differences are highlighted in the following sections.

7.1.1. To use this test you:

1. Wish to test for differences in population means.
2. Need one treatment variable and two samples.
3. Have unmatched data.
4. Have parametric data.

5a. Have at least 30 observations in each sample but the sample sizes need not be the same in the two samples. Or:

5b. The data are not normally distributed but are measured on a continuous scale and there are more than 30 observations in each sample.

6. Have two samples where the variances are similar (homogeneous).

EXAMPLE 7.1. The evolution of *Littorina littoralis* at Aberystwyth, 2002

Several years ago it was suggested that a species of periwinkle, *Littorina littoralis*, was evolving sympatrically into two species, *L. obtusata* and *L. mariae*, through niche partitioning. *L. obtusata* apparently grazes on the brown alga *Ascophyllum nodosum* on the mid shore whilst putative *L. mariae* feeds on the epiphytes growing on *Fucus serratus* on the lower shore. In a study of the sympatric evolution of *L. littoralis* a representative sample of the two groups of individuals was collected from Aberystwyth in 2002 and their shell height (mm) recorded (Table 7.1). The investigators wished to test the hypothesis that there is no difference between shell height of the two groups of periwinkles from the mid and lower shore.

In this example the criteria for using a z test are met. There is one variable (species), two samples (mid and lower shore) and 30 observations for each sample. Each periwinkle has only been measured once so the data are unmatched. The data are parametric. But are the variances similar?

One of the common major assumptions underlying many parametric tests of hypotheses is that the variance of the different samples are homogeneous (similar). A variance ratio or F test is used to examine whether this is probably true (BOX 7.1.)

BOX 7.1. How to carry out an *F* test to check for homogeneous variances before carrying out a *z* test for unmatched data

GENERAL DETAILS	THIS EXAMPLE
1. Hypotheses to be tested	**1. Hypotheses to be tested**
H$_0$: There is no difference between the variances of the two samples. The variances are homogeneous. **H$_1$**: There is a difference between the variances of the two samples.	**H$_0$**: There is no difference between the variances of the shell height of periwinkles from the lower and mid shore. **H$_1$**: There is a difference between the variances of the shell height of periwinkles from the lower and the mid shore.
2. How to work out $F_{calculated}$	**2. How to work out $F_{calculated}$**
i. Decide which is sample 1 and which is sample 2.	i. Let the periwinkles from the lower shore be sample 1 and the periwinkles from the mid shore be sample 2.
ii. For each sample calculate the variance s_1^2 and s_2^2 using the method described in BOX 3.1. Where the sums of squares of x is: $SS(x) = (\Sigma x^2) - \frac{(\Sigma x^2)}{n}$ The variance is $s^2 = SS(x)/(n-1)$	ii. Periwinkles from the lower shore $\Sigma x = 166.4$, $n = 30$, $\frac{(\Sigma x)^2}{n} = 922.96533, \Sigma x^2 = 1027.08$ $SS(x) = 1027.08 - 922.96533$ $= 104.11467$ $s^2 = 3.59016$ Periwinkles from the mid shore $\Sigma x = 224.1$, $n = 30$ $\frac{(\Sigma x)^2}{n} = 1674.027, \Sigma x^2 = 1792.85$ $SS(x) = 118.823, s^2 = 4.09734$
iii. $F_{calculated}$ is the ratio between the two variances. $F_{calculated} = \frac{\text{larger sample variance}}{\text{smaller sample variance}}$	iii. $F_{calculated} = \frac{4.09734}{3.59016}$ $= 1.14127$
3. How to find $F_{critical}$	**3. How to find $F_{critical}$**
You will come across three F tests. Each uses a different table to find $F_{critical}$. You must take great care to use the correct table. To look up the value in the F table for a **two-tailed test**, you will need to know the **degrees of freedom** (v) for each sample.	For our example, $\nu_1 = 30 - 1 = 29$ and $\nu_2 = 30 - 1 = 29$. **Interpolating** (4.3.6) from the F table: $F_{critical} = 2.10$ at $p = 0.05$.

BOX 7.1. Continued

GENERAL DETAILS	THIS EXAMPLE
Sample 1 $v_1 = n_1 - 1$ Sample 2 $v_2 = n_2 - 1$	
4. The rule If $F_{calculated}$ is less than $F_{critical}$ you do not reject the null hypothesis and may proceed with the *z* test.	**4. The rule** In this example $F_{calculated}$ (1.14.) is less than $F_{critical}$ (2.10). Therefore, you do not reject the null hypothesis.
5. What does this mean in real terms?	**5. What does this mean in real terms?** There is no significant difference ($p = 0.05$) between the variances of shell height of periwinkles from the lower and mid shores. The variances are homogeneous. You may proceed with the *z* test. If the null hypothesis is not accepted then you should consider transforming your data (3.9.) or using a non-parametric test e.g. Mann–Whitney U test (8.1.).

7.1.2. The calculation

Having satisfied ourselves that our data meet all the criteria for using this *z* test for unmatched data, we can then proceed (BOX 7.2.). Where steps have been abbreviated the full calculation is included in the Online Resource Centre.

Table 7.1. Height of shells of two putative species of periwinkles from the mid and lower shore at Aberystwyth, 2002

Shell height (mm)					
Periwinkles from the lower shore			Periwinkles from the mid shore		
5.3	4.3	6.5	11.7	9.3	11.3
8.7	8.0	6.8	4.1	7.7	7.0
5.3	10.2	5.0	8.8	6.6	6.8
5.3	5.3	5.1	9.7	9.4	8.3
5.9	7.8	4.9	5.0	8.8	4.5
2.8	7.0	3.7	6.7	5.8	6.0
8.7	6.1	2.8	6.6	5.6	7.0
2.0	5.0	3.8	7.8	7.5	6.5
5.3	5.4	3.0	7.0	6.3	6.6
6.5	5.7	4.2	7.6	12.5	5.6

BOX 7.2. How to carry out a *z* test for unmatched data

GENERAL DETAILS	THIS EXAMPLE
1. Hypotheses to be tested	**1. Hypotheses to be tested**
H_0: There is no difference between the mean values of the populations from which the samples are taken. **H_1**: There is a difference between the mean values of the populations from which the two samples are taken. If you use this test for data that are not normally distributed, then you should refer to **medians** not means in the hypotheses.	**H_0**: There is no difference between the mean shell height of the periwinkles from the lower and mid shore at Aberystwyth in 2002. **H_1**: There is a difference between the mean shell height of the periwinkles from the lower and the mid shore at Aberystwyth in 2002.
2. How to work out $z_{calculated}$	**2. How to work out $z_{calculated}$**
i. Decide which is sample 1 and which is sample 2.	i. Let the periwinkles from the lower shore be sample 1 and the periwinkles from the mid shore be sample 2.
ii. Use the variances (s^2) and n from BOX 7.1. calculated for the F test.	ii. $s_1^2 = 3.59016$, $s_2^2 = 4.09734$ $n_1 = n_2 = 30$
iii. Calculate the means ($\bar{x}$) for each sample.	iii. $\bar{x}_1 = 5.46667$, $\bar{x}_2 = 7.47$
iv. The standard error of the difference of the means (SE_D) is calculated as: $SE_D = \sqrt{\left(\frac{s_1^2}{n_1} + \frac{s_2^2}{n_2}\right)}$	iv. Standard error of the difference of the means (SE_D) $SE_D = \sqrt{\left(\frac{3.59016}{30} + \frac{4.09734}{30}\right)}$ $= 0.50621$
v. $z_{calculated} = \frac{\bar{x}_1 - \bar{x}_2}{SE_D}$	v. $z_{calculated} = \frac{5.54666 - 7.47}{0.50621}$ $= -3.79947$
3. How to find $z_{critical}$	**3. How to find $z_{critical}$**
The z distribution is unusual in that it is independent of sample size. For any one p value there is only one z value. These are given in Appendix d for a two-tailed test.	For a two-tailed test at $p = 0.05$, $z_{critical} = 1.96$
4. The rule	**4. The rule**
If the **absolute** value of $z_{calculated}$ is more than or equal to $z_{critical}$ then you may reject the null hypothesis (H_0). Samples are not likely to have come from the same statistical population.	The absolute value of $z_{calculated}$ (3.8) is more than z critical (1.96). Therefore, you may reject the null hypothesis. In fact at $p = 0.001$, $z_{critical} = 3.29$. You may therefore reject the null hypothesis at this higher level of significance.

BOX 7.2. Continued

GENERAL DETAILS	THIS EXAMPLE
5. What does this mean in real terms?	**5. What does this mean in real terms?** There is a very highly significant difference ($z = 3.8$, $p < 0.001$) between the mean shell heights of the periwinkles from the lower and mid shore at Aberystwyth. *L. mariae* is the smaller of the two apparently distinct species.

7.2. *t* test (Student's *t* test) for unmatched samples

When the sample sizes get relatively small then the distribution on which the z statistic is based changes with the sample size. This must be taken into account when sample sizes are small. This is done by using a t test, where the sample sizes are used both in the estimate of the standard error and when looking up the critical value of t in tables.

7.2.1. To use this test you:

1. Wish to test for differences in population means.
2. Have one treatment variable and two samples.
3. Have unmatched data.
4. Have parametric data.
5. Have fewer than 30 observations in each sample but the sample sizes need not be equal.
6. Have homogeneous variances.

EXAMPLE 7.2. The evolution of *Littorina littoralis* at Porthcawl, 2002

The study at Aberystwyth (2002) indicated that a species of periwinkle, *Littorina littoralis*, was evolving sympatrically into two species (Example 7.1., BOX 7.2.). To extend this study the researchers also sampled from Porthcawl in the same year. Here, however, they found far fewer periwinkles (Table 7.2.).

Table 7.2. Shell height in two putative species of periwinkles from the mid and lower shore at Porthcawl, 2002

Shell height			
Periwinkles on the lower shore		Periwinkles on the mid shore	
5.5	4.0	3.3	6.7
8.4	5.0	6.3	5.7
5.0	6.2	6.1	4.2
5.0	5.0	8.0	6.3
5.6	7.7	13.5	6.0
4.8	6.0	5.3	7.2
8.4		6.7	

These data meet the criteria for using a *t* test in that there is one treatment variable (species), two samples (mid and lower shore), and 13 observations in each sample. The data are measured on an interval scale, are parametric, and because each periwinkle has only been measured once, the data are not matched. The *F* test described in BOX 7.1. can be used to confirm that the two variances are similar. For this example $s_1^2 = 2.01$ and $s_2^2 = 5.84$, $F_{\text{calculated}} = 2.89$ with $v_1 = 12$, $v_2 = 12$ and $F_{\text{critical}} = 3.28$ at $p = 0.05$. Therefore, since $F_{\text{calculated}}$ (2.89) is less than F_{critical} (3.28) there is no significant difference between the variances of the two periwinkle samples and we may proceed with the *t* test.

7.2.2. The calculation

Having satisfied ourselves that our data meet all the criteria for using this *t* test, we can then proceed (BOX 7.3.). Where steps have been abbreviated the full calculation is included in the Online Resource Centre.

BOX 7.3. How to carry out a t test for unmatched data

GENERAL DETAILS	THIS EXAMPLE
1. Hypotheses to be tested **H$_0$**: There is no difference between the means for the two samples. **H$_1$**: There is a difference between the means for the two samples.	**1. Hypotheses to be tested** **H$_0$**: There is no difference between the mean shell height of the periwinkles from the lower and mid shore at Porthcawl in 2002. **H$_1$**: There is a difference between the mean shell height of the periwinkles from the lower and the mid shore at Porthcawl in 2002.
2. How to work out $t_{calculated}$ i. Decide which is sample 1 and which is sample 2.	**2. How to work out $t_{calculated}$** i. Let the periwinkles from the lower shore be sample 1 and the periwinkles from the mid shore be sample 2.
ii. Use the variances (s_1^2 and s_2^2) calculated for the F test, and calculate the means ($\bar{x}_1$ and $\bar{x}_2$) and record the number of observations (n) for each sample.	ii. $\bar{x}_1 = 5.89231$ $s_1^2 = 2.01244$ $n_1 = 13$ $\bar{x}_2 = 6.56154$ $s_2^2 = 5.82256$ $n_2 = 13$
iii. Determine the common variance (s_c^2) as: $s_c^2 = \frac{(n_1-1)s_1^2 + (n_2-1)s_2^2}{(n_1-1)+(n_2-1)}$	iii. $s_c^2 = \frac{[(13-1)2.01244 + (13-1)5.82256]}{(13-1)+(13-1)}$ $s_c^2 = 3.9175$
iv. Use the common variance to calculate the standard error of the difference between the means (SE$_D$) and then $t_{calculated}$. $SE_D = \sqrt{\left(\frac{s_c^1}{n_1} + \frac{s_c^2}{n_2}\right)}$ $t_{calculated} = \frac{\bar{x}_1 - \bar{x}_2}{SE_D}$	iv. $SE_D = \sqrt{\left(\frac{3.9175}{13} + \frac{3.9175}{13}\right)}$ $= 0.77633$ $t_{calculated} = \frac{5.89231 - 6.56154}{0.77633}$ $= -0.86204$
3. How to find $t_{critical}$ i. Calculate the degrees of freedom (v), where $v = (n_1-1)+(n_2-1)$ ii. Look in the t table for a two-tailed test, using $p = 0.05$ (usually the columns) and the v that you have calculated (usually the rows).	**3. How to find $t_{critical}$** $\nu = (13-1)+(13-1) = 24$ For a two-tailed test, $t_{critical} = 2.064$ at $p = 0.05$

BOX 7.3. Continued

GENERAL DETAILS	THIS EXAMPLE
4. The rule If the **absolute** value of $t_{calculated}$ is more than or equal to $t_{critical}$ you may reject the null hypothesis.	**4. The rule** The absolute value of $t_{calculated}$ ($\vert -0.862 \vert$) is less than $t_{critical}$ (2.064). Therefore you do not reject the null hypothesis. When you use an absolute value the negative sign is ignored. In this example $t_{calculated}$ is 0.862.
5. What does this mean in real terms?	**5. What does this mean in real terms?** There is no significant difference ($t = 0.862$, $p = 0.05$) in the mean shell height between the periwinkles on the lower and mid shores of Porthcawl. Therefore, these data do not support the notion that there are two distinct species.

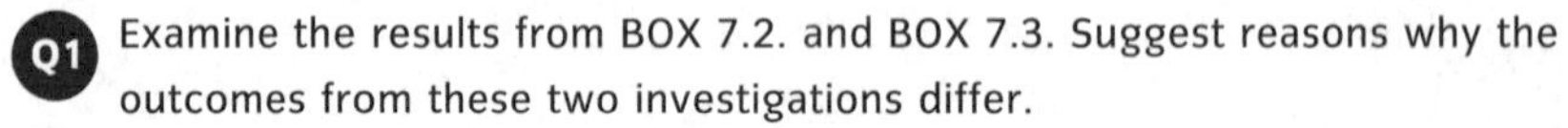

Q1 Examine the results from BOX 7.2. and BOX 7.3. Suggest reasons why the outcomes from these two investigations differ.

A1

1. There really is a difference between Porthcawl and Aberystwyth in the way these periwinkles are evolving.
2. The sampling error in the small sample from Porthcawl has masked the effect of evolution and a **Type II** error has occurred.
3. One of the assumptions was not met. For example, the data were not parametric.

7.3. *z* and *t* tests for matched data

If you are designing an investigation in which you will measure an item before a particular event or treatment and then after, this will generate matched data. The two observations for each item are not independent of each other as they relate to a single item. This needs to be allowed for in the analysis of the data. Unlike unmatched *z* and *t* tests the comparison of matched samples is based on the difference between each pair of measurements. If the samples are similar then the differences between each pair of observations will be small. The mean and standard error of these differences are used to generate $z_{calculated}$.

Matched *z* and *t* tests are virtually identical; the main difference again arises in that in the *t* test there is an allowance for the relatively small sample sizes being investigated. Since these two tests are so similar, we have amalgamated the information in one box (BOX 7.4.).

7.3.1. To use these tests you:

1. Wish to test for differences in population means.
2. Have one treatment variable and two samples.
3. Have data that are matched and therefore the sample sizes are equal.
4. Have samples in which the variances are similar (homogeneous).

5a. Have 30 or more pairs of observations and the data are parametric. In this case you may consider using the z test for matched data. Or:

5b. Have less than 30 pairs of observations and the data are parametric. In this case you may consider using a t test for matched data.

EXAMPLE 7.3. Weight loss by members of a fencing club during a 1-day competition

The members of a fencing club were aware that on days when they competed they lost weight (Table 7.3.). The weight loss was thought to be the result of dehydration. However, since not all members lost weight the team wished to know if the weight loss was significant (T. Richards, personal communication, 9.1.05).

Table 7.3. Weight loss during a 1-day fencing competition by members of a fencing club

Competitor	Weight before competition (kg)	Weight after competition (kg)	Difference between weights (kg)
1	60.00	59.55	0.45
2	59.15	58.70	0.45
3	60.20	59.80	0.40
4	62.40	61.90	0.50
5	57.20	57.20	0.00
6	60.35	59.85	0.50
7	59.80	59.40	0.40
8	60.10	60.10	0.00
9	60.20	59.90	0.30
10	59.90	59.90	0.00
11	60.00	60.00	0.00
12	61.20	60.75	0.45
13	58.50	58.05	0.45
$\bar{x}$	59.92308	59.62308	0.3
s^2	1.51656	1.35942	0.04583

This data meet the criteria for using a t test for matched data since there are less than 30 observations, there is one treatment variable (weight), and two samples (before and after the competition). The data are matched so that two observations have been recorded for each person. The test for homogeneity of variances is described in BOX 7.1. For Example 7.3. $F_{calculated} = 1.51656/1.35942 = 1.11554$, both $v = 13 - 1 = 12$ and at $p = 0.05$, $F_{critical} = 3.28$. Since $F_{calculated}$ is less than $F_{critical}$ then there is no significant difference between the two sample variances and we may proceed with the t test.

7.3.2. The calculation

Having satisfied ourselves that our data meet all the criteria for using this t test, we can then proceed (BOX 7.4.). If you are carrying out a z test for matched data the procedure is exactly the same until finding the critical values. Where steps have been abbreviated the full calculation is included in the Online Resource Centre.

BOX 7.4. How to carry out a *z* and *t* test for matched data

GENERAL DETAILS	THIS EXAMPLE
1. Hypotheses to be tested	**1. Hypotheses to be tested**
H_0: There is no difference between the means of the populations from which sample 1 and sample 2 are taken. **H_1**: There is a difference between the means of the populations from which sample 1 and sample 2 are taken.	**H_0**: There is no difference between the mean weights of individuals before and after a fencing competition. **H_1**: There is a difference between the mean weights of individuals before and after a fencing competition.
2. How to work out $z_{\text{calculated}}$ or $t_{\text{calculated}}$	**2. How to work out $t_{\text{calculated}}$**
i. Work out the difference (D) for each pair of observations. ii. Calculate the mean ($\overline{D}$) and variance (s_D^2) for these differences, where n is the number of pairs of observations. iii. The standard error of the difference of the means $\text{SE}_D = \sqrt{\frac{s_D^2}{n}}$ iv. $Z_{\text{calculated}}$ or $t_{\text{calculated}} = \frac{\overline{D}}{\text{SE}_D}$	i. The differences between each pair of observations is included in Table 7.3. (column 4). ii. The method for calculating a variance is given in BOX 3.1. The results for this example are included in Table 7.3. iii. $\text{SE}_D = \sqrt{(0.04583/13)} = 0.05938$ iv. For this example we are using a t test, therefore $t_{\text{calculated}} = (0.3/0.05938)$ $= 5.05245$
3. How to find z_{critical} or t_{critical}	**3. How to find t_{critical}**
If this is a t test for matched pairs, then t_{critical} is found for a two-tailed t test where $\nu = n - 1$ at $p = 0.05$ If this is a z test for matched pairs then z_{critical} is found from the two-tailed table of z values at $p = 0.05$.	In our example $\nu = 13 - 1 = 12$ and for a two-tailed t test at $p = 0.05$ $t_{\text{critical}} = 2.179$.
4. The rule	**4. The rule**
If the **absolute** value of $z_{\text{calculated}}$ or $t_{\text{calculated}}$ is more than or equal to z_{critical} or t_{critical} respectively then you may reject the null hypothesis.	$t_{\text{calculated}}(5.053)$ is more than t_{critical} (2.179), so we reject the null hypothesis. In fact, t_{critical} at $p = 0.001$ is 4.318, so we may reject the null hypothesis at this higher level of significance.
5. What does this mean in real terms?	**5. What does this mean in real terms?**
	There is a very highly significant difference ($t = 5.053$, $p = 0.001$) between the mean weights of individuals before and after a fencing competition.

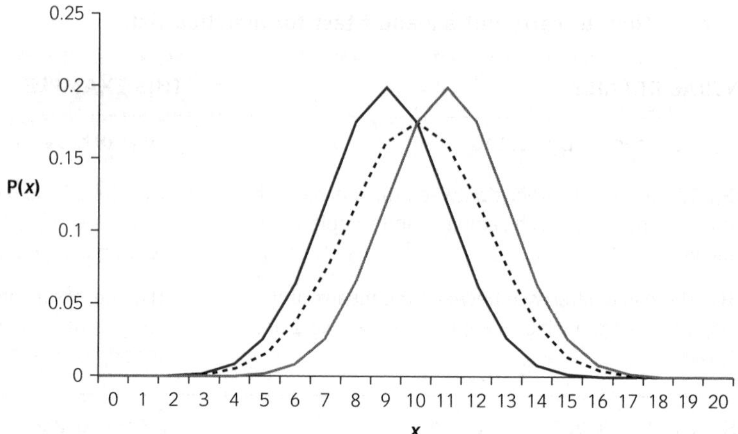

Fig. 7.1. Two samples with similar means and variances (——) which when combined (----) produce a distribution with a mean and variance similar to those of the two original samples.

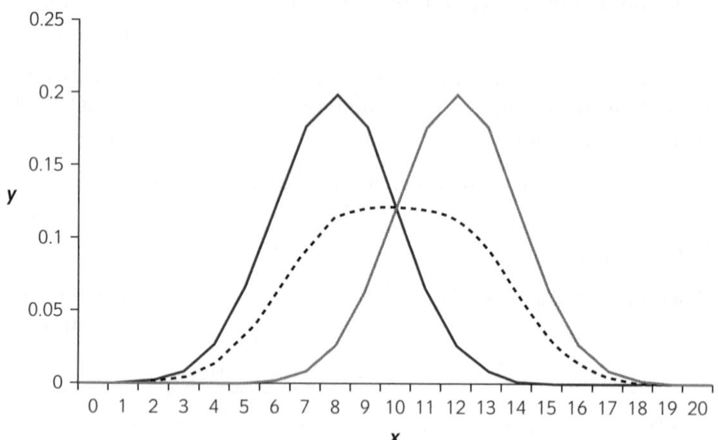

Fig. 7.2. Two samples with very different means (——) which when combined (----) produce a distribution with a variance much greater than the variance of either sample 1 or sample 2.

7.4. Introduction to parametric analysis of variance (ANOVA)

The acronym ANOVA stands for analysis of variance. Parametric ANOVAs are flexible and powerful. There are many versions of ANOVA which allow you to handle data with two or more samples, with or without replicates, with one or more treatment variables, and with or without incomplete data sets. In this chapter we examine ANOVAs suitable for one or two variables with two or more samples, with replicates. These are the ANOVAs our undergraduates most often decide to use. There are many versions of parametric ANOVAs that we do not cover. If you think that an ANOVA is likely to be the best test for your experimental design but cannot find the one you need in this chapter then we recommend Sokal & Rohlf (1981).

ANOVAs are used to test the hypothesis: there is no difference between the mean values for a given number of samples. These samples may come from different treatments or from blocks or replicates within treatments. We consider experimental designs relating to ANOVAs at the end of this Chapter. Although the hypotheses refer to sample means, the ANOVA 'looks' at this by comparing variation in the data. In Fig. 7.1. data from two samples are plotted on line graphs. The means for both samples are similar and the variances are similar and relatively small. When the two samples are combined (Fig. 7.1) there is little change either in the mean or the total variance. Compare this with Fig. 7.2. Here the two samples have very different means, although the variances for each sample are relatively small and similar to each other. If these two samples are combined the mean for this total data set is different and the combined variance is relatively large. In an analysis of the data from Fig. 7.1. you would probably not expect a significant difference between the two samples and this can be detected by comparing the total variance with the variance for the two samples. However, in the data from Fig. 7.2. the total variance is relatively large because the means of the two samples are so different from each other. In this case a comparison of the two sample variances with the total variance would suggest that these two samples will be significantly different from each other. It is this comparison of variances in the ANOVA that detects the relative differences in the means of the samples.

When asking the question 'Do samples come from the same or different populations?' it is possible to test both specific and general hypotheses (4.1.). A general hypothesis asks the question 'is there a difference?' A specific hypothesis asks a question about the nature of that difference: for example, is the mean of sample 1 significantly larger than the mean of sample 2, which is larger than the mean of sample 3. Unlike a non-parametric ANOVA, a parametric ANOVA only tests general hypotheses. Therefore, another test (e.g. Tukey's test) is used to test specific hypotheses following on from a significant parametric ANOVA.

Calculating ANOVAs by hand is straightforward but tedious. However, there are so many ANOVAs that a common mistake when using computer software is to select the wrong one. For this reason we include the five ANOVAs most frequently used by our undergraduates. We hope that this introduction will provide you with sufficient familiarity to be confident when using statistical packages.

The five ANOVAs we include are:

1. One-way ANOVA with equal sample sizes
2. Two-way ANOVA with equal sample sizes
3. Two-way nested ANOVA
4. Three-way factorial ANOVA without replicates
5. Three-way factorial ANOVA with replicates.

The ANOVAs we include show you how to analyse data from experiments with one (one-way ANOVA, 7.5.) to three (three-way ANOVA, 7.10.) variables and all but one of these designs have equal replicates. The exception is the three-way factorial ANOVA without replicates (7.11.). In the two-way ANOVAs we have a two-way ANOVA and a nested two-way ANOVA. These reflect a Model I and mixed model experimental design.

The terms Model I, Model II and mixed models refer to the variables, which can be fixed or random. In Example 7.4. the effectiveness of weaning plantlets of *Lobelia* 'Hannah' from tissue culture onto one of four composts is considered. The four composts are a consistent or fixed treatment determined by the investigator. In Example 7.5. the experiment on weaning was extended and two varieties of *Cosmos atrosaguineus* were weaned from tissue culture on one of four composts. Here, both the varieties and the composts are consistent or fixed treatments and both are determined by the investigator. An ANOVA suitable for analysing data where the treatment(s) is fixed is called a Model I ANOVA. The fixed treatments can be contrasted with Example 7.6 (p. 214). Here, soil samples were taken from three pits dug in two forests (deciduous and coniferous). The difference in this design is that the pits are 'randomized' in the forests. Pit 1 in the deciduous forest cannot be said to be the same treatment as pit 1 in the coniferous forest. In this example the forests are fixed treatments and the pits are randomized. The randomized element is called a Model II design. The overall experimental design is called a mixed model as it has a fixed treatment variable (Model I) and a randomized treatment variable (Model II) within it. The ANOVA that is used to test a mixed model design is called a nested ANOVA. The ANOVAs we include can generally be used to test a Model I or Model II ANOVA with the exception of the nested ANOVA, which is specifically for mixed models. It is therefore important that you know whether your design is Model I, Model II or a mixed model.

Q2 **Look at Example 7.7 (p. 219). There are three variables: depths, distances, and bearings from a smelter. Which of these are fixed and which random. What model is this experimental design?**

A2 All these treatments are under the control of the investigator and fixed. Therefore this is a Model I design.

In the ANOVAs that we include only the nested ANOVA is model-dependent. The other ANOVAs may be used for either pure Model I or pure Model II designs. Mixed models should be analysed using nested ANOVAs.

The method for calculating ANOVAs does not lend itself to the format we have followed elsewhere in the book. However, we include both general instructions and a specific example to help you understand the calculations. In addition there are calculation tables and ANOVA tables that illustrate both how to carry out the analysis and how to report these results.

7.5. Parametric one-way ANOVA with equal numbers of replicates

The first and simplest ANOVA we consider is one that allows you to compare samples that have been subjected to a range of conditions for one treatment, e.g. the organic content of soil in four different woodlands. Here the treatment is 'woodlands'. When there is one treatment variable the ANOVA is called a one-way ANOVA. This parametric one-way ANOVA can examine Model I or Model II designs with data that have an equal number of replicates (7.5.) or unequal replicates (Sokal & Rohlf, 1981).

7.5.1. To use this test you:

1. Wish to test for differences in population means.
2. Have one treatment variable.
3. Have parametric data.
4. Have an experimental design which means that each item is assigned at random to the samples.
5. Have samples where the variation is similar (homogeneous).
6. Have the same number of **replicates** (observations) in each sample.

EXAMPLE 7.4. The effectiveness of weaning plantlets of *Lobelia* 'Hannah' from tissue culture onto one of four composts

An undergraduate studying horticulture was interested in how plants propagated by tissue culture were then weaned onto other growing media, such as soil-based compost. She designed a series of trials that examined aspects of this weaning process. In one trial she transferred *Lobelia* 'Hannah' plantlets at random into plugs containing one of four different composts (A–D). After 8 weeks the plantlets were examined and their fresh weight recorded (Table 7.4.).

Table 7.4. The fresh weight of *Lobelia* 'Hannah' after 8 weeks weaning in one of four types of compost (A–D); calculations using these data and summary statistics

Fresh weight (g) of Lobelia 'Hannah' weaned in one of four composts (A–D)				
	A	B	C	D
	7.04 2.48	3.18 2.18	3.20 5.11	4.16 3.71
	4.86 3.03	1.52 2.12	3.93 5.32	2.94 1.60
	7.47 4.21	3.85 1.27	3.21 3.00	3.62 1.60
	4.49 6.03	1.42 2.03	3.01 2.25	2.09 2.14
	3.06 4.77	4.00 4.82	3.93 3.53	3.83 3.16
$\bar{x}$	4.744	2.639	3.649	2.885
s^2	2.84663	1.53683	0.91999	0.92307
Σx_s	47.44	26.39	36.49	28.85
Σx_s^2	250.675	83.4747	141.4319	91.5399
n_s	10	10	10	10
Σx_T		139.17		
Σx_T^2		567.1215		
N		40		

The data meet all the criteria for using this one-way parametric ANOVA in that the data are parametric and there is one treatment variable (compost). The plantlets were assigned at random to each treatment and there are 10 observations in each sample. To check whether the variances are similar an F_{max} test is carried out (BOX 7.5.). This test is very similar to the one outlined in BOX 7.1.; however, the table of critical values is not the same. You must therefore use the correct F_{max} table.

BOX 7.5. How to carry out an F_{max} test to check for homogeneous variances, before carrying out an ANOVA

GENERAL DETAILS	THIS EXAMPLE
1. Hypotheses to be tested	**1. Hypotheses to be tested**
H_0: There is no difference between the variances of the two samples. The variances are homogeneous. **H_1**: There is a difference between the variances of the two samples.	**H_0**: There is no difference between the variances of the fresh weight of plantlets grown in the four different composts. **H_1**: There is a difference between the variances of the fresh weight of plantlets grown in the four different composts.

BOX 7.5. Continued

GENERAL DETAILS	THIS EXAMPLE
2. How to work out $F_{\text{max calculated}}$ i. For each sample calculate the variance (s^2) using the method described in BOX 3.1. ii. $F_{\text{max calculated}}$ is the ratio between these two values found by dividing the largest variance by the smallest variance. $F_{\text{max calculated}} = \frac{\text{largest sample variance}(s_1^2)}{\text{smallest sample variance}(s_2^2)}$	**2. How to work out $F_{\text{max calculated}}$** i. See Table 7.4. ii. $F_{\text{max calculated}} = \frac{2.84663}{0.91999} = 3.09420$
3. How to find $F_{\text{max critical}}$ To look up the value for an F test that proceeds an ANOVA, you will need to know the number of samples (a) and the degrees of freedom ($v = n_s - 1$). n_s is the number of observations in each sample.	**3. How to find $F_{\text{max critical}}$** For our example there are 4 samples ($a = 4$). There are 10 observations in each sample so the degrees of freedom $\nu = 10 - 1 = 9$. So at $p = 0.05$, $F_{\text{max critical}} = 6.31$.
4. The rule If $F_{\text{max calculated}}$ is less than $F_{\text{max critical}}$, then you may accept the null hypothesis and proceed with the ANOVA.	**4. The rule** In this example $F_{\text{max calculated}}$ (3.09) is less than $F_{\text{max critical}}$ (6.31). Therefore, you do not reject the null hypothesis.
5. What does this mean in real terms?	**5. What does this mean in real terms?** There is no significant difference ($F_{\text{max}} = 3.09$, $p = 0.05$) between the variances of the fresh weight of the plantlets weaned on different composts. You may proceed with the ANOVA. If the null hypothesis is not accepted then youshould consider transforming your data (3.9.) or using a non-parametric test, e.g. Kruskal–Wallis (8.3.).

7.5.2. The calculation

Having satisfied ourselves that our data meet all the criteria for using this ANOVA, we can then proceed. In BOX 7.6. we have organized the calculation of $F_{\text{calculated}}$ under a number of subheadings: A. Calculate general terms; B. Calculate the sums of squares; C. Construct and complete an ANOVA calculation table. At each point we show how these steps can be applied to Example 7.4. Some of the calculation is included in Tables 7.4. and 7.5. Where steps have been abbreviated the full calculation is included in the Online Resource Centre.

Table 7.5. ANOVA table for one-way parametric ANOVA of data from Example 7.4. The fresh weight of *Lobelia* 'Hannah' after 8 weeks weaning in one of four types of compost (A–D)

Source of variation	SS	ν	s^2	F
Between composts	26.87561	3	26.87561/3 = 8.95854	5.75509
Within composts	56.03867	36	56.03867/36 = 1.55663	
Total variation	82.91428	39		

BOX 7.6. How to carry out a one-way parametric ANOVA with equal replicates

1. General hypotheses to be tested

H_0: There is no difference between the means of the populations from which the samples are taken.

H_1: There is a difference between the means of the populations from which the samples are taken.

In our example (Example 7.4.) the hypotheses we are testing are:

H_0: There is no difference between the mean fresh weight of plantlets of *Lobelia* 'Hannah' weaned on the four composts after 8 weeks.

H_1: There is a difference between the mean fresh weight of plantlets of *Lobelia* 'Hannah' weaned on the four composts after 8 weeks.

2. How to work out $F_{calculated}$

A. Calculate general terms

1. Add together all the observations in all the samples (grand total) Σx_T.
2. Square each observation in all the samples and add these together $\Sigma(x_T^2)$.
3. Add all the observations in a sample (Σx_s). This is the sample total. Do this for all samples. Square each sample total $(\Sigma x_s)^2$. Add these together. Divide this total by the number of observations in each sample (n_s).
4. Take the result from step 1, square it $(\Sigma x_T)^2$ and divide by N, where N is the total number of observations in all the samples combined.

In our example the calculations are:

1. $\Sigma x_T = 139.17$
2. $\Sigma(x_T^2) = (7.04)^2 + (4.86)^2 + \cdots\cdots (2.14)^2 + (3.16)^2 = 567.1215$
3. $\dfrac{(47.44)^2 + (26.39)^2 + (36.49)^2 + (28.85)^2}{10} = 511.8283$
4. $(139.17)^2/40 = 484.20722$

B. Calculate the sums of squares (SS)

5. SS_{total} = result from step 2 – result from step 4
6. $SS_{between^*}$ = result from step 3 – result from step 4
   * *between samples*
7. $SS_{within^{**}}$ = result from step 5 – result from step 6
   ** *within samples*

Optional check on calculations. *If you calculate the SS for each sample and add these together this figure should be the same as SS_{within}. The SS for each sample is calculated as:*

$$SS_{sample} = \Sigma(x^2) - \frac{(\Sigma x)^2}{n}$$

In our example (Table 7.4.) the calculations are:

5. $SS_{total} = 567.1215 - 484.20722 = 82.91428$
6. $SS_{between} = 511.8283 - 484.20722 = 26.87561$
7. $SS_{within} = 82.91428 - 26.87561 = 56.03867$

C. Construct and complete an ANOVA calculation table

i. Draw an ANOVA table as illustrated here.

Source of variation	SS	ν	s^2	F
Between samples	step 6			
Within samples	step 7			
Total variation	step 5			

BOX 7.6. Continued

ii. Transfer the results from the calculations for $SS_{between}$, SS_{within} and SS_{total} into the ANOVA table in column 1 (see Table 7.5.).

iii. Calculate the degrees of freedom (v).

v_{total} = total number of observations − 1 = $N-1$
$v_{between}$ = number of samples − 1 = $a-1$
v_{within} = total number of observations − number of samples = $N-a$

Enter these results in column 2.

iv. Calculate the variances (s^2) for between and within samples

Note: Variances in this context are also known as mean squares (MS).

$$s^2 = \text{MS} = \frac{\text{SS}}{v}$$

In the 'between samples' row take the value for SS and divide it by its value for v.
In the 'within samples' row take the value for SS and divide it by its value for v.

Put the results from these calculations in column 3 (Table 7.5).

v. Finally calculate the variance ratio $F_{calculated}$ where:

$$F_{calculated} = \frac{\text{'between samples' variance}}{\text{'within samples' variance}}$$

It is usual to place the results from this calculation in column 4 in the first row (Table 7.5).

3. To find $F_{critical}$

Identify the critical F value using the F tables for an ANOVA. You may need to **interpolate** (4.3.6.) these values. To find the critical value you need the degrees of freedom relating to the variances 'between samples' (v_1) and 'within samples' (v_2).

In our example 'between samples' $v_1 = 3$ and 'within samples' $v_2 = 36$. Interpolating from the F table for $p = 0.05$, $F_{critical} = 2.872$.

4. The rule

If $F_{calculated}$ is greater or equal to $F_{critical}$ then you may reject the null hypothesis.

In our example, $F_{calculated}$ (5.7551) is greater than $F_{critical}$ (2.872) so you may reject the null hypothesis.

In fact at $p = 0.01$, $F_{critical} = 4.39$, so you may reject the null hypothesis at this higher level of significance.

5. What does this mean in real terms?

There is a highly significant difference ($F = 5.76$, $p = 0.01$) between the mean fresh weight of *Lobelia* 'Hannah' when weaned on four different composts.

7.6. Tukey's test following a parametric one-way ANOVA

The one-way ANOVA described in 7.5. has allowed us to test the general hypothesis that there was no difference between the success of weaning as indicated by mean fresh weight of *Lobelia* 'Hannah' on four composts. The analysis shows that there is a significant difference (BOX 7.6.). It would be helpful to be able to be more exact in our interpretation of the data. Looking at the mean values in Table 7.4., compost A appears to be most effective in this regard, but is this significant? To make a specific evaluation of this sort you may use a multiple comparisons test. There are several of these but the one we recommend (Tukey's) is reasonably conservative, i.e. you will get fewer significant differences and are therefore able to be more discriminating in your interpretation and are less likely to make a **Type I error**.

The Tukey's test we describe is suitable for use following an ANOVA where the number of replicates is the same for all samples as in our example. If you wish to use a multiple comparison test following a one-way ANOVA with unequal replicates, refer to the Tukey–Kramer test (Heath, 1995).

7.6.1. To use this test you:

1. Wish to test specific hypotheses following a significant outcome in a parametric one-way ANOVA.
2. Should have equal numbers of observations in each sample.

To illustrate this Tukey test we will use Example 7.4. and the one-way parametric ANOVA in BOX 7.6. Therefore, we are using the Tukey's test having obtained a significant difference in the ANOVA (Table 7.5.) and there are 10 observations in each sample.

7.6.2. The calculation

In a Tukey's test each mean value for a sample is compared with every other mean value in the investigation. This difference between the pairs of means is the calculated *T* value which you then compare to a critical *T* value, which you also work out. We demonstrate this in BOX 7.7. with supporting calculations in Table 7.6.

BOX 7.7. How to carry out a Tukey's test after a significant one-way parametric ANOVA with equal replicates

GENERAL DETAILS	THIS EXAMPLE
1. Hypotheses to be tested	**1. Hypotheses to be tested**
You will be comparing two samples at a time by comparing their means. The generic hypotheses for each of these comparisons is: **H_0**: There is no difference between the means of the two samples being compared. **H_1**: There is a difference between the means of the two samples being compared.	In the Tukey's test we will be comparing six pairs of mean values. Each of these comparisons is a test of hypotheses. You could write specific hypotheses for each: for example, H_0: There is no difference between the mean fresh weight for *Lobelia* 'Hannah' grown in compost A compared to compost B, etc. However, common sense suggests that where you are making many comparisons using this test then it is sensible to report the general hypotheses only.
2. How to work out $T_{calculated}$	**2. How to work out $T_{calculated}$**
Calculate the difference between the means in all possible pair wise combinations. Each of these differences is a $T_{calculated}$ value.	See Table 7.6.

BOX 7.7. Continued

GENERAL DETAILS	THIS EXAMPLE
3. How to find $T_{critical}$	**3. How to find $T_{critical}$**
i. Examine the ANOVA calculation to find the following:	i. From Table 7.5.
n_s: the number of observations in a sample.	$n_s = 10$
a: number of samples.	$a = 4$
s^2_{within}: from the ANOVA table. For accuracy you should use the full value and not a rounded up value.	$s^2_{within} = 1.5566297$
v: degrees of freedom for the s^2_{within} from the ANOVA calculation table.	$\nu = 36$
ii. Find the q value at $p = 0.05$, in a q table for the Tukey test using a and v.	ii. To find q for $a = 4$ and $\nu = 36$ we need to interpolate from the q table: $q = 3.814$
iii. Calculate $T_{critical}$ where: $T_{critical} = q \times \sqrt{\frac{s^2_{within}}{n}}$	iii. $T_{critical} = 3.814 \times \sqrt{\frac{1.5566297}{10}}$ $= 1.50478$ at $p = 0.05$
4. The rule	**4. The rule**
Compare $T_{critical}$ to each of the differences between the means (i.e. each $T_{calculated}$). If any $T_{calculated}$ is greater than $T_{critical}$, then you may reject the null hypothesis for these two samples.	There are two $T_{calculated}$ values in Table 7.6. that are greater than $T_{critical}$ (1.505) at $p = 0.05$ and these are shown in bold in the table.
5. What does this mean in real terms?	**5. What does this mean in real terms?**
	The general significant difference detected by the ANOVA (BOX 7.6.) is accounted for by the significant difference ($p = 0.05$) in mean fresh weight of *Lobelia* 'Hannah' growing in compost A compared with compost B and compost A compared with compost D. From this information the student was able to identify which components of the compost appeared to be most important in the weaning of *Lobelia*.

Table 7.6. The difference between each pair of means from Example 7.4. The fresh weight of *Lobelia* 'Hannah' after 8 weeks weaning in one of four types of compost (A–D) (each difference is a $T_{calculated}$ value to be used in a Tukey's test)

Compost	A	B	C	D
	$\bar{x} = 4.744$	$\bar{x} = 2.639$	$\bar{x} = 3.649$	$\bar{x} = 2.885$
A $\bar{x} = 4.744$		4.744 – 2.639 = **2.105**	4.744 – 3.649 = 1.104	4.744 – 2.885 = **1.859**
B $\bar{x} = 2.639$			3.649 – 2.639 = 1.01	2.885 – 2.639 = 0.246
C $\bar{x} = 3.649$				3.649 – 2.885 = 0.764
D $\bar{x} = 2.885$				

Reporting your findings

In a multiple comparisons test you are testing many hypotheses, one for each pair of means. To save you writing each hypothesis out formally at the end an alternative method for reporting multiple comparison results is used. First, rank all your mean values from the smallest to the largest. Then underscore those not significantly different from each other, or do not underscore (by the same line) any two that are significantly different (Got that!!). It requires a bit of thought, but it does work in the end. In our example this would appear as:

B	D	C	A
$\bar{x} = 2.639$	$\bar{x} = 2.885$	$\bar{x} = 3.649$	$\bar{x} = 4.744$

This underscoring is commonly included in the table of results (Chapter 10).

An alternative, less cumbersome method is to use superscripts instead of a line. The same superscript is used for all means that are significantly different from each other, hence:

B^{b}	D^{b}	C^{ab}	A^{a}
$\bar{x} = 2.639$	$\bar{x} = 2.885$	$\bar{x} = 3.649$	$\bar{x} = 4.744$

This alternative approach is particularly useful when you are comparing many means.

7.7. Parametric two-way ANOVA with equal numbers of replicates

Parametric ANOVAs can be used to analyse data from a wide range of experimental designs. For example, if you were developing a technique for extracting DNA from bone material you might wish to compare the amount of DNA (μg) extracted across a range of temperatures and pH (Table 7.7.). This design is appropriate for the ANOVA described in this section since there are two treatment variables (temperature and pH) and each treatment, e.g. pH 3.0, is combined with all temperatures. A design like this is called orthogonal. Other features of this design are that you would have the same number of replicates for each combination of treatments and the bone samples would be allocated at random to a particular extraction method. The ANOVA described in this section is generally appropriate for a Model I or Model II ANOVA, but not for a mixed model ANOVA where a nested ANOVA is required (7.8.).

If you find you have a non-orthogonal design or unequal replicates you may use a slightly modified ANOVA (Sokal & Rohlf, 1981).

As soon as two variables are being compared in a two-way parametric ANOVA, as illustrated here, or in a two-way non-parametric ANOVA (8.5), then three pairs of hypotheses are tested. Not surprisingly the first of these three pairs of hypotheses is a general hypothesis and refers to the difference between the samples in relation to the effect of treatment 1. For example, in the design shown in Table 7.7. this would be the effect of temperature. The second pair of hypotheses refers to the effect of the second treatment, the effect of pH on DNA extraction. The two-way ANOVA also tests for an interaction between these two variables and this is the third pair of hypotheses. An interaction between treatments can be

Table 7.7. An orthogonal experimental design: the effect on DNA extraction of several temperatures is investigated in relation to a number of pHs (each category of one treatment variable is combined with each category of the second treatment variable)

pH	Temperature (°C)			
	5	10	15	20
3	✓	✓	✓	✓
7	✓	✓	✓	✓
12	✓	✓	✓	✓

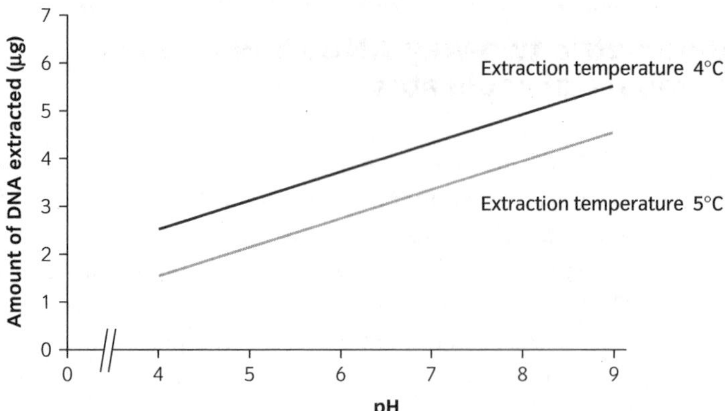

Fig. 7.3. The amount of DNA extracted increases in a similar way as pH increases, for both temperatures, indicating no interaction between pH and temperature (°C).

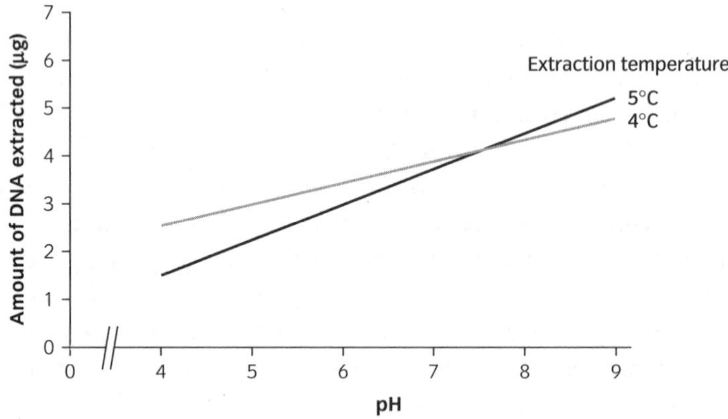

Fig. 7.4. The two temperatures do not have the same effect on the amount of DNA extracted as pH increases, indicating an interaction between temperature and pH.

predicted from a plot of the mean values (Fig. 7.3.). In Fig. 7.3. the amount of DNA extracted as the pH increases changes in a similar way for all temperatures. If there was an interaction then these lines would not be parallel (Fig. 7.4.).

EXAMPLE 7.5. The weaning of *Cosmos atrosanguineus* var. 'Pip' and var. 'Christopher' onto one of four composts following propagation by tissue culture

The undergraduate who was working on *Lobelia* 'Hannah' also examined the weaning from tissue culture of two varieties of *Cosmos atrosanguineus*, var. 'Pip' and var. 'Christopher'. As in the *Lobelia* trial (Example 7.4.), she transferred plantlets at random into plugs containing one of four different composts (A–D). After 8 weeks the plantlets were examined and the maximum height of the plants was recorded (Table 7.8.).

Table 7.8. Final height (cm) of two varieties of *Cosmos atrosanguineus* (var. 'Pip' and var. 'Christopher') grown in one of four different composts (A–D) (table includes summary statistics)

	Weaning compost							
	A		B		C		D	
Final height (cm) of *C. atrosanguineus* var. 'Pip'	20.4	22.0	19.4	22.9	18.7	18.7	20.0	20.1
	20.4	21.6	28.7	25.2	18.5	22.2	25.9	18.1
	24.9	22.7	21.1	27.7	23.2	22.9	17.0	24.1
	19.6	23.5	24.5	26.3	20.6	24.1	22.7	17.5
	19.2	25.4	22.8	18.0	22.3	22.7	25.2	25.2
Summary statistics for var. 'Pip'	$\bar{x} = 21.97$ $s^2 = 4.62011$		$\bar{x} = 23.66$ $s^2 = 12.20266$		$\bar{x} = 21.39$ $s^2 = 4.39433$		$\bar{x} = 21.58$ $s^2 = 11.87733$	
Final height (cm) of *C. atrosanguineus* var. 'Christopher'	19.5	15.7	8.9	18.1	22.9	10.8	21.6	17.8
	21.9	21.9	8.4	15.8	6.4	17.0	15.2	15.1
	21.9	19.9	15.3	16.2	14.1	21.4	11.0	14.1
	7.9	20.7	13.1	14.8	18.0	15.6	22.9	7.0
	25.5	22.6	13.8	22.6	19.3	17.6	21.0	12.2
Summary statistics for var. 'Christopher'	$\bar{x} = 9.75$ $s^2 = 23.65167$		$\bar{x} = 14.7$ $s^2 = 17.1666$		$\bar{x} = 16.31$ $s^2 = 24.13656$		$\bar{x} = 15.79$ $s^2 = 25.80767$	

7.7.1. To use this test you:

1. Wish to test for differences in population means.
2. Have two treatment variables.
3. Have parametric data.
4. Allocate all items at random to each sample.
5. Have variances that are similar (homogeneous).
6. Have the same number of replicates (observations) in each sample.
7. Have an orthogonal design.

The data from Example 7.5. meet the criteria for using a two-way parametric ANOVA with equal replicates, because the data is parametric, each plantlet was allocated at random to the treatments, there are 2 treatment variables (compost and plant variety), and there are 10 plants in each treatment. The design is orthogonal: each treatment is combined with every other treatment. To check whether the variances are similar an F_{max} test is carried out (BOX 7.5.). The variances for the data from Example 7.5. are included in Table 7.8. $F_{max\ calculated}$ is $25.80767/4.39433 = 5.8729$. At $p = 0.05$, $a = 8$ and $v = 10 - 1 - 9$, then $F_{max\ critical} = 7.87$.

$F_{max\ calculated}$ is less than $F_{max\ critical}$, so we do not reject the null hypothesis; the variances are homogeneous and we may proceed with the ANOVA. If the variances are not homogeneous you should consider **transforming** your data (3.9.) or using a non-parametric two-way ANOVA (8.5.).

7.7.2. The calculation

Having satisfied ourselves that our data meet all the criteria for using this two-way ANOVA we can then proceed. In BOX 7.8. we have organized the calculation of $F_{calculated}$ under a number of subheadings: A. Calculate general terms; B. Calculate the sums of squares; C. Construct and complete an ANOVA calculation table. At each point we show how these steps can be applied to Example 7.5. Some of the calculation is included in Table 7.9. Where steps have been abbreviated the full calculation is included in the Online Resource Centre.

BOX 7.8. How to carry out a two-way parametric ANOVA with equal replicates

1. General hypotheses to be tested

As explained in 7.7. there are three pairs of general hypotheses to be tested.

H_{0c}: There is no difference between the sample means due to treatment 1 (columns).

H_{1c}: There is a difference between the sample means due to treatment 1 (columns).

H_{0r}: There is no difference between the sample means due to treatment 2 (rows).

H_{1r}: There is a difference between the sample means due to treatment 2 (rows).

H_{0I}: There is no interaction between treatment A and treatment B in their effects on the sample means.

H_{1I}: There is an interaction between Treatment A and Treatment B in their effects on the sample means.

In our example (Example 7.5.) these hypotheses will be:

H_{0c}: There is no difference between the mean height of *C. atrosanguineus* weaned on different composts.

H_{1c}: There is a difference between the mean height of *C. atrosanguineus* weaned on different composts.

H_{0r}: There is no difference between the mean heights of *C. atrosanguineus* var. 'Pip' compared with var. 'Christopher'.

H_{1r}: There is a difference between the mean heights of *C. atrosanguineus* var. 'Pip' compared with var. 'Christopher'.

H_{0I}: There is no interaction between the weaning composts and *C. atrosanguineus* varieties in their effects on mean height.

H_{1I}: There is an interaction between the weaning composts and *C. atrosanguineus* varieties in their effects on mean height.

2. How to work out $F_{calculated}$

A. Calculate general terms

1. First add together every observation in the complete data set (grand total) Σx_T.
2. Square each and every observation and add all these squared values together Σx_T^2
3. Add all the observations in a sample (Σx_s). This is the sample total. Do this for each sample. Square each sample total $(\Sigma x_s)^2$. Add these together. Divide this by the number of observations in a sample (n_s).
4. For each column add all the observations together (Σx_c) and square the total $(\Sigma x_c)^2$. Add these squared column values together and divide this

BOX 7.8. Continued

total by the number of observations in the column (n_c).

5. For each row add all the observations (Σx_r) and square the total $(\Sigma x_r)^2$. Add these squared values together and divide this value by the number of observations in the row (n_r).
6. Square the result from step 1 and divide this by N, where N is the total number of observations in all the samples.

For this example (Table 7.9) these calculations are:

1. $\Sigma x_T = 20.4 + 20.4 + 24.9 + 19.6 + \ldots 7.0 + 21.0 + 12.2 = 1551.5$
2. $\sum x_T^2 = (20.4)^2 + (20.4)^2 + (24.9)^2 + \cdots (21.0)^2 + (12.2)^2 = 31986.69$
3. There are 8 samples in total and 10 observations in each sample. Calculate:

   $((219.7)^2 + (236.6)^2 + (213.9)^2 + (215.8)^2 + (197.5)^2 + (147.0)^2 + (163.1)^2 + (157.9)^2)/(10) = 308719.77/10 = 30871.977$
4. There are 4 columns (composts) and 20 observations in each column. Calculate:

   $((417.2)^2 + (383.6)^2 + (377.0)^2 + (373.7)^2)/20 = 602985.49/20 = 30149.2745$
5. There are 2 rows (var. 'Pip' and var. 'Christopher') and 40 observations in each row. Calculate:

   $$\frac{[(886.0)^2 + (665.5)^2]}{40} = 1227886.25/40 = 30697.15625$$
6. $(1551.5)^2/80 = 30089.40313$

B. Calculate the sums of squares (SS)

7. SS_{total} = result from step 2 – result from step 6
8. $SS_{samples}$ = result from step 3 – result from step 6
9. $SS_{treatment\ 1\ columns}$ = result from step 4 – result from step 6
10. $SS_{treatment\ 2\ rows}$ = result from step 5 – result from step 6
11. $SS_{interaction}$ = result from step 8 – result from step 9 – result from step 10
12. SS_{within} = result from 7 – result from 8

For Example 7.5., the calculations are:

7. $SS_{total} = 31986.69 - 30089.40313 = 1897.28687$
8. $SS_{samples} = 30871.977 - 30089.40313 = 782.57387$
9. $SS_{treatment\ 1\ columns} = 30149.2745 - 30089.40313 = 59.87137$
10. $SS_{treatment\ 2\ rows} = 30697.15625 - 30089.40313 = 607.75312$
11. $SS_{interaction} = 782.57387 - 59.87137 - 607.75312 = 114.94938$
12. $SS_{within} = 1897.28687 - 782.57387 = 1114.713$

C. Construct and complete an ANOVA calculation table

i. Draw an ANOVA table as illustrated here.

Source of variation	SS	ν	s^2	F	p
Treatment 1 (columns)	step 9				
Treatment 2 (rows)	step 10				
Interaction	step 11				
Within sample variation	step 12				
Total	step 7				

ii. Transfer the results from the calculations for $SS_{treatment\ 1\ columns}$, $SS_{treatment\ 2\ rows}$, SS_{within}, and SS_{total} into the ANOVA table in column 1 (see Table 7.10). (You do not use SS_{sample}.)

iii. Calculate the degrees of freedom (ν)

$\nu_{treatment\ 1\ columns}$ = number of columns – 1
$\nu_{treatment\ 2\ rows}$ = number of rows – 1
$\nu_{interaction}$ = (number of columns – 1) × (number of rows – 1)

BOX 7.8. Continued

$v_{within} = N -$ (number of rows × number of columns)

Enter these results in column 2. (Table 7.10).

iv. Calculate the variances (s^2) for the rows, columns and interaction.

Note: Variances in this context are also known as mean squares (MS).

$s^2 = \text{MS} = \text{SS}/v$

For treatment 1 (columns) take the value for SS and divide it by its value for *v*. Repeat this for treatment 2 (rows), the interaction and 'within sample' variation. Put the results from these calculations in column 3 (Table 7.10).

v. Work out the *F* ratio. Since there are three pairs of hypotheses to test, you calculate three *F* values.

Treatment 1 columns:

$$F_{calculated} = \frac{\text{treatment 1 (columns) } s^2}{\text{within samples } s^2}$$

Treatment 2 rows: $F_{calculated} = \dfrac{\text{treatment 2 (rows) } s^2}{\text{within samples } s^2}$

Interaction: $F_{calculated} = \dfrac{\text{interaction } s^2}{\text{within samples } s^2}$

Place the results from these calculations in column 4 (Table 7.10.).

3. To find $F_{critical}$

Again since you are testing three pairs of hypotheses there will be three critical *F* values to find. Using the *F* tables for an **ANOVA** and you may need to interpolate to find the values. The degrees of freedom to use are:

Treatment 1 (columns): $v_{treatment\ 1\ columns}$ and v_{within}
Treatment 2 (rows): $v_{treatment\ 2\ rows}$ and v_{within}
Interaction: $v_{interaction}$ and v_{within}

4. The rule

If $F_{calculated}$ is greater than or equal to $F_{critical}$ then you may reject the null hypothesis.

In our example when first considering the columns (composts) $F_{calculated}$ (1.2890) is less than $F_{critical}$ (2.7347) at $p = 0.05$ so we may not reject the null hypothesis. There is no significant difference (NS).

When comparing the two varieties of *C. atrosanguineus* $F_{calculated}$ (39.255) is more than $F_{critical}$ (4.076) at $p = 0.01$ so we may reject the null hypothesis.

Finally when testing for an interaction $F_{calculated}$ (2.6811) is less than $F_{critical}$ (2.4749) at $p = 0.05$ so we may not reject the null hypothesis. There is no significant interaction (NS).

5. What does this mean in real terms?

There is no significant difference ($F = 1.3$, $p = 0.05$) between the mean height of *C. atrosanguineus* weaned on different composts.

There is a highly significant difference ($F = 39.3$, $p = 0.01$) between the mean heights of *C. atrosanguineus* var. 'Pip' compared with var. 'Christopher'.

There is no significant interaction ($F = 2.5$, $p = 0.05$) between the composts and *C. atrosanguineus* varieties in their effects on mean height.

7.8. Tukey's test following a parametric two-way ANOVA

The two-way ANOVA described in 7.7. has tested three general hypotheses of which one was significant. To test specific hypotheses you need to use a multiple comparisons test, such as the Tukey's test, which allows particular samples to be compared in pairs and those contributing to the significant difference in the ANOVA can be identified. This additional testing of hypotheses allows you to make much more explicit interpretations of your data and so should always be carried out when you

Table 7.9. Calculation of general terms for two-way parametric ANOVA using data from Example 7.5. The weaning of *Cosmos atrosanguineus* var. 'Pip' and var. 'Christopher' onto one of four composts following propagation by tissue culture

		Composts used in weaning				Row totals
		A	B	C	D	
Height (cm) of *C. atrosanguineus* var. 'Pip'	Σx	219.7	236.6	213.9	215.8	886.0
	Σx^2	4868.39	5707.78	4614.87	4763.86	19954.9
	n	10	10	10	10	40
Height (cm) of *C. atrosanguineus* var. 'Christopher'	Σx	197.5	147.0	163.1	157.9	665.5
	Σx^2	4113.49	2315.4	2877.39	2725.51	442890.25
	n	10	10	10	10	40
Column totals						Grand totals
	Σx	417.2	383.6	377.0	373.7	1551.5
	Σx^2	8981.88	8023.18	7492.26	7489.37	31986.69
	n	20	20	20	20	N = 80

Table 7.10. ANOVA table for analysis of data from Example 7.5: The weaning of *Cosmos atrosanguineus* var. 'Pip' and var. 'Christopher' onto one of four composts following propagation by tissue culture, showing how some of the terms are calculated

Source of variation	SS	ν	s^2	$F_{calculated}$	$F_{critical}$	p
Between composts (columns)	59.87137	4 – 1 = 3	59.87137/3 = 19.95712	19.95712/15.48213 = 1.28904	2.7347	NS
Between varieties (rows)	607.75312	2 – 1 = 1	607.75312	607.75312/15.48213 = 39.25514	7.034	0.01
Interaction compost × varieties	114.94938	(2 – 1)(4 – 1) = 3	114.94938/3 = 38.31646	38.31646/15.48213 = 2.47488	2.7347	NS
Within sample variation	1114.713	80 – 8 = 72	1114.713/72 = 15.48213			
Total		80 – 1 = 79				

NS, not significant

have a significant result in a parametric ANOVA. However, if your interaction term is significant it is difficult to interpret the results from a Tukey's test, since the element of interaction will confound any differences between pairs of means.

The method we follow has already been described in 7.6. The general hypotheses are given in BOX 7.7. and the specific hypotheses will relate to the two means being compared at each point.

Table 7.11. $T_{calculated}$ values for Tukey test following the two-way ANOVA on the data from Example 7.5. The weaning of *Cosmos atrosanguineus* var. 'Pip' and var. 'Christopher' onto one of four composts following propagation by tissue culture

		Weaning composts used for *C. atrosaguineus* var. 'Pip'				Weaning composts used for *C. atrosaguineus* var. 'Christopher'			
		A $\bar{x}$ = 21.97	B $\bar{x}$ = 23.66	C $\bar{x}$ = 21.39	D $\bar{x}$ = 21.58	A $\bar{x}$ = 19.75	B $\bar{x}$ = 14.70	C $\bar{x}$ = 16.31	D $\bar{x}$ = 15.79
Weaning composts used for C. *atrosanguineus* var. 'Pip'	A $\bar{x}$ = 21.97		1.69	0.58	0.39	2.22	7.27	5.66	***6.18***
	B $\bar{x}$ = 23.66			2.27	2.08	3.91	**8.96**	7.35	**7.87**
	C $\bar{x}$ = 21.39				0.19	1.64	**6.69**	5.08	**5.60**
	D $\bar{x}$ = 21.58					1.83	**6.88**	5.27	**5.79**
Weaning composts used for C. *atrosanguineus* var. 'Christopher'	A $\bar{x}$ = 19.75						5.05	3.44	3.96
	B $\bar{x}$ = 14.70							1.61	1.09
	C $\bar{x}$ = 16.31								0.52

The calculated T values are the differences between the means for two samples. These differences for Example 7.5. are given in Table 7.11.

The value for $T_{critical}$ is calculated as described in BOX 7.7. For the example we are considering here, $n = 10$, $s^2_{within} = 15.482125$, $a = 8$, $v = 78$ and q when interpolated (4.3.6.) from the table of q values at $p = 0.05$ is 4.346. Therefore $T_{critical} = 4.346 \times \sqrt{(15.482125/10)} = 5.4076$.

A significant difference between the two means is indicated where the $T_{calculated}$ value exceeds the $T_{critical}$ value. These significant differences are shown in bold in Table 7.11. It shows, for example, that in the comparison between the two species the difference is less marked in compost A than in the other composts.

The Tukey's test we follow is suitable for use when each sample has the same number of replicates. If you have unequal replicates you will need to use a different two-way ANOVA (Sokal & Rohlf, 1981) and if significant you can follow this with the Tukey–Kramer test (Heath, 1995).

7.9. Two-way nested parametric ANOVA with equal replicates

A nested ANOVA is one where you do not have an orthogonal design. Instead your second factor is randomized with respect to the first. You can easily tell if you have a nested design, as no categories within your second treatment variable are quite the same. For example, you may wish to examine the effect of two different growth hormones on the growth of floral meristem explants from *Pharbitis nil* (morning glory). The tissue culture medium was made as one batch, split into two flasks and different growth hormones added to each flask. The media from these flasks were poured into tissue culture jars and left to cool. Four explants were grown in each jar of medium. Within each treatment the jars were randomized. After 4 weeks the number of shoots for each explant was recorded. The experimental design is:

Growth medium type 1			Growth medium type 2		
Jar 1	Jar 2	Jar 3	Jar 4	Jar 5	Jar 6
Explant 1	Explant 5	etc.			
Explant 2	etc.				
Explant 3					
Explant 4					

You can tell that this is a nested design since no jar will be exactly the same as any other jar. There will be slight differences in the depth of medium, location of the jar in the growth cabinet, etc. The four explants within each jar are subject to these unique factors and this grouping needs to be taken into account in the analysis. This grouping within the design is recognized by calling the growth media the *group* and the jars the *subgroup*.

Q3 **In our comparison between shoot production by explants of *Pharbitis nil* on two different tissue culture media, how many groups (a) and how many subgroups (b) are there? How many observations are in each subgroup (n_s) and how many in each group (n_g)? How many observations are there in total (N)?**

$a = 2$ (Media type 1 and media type 2.)

$b = 3$ (There are three jars in each group.)

$n_s = 4$ (There are four explants in each jar.)

$n_g = 12$ (There are three jars each with four explants in each group.)

$N = 24$ (There are 24 explants in total.)

In this design we have one main factor (the growth media) and one subordinate or nested factor (jars) and equal replicates in each subgroup. Clearly there are many other possible nested designs. There could be more than one main factor or more than one nested factor or unequal replicates. In this book we illustrate nested ANOVAs with an example with one main variable, one nested variable, and equal replicates. For informaion on other nested designs we recommend Sokal & Rohlf (1981).

EXAMPLE 7.6. Hydrogen ion concentration in deciduous and coniferous forests

A soil scientist wished to compare the hydrogen ion (H^+) concentration of soils in two different forest types, one coniferous and the other deciduous. Within each forest, three different soil pits were dug and from each pit two soil samples were taken for hydrogen ion analysis. The results were recorded as mmol H^+/l per 100 g soil (Table 7.12.).

Table 7.12. The concentration of hydrogen ions in soil (mmol H^+/l per 100 g soil) from three soil pits in two woodlands

Deciduous forest			Coniferous forest		
Pit 1	Pit 2	Pit 3	Pit 1	Pit 2	Pit 3
1.5	0.7	1.2	2.0	1.6	2.7
1.2	1.0	1.7	2.5	2.2	2.4

This experimental design is necessary because of the variable nature of soil. If only one pit was dug in each forest and multiple soil samples taken from this, there would be no assurance that the outcome of an analysis would not depend on the chance of having selected sites with soil that contained particularly high or low values of hydrogen ion concentrations. Hence, soil sampling from a number of sites within each forest is essential to ensure that any differences are due to the different type of forest and not to the location of the site. If no differences, beyond normal soil sample variability, are found between pits within the forest, any differences found can be attributed to the forest type. Hence, we would expect the difference in soil hydrogen ion concentration due to the forest type to be much greater than the differences due to location within the forests individually. Each of these aspects can be tested individually in a nested ANOVA.

7.9.1. To use this test you:

1. Wish to test for differences in population means.
2. Have two treatment variables, one of which is subordinate or nested within the main factor.
3. Have parametric data.
4. Have assigned at random each item within the subordinate variable.
5. Have samples where the variation is similar (homogeneous).
6. Have the same number of replicates (observations) in each sample.

In our example (7.6.) it is difficult to tell if the third and fifth criteria are met as there are relatively few data in each subgroup. The scale of measurement is continuous and could be parametric. However, since we are unable to confirm that these criteria are met we must acknowledge this when reporting the results from the analysis. The other criteria are met in that there are two treatment variables (forest and pits) and the pits are nested within the forests, since no one pit will be the same as any other pit. There are two replicates in each sample and these were located at random in the pits.

7.9.2. The calculation

Having satisfied ourselves as far as we can that our data meet all the criteria for using this nested ANOVA, we can then proceed. In BOX 7.9. we have organized the calculation of $F_{\text{calculated}}$ under a number of subheadings: A. Calculate general terms; B. Calculate the sums of squares;

Table 7.13. Calculation table for the parametric nested ANOVA on data from Example 7.6. Hydrogen ion concentration in a deciduous and coniferous forest

	Deciduous forest			Coniferous forest		
	Pit 1	Pit 2	Pit 3	Pit 1	Pit 2	Pit 3
Sample 1	1.5	0.7	1.2	2.0	1.6	2.7
Sample 2	1.2	1.0	1.7	2.5	2.2	2.4
Subgroup totals (Σx_{sg})	2.7	1.7	2.9	4.5	3.8	5.1
Group totals (Σx_g)		7.3			13.4	
Sum of all 12 observations (Σx_T)				20.7		

C. Construct and complete an ANOVA calculation table. At each point we show how these steps can be applied to Example 7.6. Some of the calculation is included in Table 7.13. Where steps have been abbreviated the full calculation is included in the Online Resource Centre.

Unlike the previous two-way ANOVA there are only two pairs of hypotheses to be tested. The first relates to the main treatment, in this case forests, and the second to the nested factor, the soil pits. In this analysis there is a need to distinguish different levels of grouping so the forests are groups and the soil pits are subgroups.

BOX 7.9. How to carry out a nested ANOVA with one main factor and one nested factor, with equal replicates

1. General hypotheses to be tested

H_0: There is no difference between the means of the samples.

H_1: There is a difference between the means of the samples.

In our example the two pairs of hypotheses we are testing are:

H_0: There is no difference in soil hydrogen ion concentrations due to different forest types.

H_1: There is a difference in soil hydrogen ion concentrations due to different forest types.

H_0: There is no difference in soil hydrogen ion concentrations among the different soil pits within the forest types.

H_1: There is a difference in soil hydrogen ion concentrations among the different soil pits within the forest types.

2. How to work out $F_{calculated}$

A. Calculate general terms

1. Add together all the observations in all the samples (grand total) Σx_T.
2. Square each observation in all the samples and add these together $\Sigma(x_T^2)$.
3. Add all the observations in a subgroup (Σx_{sg}). These are the subgroup totals. Do this for all subgroups. Square each subgroup total $(\Sigma x_{sg})^2$. Add these together. Divide this total by the number of observations in each subgroup (n_s).
4. Add all the observations in a group (Σx_g), square these $(\Sigma x_g)^2$ and divide by the number of observations in that group (n_g).
5. Take the result from step 1, square it $(\Sigma x_T)^2$ and divide by N, where N is the total number of observations in all the samples combined.

In our example (Table 7.13.) the calculations are:

1. $\Sigma x_T = 1.5 + 1.2 + \cdots + 2.7 + 2.4 = 20.7$
2. $\sum(x_T^2) = (1.5)^2 + (1.2)^2 + \cdots + (2.4)^2 = 40.21$
3. $\dfrac{(2.7)^2 + (1.7)^2 + (2.9)^2 + (4.5)^2 + (3.8)^2 + (5.1)^2}{2}$
$= 79.29/2 = 39.645$
4. $\dfrac{(7.3)^2 + (13.4)^2}{6} = 232.82/6 = 38.80833$
5. $(20.7)^2/12 = 35.7075$

B. Calculate the sums of squares (SS)

6. SS_{total}
= result from step 2 – result from step 5
7. SS_{groups}
= result from step 4 – result from step 5
8. $SS_{subgroups}$
= result from step 3 – result from step 4
9. SS_{within}
= result from step 2 – result from step 3

In our example the calculations are:

6. $SS_{total} = 40.21 - 35.7075 = 4.5025$

BOX 7.9. Continued

7. $SS_{groups} = 38.80833 - 35.7075 = 3.10083$
8. $SS_{subgroups} = 39.645 - 38.80833 = 0.83667$
9. $SS_{within} = 40.21 - 39.645 = 0.565$

C. Construct and complete an ANOVA calculation table

i. Draw an ANOVA table as illustrated here.

Source of variation	SS	v	s^2	F
Between groups	step 7			
Between subgroups within groups	step 8			
Within subgroups	step 9			
Total variation	step 6			

ii. Transfer the results from the calculations for SS_{groups}, $SS_{subgroups}$, and SS_{within} into the ANOVA table in column 1 (see Table 7.14.).

iii. Calculate degrees of freedom (v).

Let a = the number of groups, b = the number of subgroups, N = total number of observations, and n_{sg} = number of observations in each subgroup.

$v_{total} = N - 1$

$v_{groups} = a - 1$

$v_{subgroups} = a(b - 1)$

$v_{within} = ab(n_{sg} - 1)$

Enter these results in column 2.

iv. Calculate the variances (s^2) for the groups, subgroups and within subgroups.

Note: Variances in this context are also known as mean squares (MS).

$$s_2 = MS = \frac{SS}{v}$$

In the groups row take the value for SS and divide it by its value for v. Repeat this for the subgroups row and the within row. Put the results from these calculations in column 3 (Table 7.14.).

v. Finally to calculate the variance ratio to test the first pair of hypotheses you use the group and subgroup variances, where:

$$F_{calculated} = \frac{\text{group variance}}{\text{subgroup variance}}$$

It is usual to place the results from this calculation in column 4 in the first row (Table 7.14.).

To test the second pair of hypotheses you use the subgroups and within subgroups variances. So:

$$F_{calculated} = \frac{\text{subgroup variance}}{\text{within variance}}$$

This value is usually placed in the second row in column 4.

3. To find $F_{critical}$

Identify the critical F value using the F tables for an **ANOVA**. You may need to **interpolate** (4.3.6.) these values. To find the critical value you need the degrees of freedom relating to the two variances in the F test.

In our example to test the first pair of hypotheses the degrees of freedom are $v_1 = 1$ and $v_2 = 4$. Therefore at $p = 0.05$, $F_{critical} = 7.71$ and at $p = 0.01$, $F_{critical} = 21.20$.

To test the second pair of hypotheses, the degrees of freedom are $v_1 = 1$ and $v_2 = 6$. Therefore at $p = 0.05$, $F_{critical} = 5.99$.

4. The rule

If $F_{calculated}$ is greater than $F_{critical}$, then you may reject the null hypothesis.

In our example, when testing the first pair of hypotheses $F_{calculated}$ (14.82) is greater than $F_{critical}$ (7.71) at $p = 0.05$, but not greater than $F_{critical}$ (21.20) at $p = 0.01$, so you may reject the null hypothesis at $0.01 < p < 0.05$.

For the second pair of hypotheses, $F_{calculated}$ (2.22) is less than $F_{critical}$ (5.99) at $p = 0.05$, so you do not reject the null hypothesis.

5. What does this mean in real terms?

There is a significant difference ($F = 14.82$, $0.01 < p < 0.05$) in soil hydrogen ion concentrations due to different forest types. However, there is no difference ($p = 0.05$) in soil hydrogen ion concentrations among the different soil pits within the forest types.

Table 7.14. ANOVA table from the analysis of data from Example 7.6. Hydrogen ion concentration in a deciduous and coniferous forest, showing how some of the terms are calculated

Source of variation	SS	ν	s^2	F
Between forests (groups)	3.10083	$2-1=1$	3.10083	3.10083/0.20917 = 14.82469
Between pits (subgroups)	0.83667	$2(3-1)=4$	0.20917	3.10083/0.09417 = 2.22123
Between replicates in pits (within subgroups)	0.565	$(2\times 3)(2-1)=6$	0.09417	
Total variation	4.5025	$12-1=11$		

7.9.3. Tukey's test for nested parametric ANOVAs

It is very difficult to use a multiple comparisons test, such as Tukey's, when one variable is nested in the other and therefore any mean values for samples will be means for subgroups within groups. One circumstance where you might use a Tukey's test would be, as in our example, when the subgroups are not significantly different but the groups are. Under these circumstances you could consider all values within a group to be replicates. If you had three or more groups you could carry out a one-way ANOVA where all subgroup observations are replicates in the groups. The values from this ANOVA could then be used to test specific hypotheses. If you have only two groups, there is no point in carrying out this specific test as it will be clear from the data what the relationship is. In Example 7.6., the soil hydrogen ion concentrations are significantly greater in the coniferous forest than in the deciduous forest.

7.10. Factorial three-way parametric ANOVA

When there are more than two independent factors acting on a sample, the design is called *factorial*. It is possible to have any number of factors and test for differences arising from these in one large experiment and calculation, but in reality the experimental design and execution would become so large that it would be unmanageable. Therefore, we will limit worked examples to a demonstration of a three factor ANOVA (factors A, B, and C) with (7.11.) and without (7.10.) replicates. Here the effect of the three treatments can be tested in the usual way to determine whether there is a significant difference. We can also determine whether there is an interaction between any or all of the factors. The first three interactions, A × B, B × C, A × C, are known as first-order interactions and the interaction A × B × C is known as a second-order interaction.

There is a difference between a factorial analysis with or without replicates. If there are no replicates and for each combination of treatments there is only one observation, then the interaction term A × B × C is used in place of the SS_{within} term in the F ratio. If replicates are present then an SS_{within} term can be calculated and this is used in the F ratio. We illustrate both these processes in the following sections.

7.10.1. To use this test you:

1. Wish to test for differences in population means.
2. Have three or more treatment variables each with two or more categories.
3. Have parametric data.
4. Have an orthogonal experimental design
5. Have items assigned at random to treatments, with independent measurements and not matched.

EXAMPLE 7.7. Lead levels in soil samples taken at various depths, distances, and bearings from a smelter

Soil samples were taken to discover whether there was a significant difference in the extent of lead contamination due to the proximity of a smelter. Soil samples were taken at various bearings and distances to the smelter and at various depths of soil (Table 7.15.). Each treatment was combined with every other treatment.

Table 7.15. Summary table of the factors (3 × 4 × 4 factorial ANOVA) in the investigation of lead levels in soil samples

Factor A Distance (km)	Factor B Bearing	Factor C Soil depth (cm)
0.25	South	5
0.5	South-east	10
1.0	South-west	15
2.0		20
		30

As there are three factors, the results table would be in three dimensions. Since this cannot be represented on paper, a data table is constructed in which the third factor (C) is located within one of the other factors, in this

Table 7.16. Lead concentrations in soil (μg/g) at a number of locations (bearing and distance) and soil depths in the vicinity of a smelter

Distance (km) (Factor A)	Soil depth (cm) (Factor C)	Bearing (Factor B)		
		South	South-east	South-west
0.25	5	155	96	365
	10	102	73	345
	15	77	23	248
	20	65	12	176
	30	26	8	72
0.5	5	98	54	302
	10	65	24	238
	15	45	14	154
	20	23	12	98
	30	9	10	43
1.0	5	55	32	256
	10	36	24	189
	15	22	14	112
	20	12	8	54
	30	9	5	23
2.0	5	25	18	167
	10	18	14	98
	15	15	13	43
	20	8	6	22
	30	7	4	17

case A. This does not imply that factor C is in any way dependent on factor A (Table 7.16.).

The data from this example meet the criteria for using a factorial ANOVA with no replicates as there are three treatment variables (soil depth, bearing, and distance from the smelter), each treatment variable has three or four categories and the design is orthogonal. Each soil sample is unique to that set of treatments, so the data are not matched. As there are only single observations for each particular set of treatments it is difficult to tell whether the data are parametric other than by criterion 1 (BOX 3.2.). This criterion would support the idea that the data are parametric.

7.10.2. The calculation

In this calculation we will examine the effects of the treatments (factor A, factor B and factor C) and the first-order interactions (A × B, A × C, B × C). The second-order interaction (A × B × C) is used in place of the $\text{variance}_{\text{within}}$ term. Therefore, we are testing six pairs of hypotheses and carrying out an F test for each. The calculation of these variance ratios have been organized into a number of steps: A. Calculate the general terms; B. Calculate the sums of squares; C. Construct and complete an

ANOVA table. We illustrate this calculation using the data from Table 7.16. Where steps have been abbreviated the full calculation is included in the Online Resource Centre.

BOX 7.10. How to carry out a three-way factorial parametric ANOVA without replicates

1. General hypotheses to be tested

Since there are six pairs of hypotheses we have not included general hypotheses but only those from our example.

H_0: There is no difference between the mean lead concentration in the soil samples due to distance from the smelter (factor A).

H_{1A}: There is a difference between the mean lead concentration in the soil samples due to distance from the smelter (factor A).

H_{0B}: There is no difference between the mean lead concentration in the soil samples due to the bearing from the smelter (factor B).

H_{1B}: There is a difference between the mean lead concentration in the soil samples due to the bearing from the smelter (factor B).

H_{0C}: There is no difference between the mean lead concentration in the soil samples due to the depth at which the soil was sampled (factor C).

H_{1C}: There is a difference between the mean lead concentration in the soil samples due to the depth at which the soil was sampled (factor C).

$H_{0(A \times B)}$: There is no interaction between distance (factor A) and bearing (factor B) in their effects on the mean lead concentration in the soil.

$H_{1(A \times B)}$: There is an interaction between distance (factor A) and bearing (factor B) in their effects on the mean lead concentration in the soil.

$H_{0(A \times C)}$: There is no interaction between distance (factor A) and soil sample depth (factor C) in their effects on the mean lead concentration in the soil.

$H_{1(A \times C)}$: There is an interaction between distance (factor A) and soil sample depth (factor C) in their effects on the mean lead concentration in the soil.

$H_{0(B \times C)}$: There is no interaction between bearing (factor B) and depth of sampling (factor C) in their effects on the mean lead concentration in the soil.

$H_{1(B \times C)}$: There is an interaction between bearing (factor B) and depth of sampling (factor C) in their effects on the mean lead concentration in the soil.

2. How to work out $F_{\text{calculated}}$

A. Calculate general terms

1. Calculate the grand total (Σx_T), by adding together all the observations in the data set. Note the total number of observations (N).

   $\Sigma x_T = 155 + 96 + 365 + \ldots\ldots 7 + 4 + 17$
   $= 4358.0$
   $N = 60$

2. Square each and every observation and add these together (Σx_T^2).

   $\Sigma x_T^2 = (155)^2 + (96)^2 + (365)^2 + \cdots\cdots (7)^2 + (4)^2 + (17)^2 = 780172.0$

3. Summarize the data from the three-way table (Table 7.16.) into three two-way tables.

   B × C table
   For the B × C table (Table 7.17.) total the four observations for bearing South and 5 cm soil depth. This total goes into the first row, first column of the B × C table. Add the four observations for bearing South-east and 5 cm soil depth. This total goes into the first column second row of the A × C table, etc. The number of observations being counted $= a = 4$.

   A × C table
   For the A × C two-way table (Table 7.18.), add together the three observations for 5 cm soil depth and distance 0.25 km. This total goes into the first row first column of the A × C table. Add together the three observations for 5 cm soil depth and distance 0.5 km. This total goes in the first column second row, etc.
   The number of observations being counted $= b = 3$

   A × B table
   For the A × B table (distance × bearing) first add the five values for 0.25 km distance and bearing South. The total goes into the first cell (first row first column) of the two way table A × B. Add the five values for distance 0.5 km, bearing South. This total goes into the second cell (second row first column) of the A × B table, etc.
   The number of observations being counted $= c = 5$.

BOX 7.10. Continued

Table 7.17. B × C two-way table (bearing × soil sample depth (cm)) for lead levels in soil samples (μg/g)

Factor B Bearing	Factor C Depth at which soil samples taken (cm)					Row totals
	5	10	15	20	30	
South	333	221	159	108	51	872
South-east	200	135	64	38	27	464
South-west	1090	870	557	350	155	3022
Column totals	1623	1226	780	496	233	4358

Table 7.18. A × C two-way table (distance km × soil depth (cm)) for lead levels in soil samples (μg/g)

Factor A Distance from smelter (km)	Factor C Depth at which soil samples taken (cm)					Row totals
	5	10	15	20	30	
0.25	616	520	348	253	106	1843
0.5	454	327	213	133	62	1189
1.0	343	249	148	74	37	851
2.0	210	130	71	36	28	475
Column totals	1623	1226	780	496	233	Grand total 4358

Table 7.19. A × B two-way table (distance (km) × bearing) for lead levels in soil samples (μg/g)

Factor A Distance from smelter (km)	Factor B Bearing from smelter			Row totals
	South	South-east	South-west	
0.25	425	212	1206	1843
0.5	240	114	835	1189
1.0	134	83	634	851
2.0	73	55	347	475
Column totals	872	464	3022	Grand total 4358

4. Square each and every row total from the A × C two-way table (Table 7.18.). Add these squared values together and divide by $(b \times c)$:

$$= \frac{(1843)^2 + (1189)^2 + (851)^2 + (475)^2}{3 \times 5}$$
$$= 384013.0667$$

5. Square each and every row total from the B × C two-way table (Table 7.17.). Add these squared values together and divide by $(a \times c)$:

$$= \frac{(872)^2 + (464)^2 + (3022)^2}{4 \times 5} = 505408.2$$

6. Square each and every column total from the A × C two-way table (Table 7.18.). Add these squared values together and divide by $(a \times b)$:

$$= \frac{(1623)^2 + (1226)^2 + (780)^2 + (496)^2 + (233)^2}{4 \times 3}$$
$$= 420492.5$$

7. Square each value in the A × B two-way table (Table 7.19.), add these squared values together and divide by c:

$$= \frac{(425)^2 + (212)^2 + (1206)^2 + \ldots\ldots(347)^2}{5}$$
$$= 600678.0$$

8. Square each value in the A × C two-way table (Table 7.18.), add these squared values together and divide by b:

$$= \frac{(616)^2 + (520)^2 + \ldots\ldots(36)^2(28)^2}{3}$$
$$= 500890.6667$$

9. Square each value in the B × C two-way table (Table 7.17.), add these squared values together and divide by a:

$$= \frac{(333)^2 + (221)^2 + \ldots\ldots(350)^2 = (155)^2}{4}$$
$$= 666386.0$$

10. Square the grand total $(\Sigma x_T)^2$ from step **1** and divide by the total number of observations (N): $= (4358)^2/60 = 316536.0667$

BOX 7.10. Continued

B. Calculate the sums of squares (SS)

11. SS_A = result from step **4** – result from step **10**
$= 384013.0667 - 316536.0667 = 67477.0$

12. SS_B = result from step **5** – result from step **10**
$= 505408.2 - 316536.0667 = 188872.1333$

13. SS_C = result from step **6** – result from step **10**
$= 420492.5 - 316536.0667$
$= 103956.4333$

14. SS_{AB} = result from step **7** + result from step **10** – result from step **4** – result from step **5**
$= 600678.0 + 316536.0667 - 384013.0667 - 505408.2 = 27792.8$

15. SS_{AC} = result from step **8** + result from step **10** – result from step **4** – result from step **6**
$= 500890.6667 + 316536.0667 - 384013.0667 - 420492.5$
$= 12921.1667$

16. SS_{BC} = result from step **9** + result from step **10** – result from step **5** – result from step **6**
$= 666386.0 + 316536.0667 - 505408.2 - 420492.5 = 57021.3667$

17. SS_{ABC} = (result from step **2** + results from steps **4** + **5** + **6**) – (results from steps **7** + **8** + **9** + **10**)
$= (780172.0 + 384013.0667 + 505408.2 + 420492.5) - (600678.0 + 500890.6667 + 666386.0 + 316536.0667)$
$= 5595.0333$

C. Construct and complete an ANOVA calculation table

i. Construct and complete an ANOVA table as shown in Table 7.20.

ii. Transfer the results from the calculations for sums of squares (SS) into the table in column 1 (Table 7.20.).

iii. Calculate the degrees of freedom (v) where:

$v_A = a - 1$ = number of samples – 1
$v_B = b - 1$ = number of samples – 1
$v_C = c - 1$ = number of samples – 1
$v_{A \times B} = (a-1)(b-1)$
$v_{A \times C} = (a-1)(c-1)$
$v_{B \times C} = (b-1)(c-1)$
$v_{A \times B \times C} = (a-1)(b-1)(c-1)$

Enter these results in column 2 (Table 7.20.).

iii. Calculate the variances (s^2)
Note: Variances in this context are also known as mean squares (MS).

$$s^2 = \text{MS} = \frac{\text{SS}}{v}$$

In the Factor A row take the value for SS_A and divide it by the value for v_A.

In the Factor B row take the value for SS_B and divide it by the value for v_B, and so on. Put the results from these calculations in column 3 (Table 7.20.).

iv. How to work out $F_{calculated}$
As in the F_{max} test, $F_{calculated}$ is a ratio. The F value for each pair of hypotheses to be tested is calculated by the ratio of its variance over the variance of A × B × C (which represents the 'error' variance in a factorial with no replicates and assuming that the third factor interaction is zero), e.g:

$$F_{calculated} = \frac{s^2_A}{s^2_{ABC}}$$

The results from these calculations are placed in column 4 in the first row and so on (Table 7.20.).

3. To find $F_{critical}$

Identify the critical F value using the F tables for an ANOVA. You may need to interpolate (4.3.6.) these values, where the degrees of freedom are those relating to the numerator (v_1) and denominator which is $s^2_{ABC}(v_2)$. Since you are testing six pairs of hypotheses you will need to find six values for $F_{critical}$.

4. The rule

The general rule is that if $F_{calculated}$ is greater then or equal to $F_{critical}$ then you may reject the null hypothesis.

In this factorial analysis you should examine the first-order interaction terms first (i.e., A × B, A × C, and B × C). If these are significant then it is not meaningful to test the main effects of factors A, B, and C as these elements have been shown to not be contributing separately to the overall outcome.

BOX 7.10. Continued

If the interaction terms are not significant then you may proceed to test the main effects.

In our example the interaction terms are all significantly different (Table 7.20.).

5. Relate your findings back to your investigations

The distance × bearing, distance × soil sample depth and bearing × soil sample depth are all highly significant interactions ($p = 0.01$). Therefore, the effect of distance from the smelter on the lead concentrations in the soil samples depends on the bearing and soil sample depth. In addition, the effect of the soil sample depth on the lead concentrations depends on the bearing from the smelter.

Table 7.20. ANOVA table for analysis of data from Example 7.7. Lead concentrations in soil (μg/g) at a number of locations (bearing and distance) and soil depths from a smelter

Source of variation	SS	ν	s^2	$F_{calculated}$	$F_{critical}$ at $p = 0.01$
Factor A Distance (km)	67477.0	3	22492.3	96.48134	
Factor B Bearing	188872.1333	2	94436.06	405.08551	
Factor C Soil sample depth (cm)	103956.43	4	25989.11	111.48083	
Interaction A × B Distance (km) × bearing	27792.80	6	4632.13	19.86963	3.67
Interaction A × C Distance (km) × Soil sample depth (cm)	12921.1667	12	1076.76	4.61880	3.03
Interaction B × C Bearing × Soil sample depth	57021.37	8	7127.67	30.5743	3.36
Interaction A × B × C	5595.03	24	233.13		
SS_{within} (Box 7.11)	result from step **19**	$\nu_{within} = abc(n-1)$	$s^2_{within} = SS_{within}/\nu_{within}$		

7.10.3. Tukey's test and a parametric three-way ANOVA

As we explained in 7.8. any significant interactions between variables will make it very difficult to interpret any multiple comparisons between pairs of means. Therefore, you should restrict your use of Tukey's test to experiments where only the main factors are significant and not the interaction terms. In these cases you apply the Tukey's test as we describe in BOX 7.7. and section 7.8.

7.11. Factorial three-way parametric ANOVA with replicates

Carrying out a factorial ANOVA where you have replicates, i.e. more than one observation for anyone particular combination of treatments, is very similar to the method described in 7.10. In the present section we therefore only highlight the differences that you would need to take into account to allow for replication within the design.

7.11.1. To use this test you:

Need to satisfy the criteria listed in 7.10.1. with two additional criteria:

6. Have samples with homogeneous variances.
7. Have equal numbers of replicates.

With more data you will be able to be more exact as to whether your data are parametric. To test criterion 6 you should carry out the F_{max} test (BOX 7.5.). If you do not have equal numbers of replicates you should refer to Sokal & Rohlf (1981).

7.11.2. The calculation

The key difference between this test and the one outlined in 7.10. is that you will have an error term SS_{within}. The changes required to allow for this are shown in BOX 7.11.

BOX 7.11. How to carry out a three-way factorial parametric ANOVA, with replicates. An extension of the method outlined in BOX 7.10.

1. General hypotheses to be tested

Apart from the general hypotheses included in BOX 7.10. you will have a further pair of hypotheses relating to the second order interaction A × B × C.

$H_{0(A \times B \times C)}$: There is no interaction between factor A, factor B, and factor C in their effects on the sample means.

$H_{1(A \times B \times C)}$: There is an interaction between factor A, factor B, and factor C in their effects on the sample means.

2. How to work out $F_{calculated}$

A. Calculate general terms

18. Using the data from the original three-way table of results sum each cell, square this value, add all these squared values together and divide by n (the number of observations in each cell).

BOX 7.11. (Continued)

For example a small extract from a results table might look like:

Distance (km)	Soil sample depth (cm)	Bearing	
		South	South-east
0.25	5	156	96
		158	98
		155	94
	10	101	
		103	etc.
		102	

Step 18 would then be:

$((156+158+155)^2+(96+98+94)^2+(101+103+102)^2+\cdots)/3$

B. Calculate the sum of squares (SS)

19. SS_{within} = result from step 2 − result from step 18.

C. Construct and complete an ANOVA calculation table

An additional row is added to the bottom of the ANOVA table (Table 7.20):

s^2_{within} is now used as the denominator in working out the $F_{calculated}$ values instead of the s^2_{ABC} as in BOX 7.8. In addition, it is possible to test hypotheses relating to the second-order interaction A × B × C, so $F_{calculated}$ for this is s^2_{ABC}/s^2_{within}.

Again it is best to consider the interactions first before the main treatment effects. If any of the interactions are significant then it is not sensible to use a Tukey's test on these data.

7.12. Experimental design and parametric statistics

There are a number of very different statistical tests covered in this chapter which reflect a number of different experimental designs.

7.12.1. One variable

In experimental designs which seek to examine the effect of one variable we consider investigations where you will collect unmatched data or matched data. We also discuss a specific design (Latin square) for reducing the effect of non-treatment variables.

i. Unmatched

We have considered three tests that may be used to analyse parametric data with one variable. The first two (z and t tests) can be used when you have only two categories for your treatment variable and the one-way ANOVA can be used when you have two or more categories. If you have two categories it is common to design the investigation so that a t or z test may be used, because the analysis for these tests is thought to be simpler. This general design is shown in Table 7.21.

Table 7.21. Experimental design with one treatment variable may be tested by a z or t test for unmatched pairs or one-way ANOVA if all other criteria are met

<table>
<tr><th>Category 1</th><th>Category 2</th><th>Category 3 etc.</th></tr>
<tr><td>observation 1</td><td>observation 1</td><td>observation 1 etc.</td></tr>
<tr><td>observation 2</td><td>observation 2</td><td></td></tr>
<tr><td>observation 3 etc.</td><td>observation 3, etc.</td><td></td></tr>
<tr><td colspan="2">If two categories only you may use a t or z test for unmatched data, if all other criteria are met</td><td></td></tr>
<tr><td colspan="3">If three or more categories consider a one-way ANOVA, if all other criteria are met</td></tr>
</table>

The number of observations in each category is the second key factor that determines the statistical test you are likely to use and must therefore be considered before you carry out your investigation. For example, if you decided to use a z test you will need more than 30 observations in each sample and to use a t test you have under 30 observations in each sample. If this design is to be analysed by an ANOVA the number of observations are usually determined by practical constraints. These tests may be used if you have unequal numbers of observations in your categories (samples).

This design is known as a completely randomized block design as each **item** is assigned at random to a category. The term 'block' is often used in relation to experimental designs and ANOVAs and in this instance each block is all the observations in a category.

If you have only two categories one could be from a **control** treatment and these tests would be an effective way of testing whether your treatment makes a significant impact on your experimental system compared with the control. However, you must not use the same item for the control and the treatment: all observations must be independent measurements on different items. Each observation within a sample is in a sense a replicate, although they are not usually referred to as such.

ii. Matched

If you have one treatment variable where your items are examined before and after the treatment then these are matched observations. The design in this case (Table 7.22.) looks similar to those suitable for a correlation or regression analysis, but you are only examining one variable.

Table 7.22. Experimental design with one treatment variable and for matched data

Item	First measure	Second measure
1	observation 1	observation 1
2	observation 2 etc.	observation 2
3		
4 etc.		

Again the number of observations in a sample is critical to the statistical test you use and the statistical test you are planning to use will determine the sample size. For example, if you are planning to use a z test then you need more than 30 items in your experiment. One of the samples could be a null state and could therefore be a control. Each observation within the sample is in a sense a replicate, although rarely referred to as such.

iii. Latin square

When designing experiments you will seek to identify and control or minimize the effects of non-treatment variables. For example, you can minimize the effect of non-treatment variables by the randomization of items to **treatments** (2.2.6.). However, there is one specific experimental design called a Latin square that is structured to minimize the effect when there are two known non-treatment variables that are graded across the area in which you are working. For example, you may need to organize plants in a greenhouse and know that the light levels drop off as you move away from the windows; similarly, you may have a door at one end and know that there is therefore a temperature gradient along the bench caused by a draught from the door. In these circumstances you can organize your items or blocks in a non-random arrangement. In our example of a Latin square (Fig. 7.5.) you can see that there are four categories (A, B, C, and D) for a single variable. Each category is present in only one row and one column. The ANOVA that is used to analyse the results from a design of this nature partitions the variation into that due to the non-treatment variable 1 + variation due to non-treatment variable 2 + variation due to the treatment + sampling error. We have not included a Latin square ANOVA in this book; instead we refer you to Fowler *et al.* (1998).

Gradient for non-treatment variable 1 →

Gradient for non-treatment variable 2 ↓

A	D	C	B
C	B	A	D
B	C	D	A
D	A	B	C

Fig. 7.5. A Latin square for two non-treatment variables and one treatment variable with four categories.

7.12.2. **Two variables**

When investigations are extended to two treatment variables you have a number of other factors to consider including whether the variables are fixed or randomized in relation to each other.

i. Fixed effects

A two-way ANOVA is the simplest test suitable for the experimental design with two treatment variables (Table 7.23.) This design is orthogonal, where each category of treatment 1 is tested against each category of treatment 2. The items are allocated at random to a category which contrasts with the designs suitable for analysis by correlation and regression tests where two observations are recorded for each item, one observation for each treatment variable (6.9.).

Table 7.23. Experimental design for a two-way parametric ANOVA

	Variable 1	
Variable 2	**Category 1**	**Category 2, etc.**
Category 1	*BLOCK 1, 1*	*BLOCK 1, 2*
	observation 1	observation 1
	observation 2	observation 2
	observation 3 etc.	observation 3 etc.
Category 2, etc.	*BLOCK 2, 1*	*BLOCK 2, 2*
	observation 1	observation 1
	observation 2	observation 2
	observation 3 etc.	observation 3 etc.

A two-way ANOVA will allow you to examine three pairs of hypotheses relating to two treatment variables and their possible interaction. Again, ANOVAs are very flexible and can be modified for designs with or without replicates and where the number of replicates in each category are not equal. The two-way ANOVA we have included in this chapter is a two-way parametric ANOVA for an equal number of replicates. The generic design is known as the randomized block design as the items are randomized to a block and the blocks are randomized to the treatments. As we explained in 2.2.6. this randomization of items within treatments is a way of minimizing the impact of non-treatment variation and is therefore a powerful tool that can be used when designing an experiment.

ii. Nested

With two or more variables you need to consider whether they are fixed or random. Table 7.23. is an example of fixed variables or a Model I design and may be contrasted with Table 7.24., which illustrates a mixed model where one variable is fixed (main effect) and the subordinate variable is randomized within the categories of the first variable (Model II). We saw this in Example 7.6. where soil samples had been taken from a number of pits dug in two forests. Clearly, pit 1 in the coniferous forest is not the same as pit 1 in the deciduous forest: the pits as a treatment are randomized within the forests. This design is not orthogonal but is nested. A mixed model design should be analysed using a nested ANOVA. Again, there are many versions of mixed models. In this chapter we include a mixed model with one fixed (main) variable and one random variable with equal replicates (Example 7.6.). For other designs, including how to deal with missing data, we refer you to Sokal & Rohlf (1981).

Table 7.24. Mixed model experimental design with one fixed main variable and one random, nested variable, with equal replicates

	Variable 1			
	Category 1		Category 2	
Variable 2	Category 1	Category 2	Category 3	Category 4
	BLOCK 1	BLOCK 2	BLOCK 1	BLOCK 2
	observation 1	observation 1	observation 1	observation 1
	observation 2	observation 2	observation 2	observation 2
	observation 3 etc.	observation 3 etc.	observation 3 etc.	observation 3 etc.

7.12.3. **Three variables**

The principles behind the randomized block design (two-way ANOVA) and mixed model design (two-way nested ANOVA) can, as we saw in 7.10. and 7.11., be extended to experiments involving three variables. The advantage of a comprehensive design examining many variables at one go rather than several small experiments examining one or two variables at a time is that you reduce the likelihood of a Type I error occurring. You are also more likely to detect significant interactions between the variables. The analysis of comprehensive designs is relatively more complex and the detection of significant interactions makes it more difficult to interpret your data.

The choice of statistical test for experimental designs that examine three of more treatment variables is dependent, as in the one variable and two variable designs on the model (pure Model I, pure Model II, or mixed model) and the numbers of replicates. Some of these alternate designs are considered in other texts, including Sokal & Rohlf (1981).

7.12.4. **General and specific hypotheses**

In this chapter we introduced you to Tukey's test, which, unlike other tests we have considered in this book so far, allows you to test specific hypotheses and so gain more explicit information about your experimental system. Tukey's can be used after the ANOVAs included in this book, with some restrictions. For example, Tukey's can only be used if you have a significant ANOVA. When testing experiments with two or more variables significant interactions can make it difficult to interpret any results from a Tukey's test. However, the value of these specific comparisons should not be underestimated and where possible you should design investigations where you may be able to test general and specific hypotheses.

In all of the designs we have outlined in this chapter controls may be included. Testing specific hypotheses using the Tukey's test can allow you to make more explicit comparisons between any controls and the effects of the treatments.

Summary of Chapter 7

- This chapter draws on your understanding of distributions and in particular the normal distribution (3.2.1.). The normal distribution is used as the basis for parametric statistics. Data that are parametric (BOX 3.2.) or have been normalized following transformation (3.9.) may be analysed using parametric statistics.
- The basic elements of parametric statistics are the sum of squares and variance which we introduced in Chapter 3 (BOX 3.1.).
- Parametric statistics are flexible and powerful. In 7.12. we outline the range of experimental designs that may be suitable for analysis by parametric tests, if all other criteria are met.
- We consider three groups of parametric tests in this chapter: *t* tests for matched and unmatched data (7.1. and 7.2.), *z* tests for matched and unmatched data (7.3.), and parametric ANOVAS with Tukey's test for one, two, or three variables (7.4–7.11.).
- The ANOVAs followed by a Tukey's test allows you to test both general and specific hypotheses. This provides you with considerably more explicit information about the effects of your treatment variables and is a design that should be used whenever possible (e.g. 7.6. and 7.8.).
- In Chapter 8 we consider the non-parametric equivalents to the tests covered in this chapter. Therefore, if your data are not parametric (3.8.) or cannot be normalized by transforming your data (3.9.), you should refer to Chapter 8.
- The Online Resource Centre includes interactive exercises that test your understanding of this chapter with other topics, particularly those considered in Chapters 2, 5–8, and 10.

8

Hypothesis testing: Do my samples come from the same population? Non-parametric data

When testing hypotheses almost the first step you take is to determine whether your data are **parametric** or **non-parametric** (BOX 3.2.). Parametric tests are more **powerful** so where possible you should use these tests. If your data are not parametric you may be able to **transform** them to 'normalize' the data and so enable you to use a parametric test (3.9.). Failing that, you should consider the tests described in this chapter.

This chapter tells you how you may analyse non-parametric data to test the **hypothesis** that two or more **samples** come from the same **statistical population**. This is achieved by contrasting **medians**. Five tests are covered. Worked examples are given for each test. If this is the first time you have used these tests you should cover up these worked examples and use the general information and the data provided to work through these examples. Then check your answers before using the test on your own data. If your answer differs considerably from that given you should check your calculation by going to the Online Resource Centre. We would expect it to take you about 3 hours to complete these worked examples.

The tests described in this chapter are known as **ranking** tests. They are 'distribution free' and the data therefore do not have to be normally distributed. Look at the data below. Each sample has been organized into numerical order and you can see that there is a small overlap between the two data sets.

Sample 1 1 3 5 6 7 7 9 10 10 11

Sample 2 10 11 11 12 13 13 13

If the two data sets are from different statistical populations then you would expect the overlap to be small. If the two samples are from the same population then you would expect the overlap to be greater. The ranking tests provide a measure of this overlap and an estimate of the probability that this overlap could occur by chance. If you are not familiar with the process of ranking data you should refer to BOX 3.3. where we give examples.

How to choose the correct test

Each test has several requirements that must be met and these details are given at the start of each section. The following guide takes you to the most likely test for your data. It is assumed that you have non-parametric data.

You have one treatment **variable**. You are going to compare two samples. The data are **unmatched**. You have 20 **observations** or less in each sample.	Mann–Whitney U test (8.1.)
You have one treatment variable. You are going to compare two samples. The data are unmatched. The data are measured on a continuous scale and you have more than 30 observations in each sample.	*z* test for unmatched data (Chapter 7 (7.1.))
You have one treatment **variable**. You are going to compare two samples. The data are **unmatched**. You have more than 20 **observations** in each sample.	Sokal & Rohlf, 1981
You have one treatment variable. You are going to compare two samples. The data are **matched**. You have less than 30 pairs of observations.	Wilcoxon's rank paired test (8.2.)
You have one treatment variable. You are going to compare two samples. The data are matched. You have more than 30 pairs of observations.	*z* test for matched data (Chapter 7 (7.2))
You have one treatment variable. You are going to compare two or more samples. You wish to test **general** and **specific** hypotheses.	One-way ANOVA (Kruskal–Wallis test) (8.3. and 8.4)

You have more than one treatment variable. You are going to compare two or more samples. You wish to test general and specific hypotheses. You will be using a calculator.	Two-way non parametric ANOVA (8.5. and 8.6)
You have more than one treatment variable. You are going to compare two or more samples. You wish to test general hypotheses. You want to use a computer.	Scheirer–Ray–Hare test (8.7.)

Q1 **The impact of consuming 1.5 units of alcohol was tested on reaction times in six men and ten women.**

What information do you know from this brief description that can help you identify which might be the most suitable statistical test to use? What more do you need to find out before you can make a definite decision about which test to use?

A1 There is one treatment variable (consumption of alcohol). The data are matched with observations on a total of 16 volunteers before and after consumption of alcohol. The observations (time taken) are measured on a continuous scale and the data may therefore be parametric. Gender is a possible non-treatment variable.

To decide which test to use and if the experiment needs to be modified you need to decide if the data are parametric. You cannot confirm this until you have carried out the investigation or a pilot experiment. If the data are parametric, you may consider the *t* test for matched pairs. If the data are not parametric, then you may consider using the Wilcoxon's matched pairs test. It would be worth analysing the male and female results separately to obtain some information about this non-treatment variable.

8.1. Mann–Whitney U test

We have already introduced you to the idea of ranking observations (BOX 3.3.) and applied this idea in Chapter 6 when testing the hypotheses: there is no association between two variables and using the Spearman's rank test. The Mann–Whitney U test uses the rank value of observations and the sum of these ranks for each sample to establish whether the median of two samples is significantly different.

8.1.1. **To use this test you**

1. Wish to test for differences in population medians.
2. Have one treatment variable and two samples.
3. Have data that are non-parametric and unmatched.
4. Have data that can be ranked (3.1. and 3.8.2.).
5. Have two samples which both have a similar shaped distribution. For example, if one distribution is skewed to the left and the other to the right (3.4.4.) then you should not use this test. (If this does arise you could try transforming the data (3.9.).)
6. Should not use this test if one sample has only one observation or if both samples have less than five observations each.
7. Need not have equal sample sizes.

We will illustrate the Mann–Whitney U test using Example 7.2. The evolution of *Littorina littoralis* at Porthcawl, where we know there is some doubt as to whether this data meet all the criteria for a parametric test. In this Example investigators used a **systematic random** sampling method to collect periwinkles from the mid shore and lower shore and measured the shell height (mm) (Table 7.2.). The investigators wished to test the hypothesis that there was a difference between these groups of periwinkles as had been seen in an earlier study carried out at Aberystwyth (Example 7.1.).

This data set meets all the criteria for using a Mann–Whitney U test since there is one variable (periwinkles) and two samples (mid shore and lower shore periwinkles). The data are unmatched as each periwinkle was measured only once. The data can be ranked but since there are less than

Table 7.2. Shell height in two groups of periwinkles from the mid and lower shore at Porthcawl, 2002

Shell heights (mm)			
Periwinkles on the lower shore		Periwinkles on the mid shore	
5.5	4.0	3.3	6.7
8.4	5.0	6.3	5.7
5.0	6.2	6.1	4.2
5.0	5.0	8.0	6.3
5.6	7.7	13.5	6.0
4.8	6.0	5.3	7.2
8.4		6.7	

30 observations a z test should not be used. The distributions are similar. Both samples have more than five observations.

8.1.2. The calculation

We show you the general process and a specific example of a Mann–Whitney U test in BOX 8.1. In addition you will need to refer to the calculation table (8.1.). Where steps have been abbreviated the full calculation is included in the Online Resource Centre.

BOX 8.1. How to carry out a Mann–Whitney U test

GENERAL DETAILS	THIS EXAMPLE
1. Hypotheses to be tested	**1. Hypotheses to be tested**
H_0: There is no difference between the median values in sample 1 and sample 2. **H_1:** There is a difference between the median values in sample 1 and sample 2.	**H_0:** There is no difference between the median shell heights of the periwinkles from the mid and lower shore at Porthcawl. **H_1:** There is a difference between the median shell heights of periwinkles from the mid and lower shore at Porthcawl.
2. To work out $U_{calculated}$	**2. To work out $U_{calculated}$**
i. Decide which of the samples is sample 1 and which sample 2.	i. Let the lower shore periwinkles be sample 1 and the mid shore periwinkles be sample 2.
ii. Combine all observations (sample 1 and sample 2) and arrange in numerical order.	ii. See Table 8.1.
iii. Assign ranks to these values. Where there are tied values take the middle (average) rank for these values (BOX 3.3.).	iii. See Table 8.1.
iv. Add up the ranks ($\Sigma r_1 = R_1$) for sample 1. Repeat for sample 2 ($\Sigma r_2 = R_2$).	iv. See Table 8.1. $R_1 = 153.5$ and $R_2 = 197.5$.
v. Record n_1, the number of observations in sample 1 and n_2, the number of observations in sample 2.	v. In this example $n_1 = 13$, $n_2 = 13$.
vi. You now calculate two U terms, U_1 and U_2. Watch the formulae closely. A common error is to use R_1 when calculating U_1. $U_1 = n_1 n_2 + \frac{n_2(n_2+1)}{2} - R_2$ $U_2 = n_1 n_2 + \frac{n_1(n_1+1)}{2} - R_1$	vi. Taking each calculation in stages: $n_1 n_2 = 13 \times 13 = 169$ $n_2(n_2+1) = 13(13+1) = 13 \times 14 = 182$ Therefore: $U_1 = 169 + \frac{182}{2} - 197.5 = 62.5$ $U_2 = 169 + \frac{182}{2} - 153.5 = 106.5$

BOX 8.1. Continued

GENERAL DETAILS	THIS EXAMPLE
vii. Examine the two U values. The smallest U value is $U_{\text{calculated}}$.	vii. Therefore $U_{\text{calculated}} = 62.5$
viii. You can check your maths at this point since $U_1 + U_2 = n_1 \times n_2$.	viii. $(62.5 + 106.5) = (13 \times 13)$ correct.
3. To find U_{critical}	**3. To find U_{critical}**
To find U critical using the table of critical values you need to know n_1 and n_2. Where the n_1 row intersects the n_2 column for $p = 0.05$ this is the critical value. This is a two-tailed test.	$n_1 = n_2 = 13$ and at $p = 0.05$, $U_{\text{critical}} = 45$
4. The rule	**4. The rule**
If the calculated value of U is less than the critical value of U then you may reject the null hypothesis (H_0).	In our example $U_{\text{calculated}}$ (62.5) is more than U_{critical} (45) and therefore you do not reject the null hypothesis.
5. What does this mean in real terms?	**5. What does this mean in real terms?**
	There is no significant difference ($U = 62.5$, $p = 0.05$) between the median shell height of the periwinkles from the mid and lower shore at Porthcawl. The results from this investigation do not support the notion that sympatric speciation is occurring. This concurs with the analysis of the data using a t test (BOX 7.3.).

8.2. Wilcoxon's matched pairs test

Occasionally you may design an investigation in which you observe an item before and after a particular event or treatment. This type of data is called matched and has to be handled in a different way to that described in 8.1., because the 'before' measure is not independent of the 'after' measure. In Example 7.2. none of the periwinkles was measured more than once. However, in Example 8.1. each item is observed twice (before and after the treatment). In the Wilcoxon's matched pairs test this lack of independence is recognized and allowed for.

Table 8.1. Calculation of ranks for the height of shells in two groups of periwinkles from the mid and lower shore at Porthcawl, 2002

Height of shells (mm)			
Periwinkles from the lower shore		Periwinkles from the mid shore	
Observations in numerical order	Rank	Observations in numerical order	Rank
		3.3	1
4.0	2		
		4.2	3
4.8	4		
5.0	6.5		
5.0	6.5		
5.0	6.5		
5.0	6.5		
		5.3	9
5.5	10		
5.6	11		
		5.7	12
		6.0	13.5
6.0	13.5		
		6.1	15
6.2	16		
		6.3	17.5
		6.3	17.5
		6.7	19.5
		6.7	19.5
		7.2	21
7.7	22		
		8.0	23
8.4	24.5		
8.4	24.5		
		13.5	26
	$R_1 = 153.5$		$R_2 = 197.5$

8.2.1. To use this test you:

1. Wish to test for differences in population medians.
2. Have one treatment variable with two samples.
3. Have non-parametric data.
4. Have matched data.
5. Have data that can be ranked and are **quantitative.**
6. Have at least six pairs of observations where the difference between the observations is more than zero.
7. Have two samples which both have a similar shaped distribution. For example, if one distribution is skewed to the left and the other to the right then you should not use this test.

EXAMPLE 8.1. Enjoyment of consuming chocolate at two times in the day

A small randomized study was carried out where undergraduates were asked to eat a particular chocolate bar at 7 a.m. and at 6 p.m. On each occasion the students were asked to rate their enjoyment on a scale of 1 (low) to 10 (high). The first three columns of Table 8.2. show the results.

The data meet the criteria for using a Wilcoxon's matched pairs test in that there is one variable (time of day) and two samples (7 a.m. and 6 p.m.). Two measurements are made for each student. These two measurements are not independent of each other and the data are therefore 'matched'. There are less than 30 pairs of observations so a matched z test may not be

Table 8.2. Enjoyment rating for consumption of chocolate bars at two times in the day with calculations for the Wilcoxon's matched pairs test

Student	Enjoyment rating at 7 a.m.	Enjoyment rating at 6 p.m.	Difference (d)	Absolute rank for d	Signed rank for d
1	2	2	0		
2	2	5	3	3	3
3	6	2	−4	5	−5
4	2	5	3	3	3
5	4	2	−2	1	−2
6	1	4	3	3	3
7	2	8	6	6	6

used. The observations are measured on a scale of 1–10 and are therefore rankable, quantitative, but non-parametric. There are six pairs of observations where the difference between the observations is more than 0. The sample size is small so that it is difficult to be certain if criterion 7 is met; however, **histograms** for the samples indicate that they are similar in shape.

8.2.2. The calculation

In this test the calculation is dependent on the difference between each pair of observations and the sum of these differences. The general principles and a worked example are shown in BOX 8.2. Some steps in the calculation are included in Table 8.2., columns 4–6. Where steps have been abbreviated the full calculation is included in the Online Resource Centre.

8.3. One-way non-parametric ANOVA (Kruskal–Wallis test): General hypotheses

The acronym ANOVA stands for analysis of variance. Parametric ANOVAs are more powerful and flexible (Chapter 7). However, it is common in science to find yourself dealing with data that do not fit the criteria for a parametric ANOVA and cannot be normalized by transforming them (3.9.). Under these circumstances non-parametric tests may be used. In these tests you will be comparing the medians of the samples by using ranks and sums of ranks. The one-way non-parametric ANOVA used to test general hypotheses was first described by Kruskal and Wallis

BOX 8.2. How to carry out a Wilcoxon's matched pairs test

GENERAL DETAILS	THIS EXAMPLE
1. Hypotheses to be tested **H_0:** There is no difference between the median of sample 1 and the median of sample 2. **H_1:** There is a difference between the median of sample 1 and the median of sample 2.	**1. Hypotheses to be tested** **H_0:** There is no difference between the median chocolate enjoyment scores at 7 a.m. and at 6 p.m. in a small group of undergraduates. **H_1:** There is a difference between the median chocolate enjoyment scores at 7 a.m. and at 6 p.m. in a small group of undergraduates.
2. To work out $T_{calculated}$	**2. To work out $T_{calculated}$**
i. Decide which is sample 1 and which is sample 2.	i. Let the enjoyment scores recorded at 7 a.m. be sample 1 and the enjoyment scores recorded at 6 p.m. be sample 2.
ii. Construct a calculation table.	ii. See Table 8.2.
iii. Calculate the difference (d) between each pair of observations. Where the observation for sample 2 is larger than the observation for sample 1, the difference will be negative.	iii. See Table 8.2., column 4.
iv. Rank the **absolute** d values (i.e. ignore the negative signs). Also ignore any $d=0$, since this will not contribute to the evaluation of the data in this test.	iv. See Table 8.2, column 5. Ignore the 0 values. So $d=2$ is the lowest value and has the rank 1; there are three values of $d=3$ so these share the average rank 3, etc.
v. Assign to each rank the positive or negative signs from the d value.	v. See Table 8.2, column 6.
vi. Sum all the ranks that are negative in sign. The absolute total is used.	vi. The sum of negative ranks $=1+5=6$
vii. Sum all the ranks that are positive in sign.	vii. The sum of positive ranks $=3+3+3+6=15$
viii. Examine the two values. The smaller value is the calculated value of T.	viii. $T_{calculated}=6$
3. To find $T_{critical}$ This is a **two-tailed test.** Use a T table of critical values where N is the number of pairs of observations used to provide ranks (i.e. not those where $d=0$.)	**3. To find $T_{critical}$** For this example one difference is zero therefore $N=6$ and $T_{critical}=2$ at $p=0.05$.
4. The rule When $T_{calculated}$ is equal to or less than $T_{critical}$ then you may reject the null hypothesis (H_0).	**4. The rule** In this example $T_{calculated}$ (6) is more than $T_{critical}$ (2), therefore you do not reject the null hypothesis.
5. What does this mean in real terms?	**5. What does this mean in real terms?** There is no significant difference ($T=6$, $p=0.05$) between the median chocolate enjoyment scores at 7 a.m. compared with 6 p.m. in a small group of undergraduates.

and is usually called the Kruskal–Wallis test. This test has been extended by Meddis (1984) and Barnard *et al.* (2001) to allow testing of specific hypotheses. Testing general and specific hypotheses in this non-parametric ANOVA is the equivalent of a one-way parametric ANOVA in combination with a Tukey's test. To highlight the similarity of use of these two tests to the ANOVAs described in Chapter 7 we refer to them as the one-way non-parametric ANOVAs. Unlike the comparable parametric tests you do not need to obtain a significant difference when testing the general hypotheses before testing specific hypotheses. We illustrate the use of this one-way non-parametric ANOVA using Example 8.2. (Table 8.3.)

EXAMPLE 8.2. The density of *Bellis perennis* (daisy plants) at four different locations on the University of Worcester campus, 2002

Students in their first year at the University of Worcester randomly sampled four locations on campus using a 0.5 m^2 quadat (G. Davis, personal communication, 20.8.03). The number of *B. perennis* in each of eight quadrats in the four areas was recorded (Table 8.3.). Is there a significant difference between these four areas?

Table 8.3. The number of *Bellis perennis* growing at four locations on the campus of the University of Worcester (G. Davis, personnal communication, 20.8.01)

Number of *B. perennis* on the Cricket Pitch (CP)	Number of *B. perennis* on the Lawn (L)	Number of *B. perennis* on the Quadrangle (Q)	Number of *B. perennis* on the Rugby Pitch (RP)
8	15	10	15
9	13	16	10
9	15	18	15
12	18	13	12
4	11	12	8
5	12	16	5
5	13	8	10
7	13	12	10
Median = 7.5	13	12.5	10

8.3.1. To use this test you:

1. Wish to test for differences in population medians.
2. Have one treatment variable and three or more samples.

3. Have data that is non-parametric but can be ranked.
4. Do not need equal sample sizes.
5. Must, if there are only three samples, have more than five observations per sample.

Q2 **Do the data from Example 8.2. meet the criteria for a non-parametric one-way ANOVA (Kruskal–Wallis test)?**

A2 Yes. The data from Example 8.2. are suitable for analysis by a non-parametric ANOVA as the unit of measurement is 'number of plants', a non-parametric measure which can be ranked. There is one variable (location on campus); four samples (cricket pitch (CP), lawn (L), quadrangle (Q), and rugby pitch (RP)), and eight replicates (observations) in each sample.

8.3.2. The calculation

The method for testing general hypotheses (the Kruskal–Wallis test) is outlined in BOX 8.3. and the calculation table (8.4.). To test specific hypotheses you should refer to BOX 8.4. Where steps have been abbreviated the full calculation is included in the Online Resource Centre.

BOX 8.3. How to carry out a one-way non-parametric ANOVA (Kruskal–Wallis test): general hypotheses

GENERAL DETAILS	THIS EXAMPLE
1. Hypotheses to be tested **H_0:** There is no difference between the medians of the samples. **H_1:** There is a difference between the medians of the samples.	**1. Hypotheses to be tested** **H_0:** There is no difference between the median density of *Bellis perennis* at the four locations. **H_1:** There is a difference in the median density of *Bellis perennis* at the four locations.
2. To work out $K_{calculated}$ i. Combine all the observations and arrange them in numerical order. Assign the appropriate rank to each observation. ii. Sum the ranks for each sample, $\Sigma r = R_s$ iii. Record n_s, the number of observations in each sample. iv. Square each R value (R^2). Divide each R^2 by its n_s value. $= R^2/n_s$ v. Add all the R^2/n_s values together $= \Sigma(R^2/n_s)$ vi. Add all the n_s values together, $\Sigma n_s = N$ vii. The test statistic K is calculated as: $$K_{calculated} = \left(\Sigma\left(\frac{R^2}{n_s}\right) \times \frac{12}{N(N+1)}\right) - 3(N+1)$$	**2. To work out $K_{calculated}$** i, ii, iii, iv. See Table 8.4 (ranks). v. $\Sigma\frac{R^2}{n_s} = 392 + 4278.125 + 3549.0313 + 1755.2813$ $= 9974.4376$ vi. $\Sigma n_s = N = 8 + 8 + 8 + 8 = 32$ vii. $K_{calculated}$ $$= \left(9974.4376 \times \frac{12}{32(32+1)}\right) - 3(32+1) = 14.345882$$
3. To find $K_{critical}$ Using a table of critical values for χ^2 and $p = 0.05$. The **degrees of freedom** (v) = number of samples − 1. This is a two-tailed test.	**3. To find $K_{critical}$** $v = 4 - 1 = 3$ $K_{critical}$ at $p = 0.05 = 7.81$
4. The rule If the calculated value of K is more than the critical value of K then you may reject the null hypothesis.	**4. The rule** $K_{calculated}$ (14.35) is more than $K_{critical}$ (7.81) and we may therefore reject the null hypothesis. In fact at $p = 0.01$, $K_{critical} = 11.34$ and at $p = 0.001$, $K_{critical} = 16.27$. Therefore, we can reject the null hypothesis at $p = 0.01$, but not at $p = 0.001$.
5. What does this mean in real terms?	**5. What does this mean in real terms?** There is a highly significant difference ($K = 14.35$, $0.01 > p > 0.001$) between the median densities of *Bellis perennis* at the four locations.

Table 8.4. Calculation for one-way non-parametric ANOVA: General and specific hypotheses examining the number of *Bellis perennis* growing at four sites on the campus of the University of Worcester (G. Davis, personal communication, 20.8.02)

Cricket Pitch (CP)		Lawn (L)		Quadrangle (Q)		Rugby Pitch (RP)	
Number	Rank	Number	Rank	Number	Rank	Number	Rank
8	7	15	26.5	10	12.5	15	26.5
9	9.5	13	22.5	16	29.5	10	12.5
9	9.5	15	26.5	18	31.5	15	26.5
12	18	18	31.5	13	22.5	12	18
4	1	11	15	12	18	8	7
5	3	12	18	16	29.5	5	3
5	3	13	22.5	8	7	10	12.5
7	5	13	22.5	12	18	10	12.5
For BOX 8.3.	R 56		R 185		R 168.5		R 118.5
	n 8		n 8		n 8		n 8
	R^2 3136		R^2 34225		R^2 28392.25		R^2 14042.25
	$R^2/n = \frac{3136}{8} = 392$		$R^2/n = 4278.125$		$R^2/n = 3549.0313$		$R^2/n = 1755.2813$
For BOX 8.4.	λ 1		λ 4		λ 3		λ 2
	λ^2 1		λ^2 16		λ^2 9		λ^2 4
	λR 56		λR 740		λR 505.5		λR 237
	λn 8		λn 32		λn 24		λn 16
	$\lambda^2 n$ 8		$\lambda^2 n$ 128		$\lambda^2 n$ 72		$\lambda^2 n$ 32

BOX 8.4. How to carry out a one-way non-parametric ANOVA: specific hypotheses

GENERAL DETAILS	THIS EXAMPLE
1. Hypotheses to be tested	**1. Hypotheses to be tested**
H_0: The medians of the samples do not follow the predicted order. **H_1:** The medians of the samples do follow the predicted order.	**H_0:** The rank order for the median density of *Bellis perennis* does not follow the predicted order. **H_1:** The median density of *Bellis perennis* follows the rank order CP < RP < Q < L.
2. To work out $z_{calculated}$	**2. To work out $z_{calculated}$**
i. Each sample is given a weighting (λ) that reflects the predicted order.	i. For our specific hypothesis the cricket pitch has the lowest density of daisies so we will give this a weighting (λ) of 1. We will increase each weighting by 1 to reflect their apparent relationships to each other. CP<RP<Q<L λ 1 2 3 4

BOX 8.4. Continued

GENERAL DETAILS	THIS EXAMPLE
ii. For each sample calculate λ^2, λR, λn, $\lambda^2 n$	ii. Table 8.4.
iii. Sum these across samples, to calculate $N = \Sigma n$; $\Sigma(\lambda R)$; $\Sigma(\lambda n)$; $\Sigma(\lambda^2 n)$	iii. $N = \Sigma n = 8 + 8 + 8 + 8 = 32$ $\Sigma(\lambda R) = 56 + 740 + 505.5 + 237 = 1538.5$ $\Sigma(\lambda n) = 8 + 32 + 24 + 16 = 80$ $\Sigma(\lambda^2 n) = 8 + 128 + 72 + 32 = 240$
iv. Calculate the terms L, E, and V, where: $L = \Sigma(\lambda R)$ $E = (N+1)\dfrac{\Sigma \lambda n}{2}$ $V = (N+1) \times \dfrac{[N(\Sigma \lambda^2 n) - (\Sigma \lambda n)^2]}{12}$	iv. $L = 1538.5$ $E = (32+1)\dfrac{(80)}{2} = 1320$ $V = (32+1) \times \dfrac{[32(240) - (80)^2]}{12}$ $= 3519.99999$
v. Calculate the test statistic z $z_{\text{calculated}} = \dfrac{(L-E)}{\sqrt{V}}$	v. $z_{\text{calculated}} = \dfrac{1538.5 - 1320}{\sqrt{3520}}$ $= 3.68282$
3. To find z_{critical} Unusually this is a **one-tailed test** as we are considering a specific prediction. z is independent of sample size and therefore you do not need to know either the sample number or the degrees of freedom.	**3. To find z_{critical}** For a one-tailed test, at $p = 0.05$, $z_{\text{critical}} = 1.645$.
4. The rule If $z_{\text{calculated}}$ is more than z_{critical} then you may reject the null hypothesis (H_0).	**4. The rule** Since $z_{\text{calculated}}$ (3.68) is more than z_{critical} (1.64) at $p = 0.05$ then you may reject the null hypothesis (H_0). In fact at $p = 0.001$ for a one-tailed test $z_{\text{critical}} = 3.10$. Therefore, you may reject the null hypothesis at this higher level of significance.
5. What does this mean in real terms?	**5. What does this mean in real terms?** The median density of *Bellis perennis* is not significantly different ($z = 3.68$, $p = 0.001$) from the predicted order CP < RP < Q < L.

8.4. One-way non-parametric ANOVA: Specific hypotheses

In the Tukey's test that we used to test specific hypotheses following a significant one-way parametric ANOVA, the specific hypotheses were the differences between pairs of means. In this non-parametric test the specific hypotheses are more flexible, dependent on the trends you observe in your data, and unlike the parametric ANOVAs you may test specific hypotheses without first confirming a significant difference in the general hypothesis testing. Look at the medians for Example 8.2. (Table 8.3.). From this we may propose that the median density of daisies can be ranked as follows: the cricket pitch has the lowest daisy density, followed by the rugby pitch, then the quadrangle, and finally the lawn has the highest density. This specific proposal can now be tested statistically.

Central to this test is a term denoted by the symbol λ. This is a value that is used to weight the sums of ranks and reflect the specific hypotheses that you are testing. In our example, (Box 8.4 and Table 8.4) we only examine one pair of specific hypotheses and therefore only show you one example of selecting appropriate values for λ. For other examples see 8.6. and Barnard *et al.* (2001).

8.5. Two-way non-parametric ANOVA: General hypotheses

This test is an extension of the one-way ANOVA but allows you to examine two treatment variables and their **interaction**. As explained in Chapter 7 in relation to the comparable two-way parametric ANOVA, when you compare two treatment variables in this way you will test three pairs of hypotheses. The first test examines the effect of treatment 1, the second the effect of treatment 2, and the third the interaction between treatment 1 and treatment 2. All these tests are carried out within one calculation and so you reduce the likelihood of generating a **type I** error. We illustrate the two-way non-parametric ANOVA using data from Example 8.3. (Table 8.5.).

8.5.1. To use this test you:

1. Wish to test for differences in population medians.
2. Have two treatment variables each with at least two categories.

3. Have an **orthogonal** design.
4. Have non-parametric data that can be ranked.
5. Can test both general and specific predictions if there are equal numbers of observations in each sample. If there are not equal numbers of observations in each sample only specific predictions can be tested.

EXAMPLE 8.3. The effect of fertilizer treatment on the density of *Bellis perennis* (daisy) at two locations on the University of Worcester campus, 2003

At the University of Worcester the lawn and cricket pitch were divided and half of each plot was treated with fertilizer. Students extended their original study (Example 8.2.) and using stratified random sampling with quadrats recorded the numbers of *Bellis perennis* in these four locations (the cricket pitch with fertilizer, the cricket pitch without fertilizer, the lawn with fertilizer, and the lawn without fertilizer) (G. Davis, personal communication, 20.8.03) (Table 8.5.). The data are organized into blocks and these will be referred to as block A–block D.

Table 8.5. The density of *Bellis perennis* (daisies) in two locations, with or without fertilizer treatments at the University of Worcester, 2003 (G. Davis, personal communication, 20.8.03)

	Number of *B. perennis* in each quadrat	
	Lawn (L)	Cricket pitch (CP)
Fertilizer	8	15
	9 (A)	13 (B)
	9	15
	12	18
Median	9	15
No fertilizer	4	11
	5	12
	5 (C)	13 (D)
	7	13
Median	5	12

The data from Example 8.3. meet the criteria for using this test since the unit of measurement is 'number of daisy plants', which is a non-parametric measure which can be ranked. There are two variables (location and fertilizer treatment) each with two categories. Each sample has four replicates (observations) and therefore both general and specific hypotheses can be tested.

8.5.2. The calculation

As in the two-way parametric ANOVA for an orthogonal design three pairs of hypotheses are tested: the differences between the medians due to treatment 1, the differences between the medians due to treatment 2, and the interaction between the two treatment variables. The idea of an interaction is explained in Chapter 7 (7.7.). The following set of BOXES and calculation tables are arranged to take you through testing one pair of hypotheses at a time:

a. General hypothesis: Is there a difference between the columns (i.e. locations) – go to BOX 8.5.A. and Table 8.6.
b. General hypothesis: Is there a difference between the rows (i.e. fertilizer treatment) – go to BOX 8.5.B. and Table 8.7.
c. General hypothesis: Is there an interaction between variable 1 and variable 2, i.e. Is there an interaction between the locations and the effect of the fertilizer – go to BOX 8.5.C. and Table 8.8.

Our analysis of these data shows us that only the location is a significant factor in the density of daisies in this experiment. Neither the fertilizer nor the interaction between fertilizer and location as treatments had a significant effect on the median density of daisies. However, this has only been a test of general hypotheses. To be able to draw more explicit conclusions about the trends in the data we should test specific hypotheses (8.6.).

Table 8.6. Calculation table for two-way non-parametric ANOVA test: General hypotheses (columns) examining the number of daisies per quadrat in two locations with or without fertilizer treatments

	Cricket pitch		Lawn	
	Number	Rank	Number	Rank
Fertilizer	8	5	15	14.5
	9	6.5	13	12
	9	6.5	15	14.5
	12	9.5	18	16
No fertilizer	4	1	11	8
	5	2.5	12	9.5
	5	2.5	13	12
	7	4	13	12
Totals for the columns	n	8	n	8
	R	37.5	R	98.5
	R^2	1406.25	R^2	9702.25
	$\frac{R_{c1}^2}{n_{c1}}$	175.78125	$\frac{R_{c2}^2}{n_{c2}}$	1212.7813

BOX 8.5.A. How to carry out a two-way non-parametric ANOVA test: general hypotheses (columns)

GENERAL DETAILS	THIS EXAMPLE
1. Hypotheses to be tested **H_0:** There is no difference between the median values for samples given treatment 1 (columns). **H_1:** There is a difference between the median values for samples given treatment 1 (columns).	**1. Hypotheses to be tested** **H_0:** There is no difference between the median density of *Bellis perennis* on the lawn and on the cricket pitch. **H_1:** There is a difference between the median density of *Bellis perennis* on the lawn and on the cricket pitch.
2. To work out $K_{calculated}$	**2. To work out $K_{calculated}$**
i. Combine all the observations and arrange them in numerical order. Assign the appropriate rank to each observation.	i., ii. and iii. Table 8.6.
ii. Sum the ranks for column 1 of treatment 1 (R_{c1}) and column 2 of treatment 1 (R_{c2}) and continue for all columns of data.	There are two columns, 'cricket pitch' and 'lawn', and eight observations in each column.
iii. Sum the number of observations for column 1 of treatment 1 (n_{c1}) and column 2 of treatment 1 (n_{c2}), etc. (Remember to test general hypotheses all n values should be the same.)	
iv. Using the column totals. For column 1 square the sum of ranks (R^2_{c1}) and divide by the number of observations in that first column (R^2_{c1}/n_{c1})	iv. $\frac{R^2_{c1}}{n_{c1}} = 175.78125$
In a similar way calculate R^2_{c2}/n_{c2} for column 2.	$\frac{R^2_{c2}}{n_{c2}} = 1212.7813$
(Continue for each column of data.)	$\Sigma(R^2/n) = 175.78125 + 1212.7813 = 1388.5626$
Add together all these R^2/n values ($\Sigma R^2/n$).	$N = 8 + 8 = 16$
Add all the n_c values together $= N$.	
v. The test statistic K is calculated as before: $K_{\text{calculated columns}} = \left(\Sigma\left(\frac{R^2}{n_s}\right) \times \frac{12}{N(N+1)}\right) - 3(N+1)$	v. $K_{\text{calculated columns}} =$ $\left(1388.5626 \times \frac{12}{16(16+1)}\right) - 3(16+1) = 10.26012$
3. To find $K_{critical}$ Using a table of critical values for χ^2 and $p = 0.05$. The degrees of freedom (v) = number of columns − 1. This is a **two-tailed test.**	**3. To find $K_{critical}$** Since $v = 2 - 1 = 1$ $K_{\text{critical columns}} = 3.84$ at $p = 0.05$

BOX 8.5.A. Continued

GENERAL DETAILS	THIS EXAMPLE
4. The rule If the calculated value of K is more than the critical value of K then you may reject the null hypothesis.	**4. The rule** $K_{\text{calculated columns}}$ (10.26) is more than K_{critical} (3.84). You may therefore reject the null hypothesis (H_0). In fact at $p=0.01$, $K_{\text{critical}}=6.64$ and at $p=0.001$, $K_{\text{critical}}=10.83$. Therefore, we can reject the null hypothesis at $p=0.01$, but not at $p=0.001$.
5. What does this mean in real terms? NOW GO ON TO BOX 8.5.B.	**5. What does this mean in real terms?** There is a highly significant difference ($K=10.26$, $0.01>p>0.001$) in the median density of *Bellis perennis* on the lawn and on the cricket pitch.

Table 8.7. Calculation table for two-way non-parametric ANOVA test: General hypothesis (rows) examining the number of daisies per quadrat in two locations with or without fertilizer treatments

	Cricket pitch		Lawn		Total for rows	
	Number	Rank	Number	Rank		
Fertilizer	8	5	15	14.5	n	8
	9	6.5	13	12	R	84.5
	9	6.5	15	14.5	R^2	7140.25
	12	9.5	18	16		
					$\frac{R_{r1}^2}{n_1}$	$= 892.53125$
No fertilizer	4	1	11	8	n	8
	5	2.5	12	9.5	R	51.5
	5	2.5	13	12	R^2	2652.25
	7	4	13	12		
					$\frac{R_{r2}^2}{n_2}$	$= 331.53125$

BOX 8.5.B. How to carry out a two-way non-parametric ANOVA test: general hypotheses (rows)

GENERAL DETAILS	THIS EXAMPLE
1. Hypotheses to be tested **H_0:** There is no difference in the median values for samples in response to treatment 2 (rows). **H_1:** There is a difference in the median values for samples in response to treatment 2 (rows).	**1. Hypotheses to be tested** **H_0:** There is no difference in the median density of *Bellis perennis* between the areas treated with fertilizer and those not so treated. **H_1:** There is a difference in the median density of *Bellis perennis* between the areas treated with fertilizer and those not so treated.
2. To work out $K_{calculated}$ i. Sum the ranks for row 1 of treatment 1 (R_{r1}) and row 2 of treatment 1 (R_{r2}). Continue for all rows. ii. Sum the number of observations for row 1 of treatment 1 (n_{r1}) and row 2 of treatment 1 (n_{r2}). Continue for all rows. iii. From the row totals. For row 1 square the sum of ranks (R^2_{r1}) and divide by the number of observations in that first row $\left(\frac{R^2_{r1}}{n_{r1}}\right)$. In a similar way calculate $\frac{R^2_{r2}}{n_{r2}}$ for row 2. (Continue for each row of data.) Add together all these R^2/n values ($\Sigma R^2/n$). Add all the n_r values together $= N$. iv. The test statistic K is calculated as before. $K_{\text{calculated rows}} = \left(\Sigma\left(\frac{R^2}{n_s}\right) \times \frac{12}{N(N+1)}\right) - 3(N+1)$	**2. To work out $K_{calculated}$** i., ii. Table 8.7.There are two rows, 'fertilizer' and 'no fertilizer', each with eight observations in the row. iii. $\frac{R^2_{r1}}{n_{r1}} = 893.53125$ $\frac{R^2_{r2}}{n_{r2}} = 331.53125$ $\Sigma\frac{R^2}{n} = 892.53125 + 331.53125$ $= 1224.0625$ $\Sigma n = N = 8 + 8 = 16$ iv. $K_{\text{calculated rows}}$ $= \left(1224.0625 \times \frac{12}{16(16+1)}\right) - 3(16+1)$ $= 3.00260$
3. To find $K_{critical}$ Using a table of critical values for χ^2 and $p = 0.05$. The degrees of freedom (v) = number of rows − 1. This is a two-tailed test.	**3. To find $K_{critical}$** Since $\nu = 2 - 1 = 1$, $K_{\text{critical rows}} = 3.84$ at $p = 0.05$
4. The rule If the calculated value of K is more than the critical value of K then you may reject the null hypothesis (H_0).	**4. The rule** $K_{\text{calculated rows}}$ (3.00) is less than $K_{critical}$ (3.84). You may not reject the null hypothesis (H_0).

BOX 8.5.B. Continued

GENERAL DETAILS	THIS EXAMPLE
5. What does this mean in real terms? NOW GO ON TO BOX 8.5.C.	**5. What does this mean in real terms?** There is no significant difference ($K=3.00$, $p=0.05$) in the median density of *Bellis perennis* in areas treated by fertilizer and not so treated.

Table 8.8. Calculation table for two-way non-parametric ANOVA test: General hypothesis: Interaction. Examining the number of daisies per quadrat in two locations with or without fertilizer treatments

Daisy density per quadrat in area where fertilizer added	Cricket pitch		Lawn	
	Number	Rank	Number	Rank
	BLOCK A		BLOCK B	
	8	5	15	14.5
	9	6.5	13	12
	9	6.5	15	14.5
	12	9.5	18	16
For BOX 8.5.3.	n	4	n	4
	R	27.5	R	57
	R^2	756.25	R^2	3249
	R^2/n	189.0625	R^2/n	812.25
For BOX 8.6.	λ	1	λ	−1
	λ^2	1	λ^2	1
	λR	27.5	λR	−57
	λn	4	λn	−4
	$\lambda^2 n$	4	$\lambda^2 n$	4
Daisy density in area where fertilizer is not added	Number	Rank	Number	Rank
	BLOCK C		BLOCK D	
	4	1	11	8
	5	2.5	12	9.5
	5	2.5	13	12
	7	4	13	12

Table 8.8. Continued

Daisy density per quadrat in area where fertilizer added	Cricket pitch		Lawn	
	Number	Rank	Number	Rank
For BOX 8.5.3.	n	4	n	4
	R	10	R	41.5
	R^2	100	R^2	1722.25
	R^2/n	25	R^2/n	430.5625
For BOX 8.6.	λ	−1	λ	1
	λ^2	1	λ^2	1
	λR	−10	λR	41.5
	λn	−4	λn	4
	$\lambda^2 n$	4	$\lambda^2 n$	4

BOX 8.5.C. How to carry out a two-way non-parametric ANOVA test: general hypotheses (interaction)

GENERAL DETAILS	THIS EXAMPLE
1. Hypotheses to be tested	**1. Hypotheses to be tested**
H_0: There is no difference between the median values of all samples due to an interaction between treatment variables 1 and 2.	**H_0:** There is no difference between the median density of *Bellis perennis* due to an interaction between location and fertilizer treatment.
H_1: There is a difference between the median values of all samples due to an interaction between treatment variables 1 and 2.	**H_1:** There is a difference between the median density of *Bellis perennis* due to an interaction between location and fertilizer treatment.
2. To work out $K_{calculated}$	**2. To work out $K_{calculated}$**
i. For each block calculate R, R^2, n, R^2/n.	i. Table 8.8. There are four blocks A–D, each with four observations.
ii. Add all the R^2/n values together: $\Sigma(R^2/n)$.	ii. $\Sigma(R^2/n)$ $=189.0625+821.25+25+430.5625=1456.875$
iii. Add all the n values from the blocks. $\Sigma n=N$	iii. $N=4+4+4+4=16$
iv. $K_{calculated\ total}$ is calculated as before. $\left(\Sigma\left(\frac{R^2}{n_s}\right)\times\frac{12}{N(N+1)}\right)-3(N+1)$	iv. $K_{calculated\ total}=\left(1456.875\times\frac{12}{16(16+1)}\right)-3(16+1)$ $=13.27380$

BOX 8.5.C. Continued

GENERAL DETAILS	THIS EXAMPLE
v. You now have three values for K: $K_{columns}$ (BOX 7.4.1.), K_{rows} (BOX 7.4.2), and K_{total} (step iv. above) Using these values calculate $K_{interaction}$: $K_{interaction} = K_{total} - K_{rows} - K_{columns}$	v. $K_{interaction}$ $K_{columns} = 10.26012$ $K_{rows} = 3.00260$ $K_{total} = 13.27380$ $K_{interaction} = 13.2738 - 3.00276 - 10.26012$ $= 0.0110246$
3. To find $K_{critical}$ The degrees of freedom (v) are v for K_{total} = no. blocks − 1 v for K_{rows} = no. rows − 1 v for $K_{columns}$ = no. columns − 1 v for $K_{interaction} = v_{total} - v_{rows} - v_{columns}$ Use the v for $K_{interaction}$ and $p = 0.05$ to locate the critical value of K in a χ^2 table as a two-tailed test.	**3. To find $K_{critical}$** v for $K_{total} = 4 - 1 = 3$ v for $K_{fertilizer} = 2 - 1 = 1$ v for $K_{location} = 2 - 1 = 1$ v for $K_{interaction} = 3 - 1 - 1 = 1$ Therefore the critical value for $K_{interaction}$ at $p = 0.05$ and $\nu = 1$ is 3.84.
4. The rule If the calculated value of K is more than the critical value of K then you may reject the null hypothesis (H_0).	**4. The rule** $K_{calculated\ interaction}$ (0.01) is less than $K_{critical}$ (3.84) so you may not reject the null hypothesis.
5. What does this mean in real terms?	**5. What does this mean in real terms?** There is no significant interaction ($K = 0.01$, $p = 0.05$) between the fertilizer treatment and location in the median density of *Bellis perennis* at the University of Worcester.

8.6. Two-way non-parametric ANOVA: Specific hypotheses

In 8.5. we tested the general hypothesis 'there is a difference ...'. From the analysis of Example 8.3. it was clear that there was a significant difference due to the location and not due to the fertilizer treatment or any interaction. It would help our understanding of this experimental system if we could test more specific hypotheses. The details of the specific hypotheses will depend on the data. You construct one or more hypothesis based on the relationships that you think may be real and supported by your data. Look at the medians given for the data in Table 8.5. The specific hypothesis that we will test is that the effect of the fertilizer will be greater on the cricket pitch than on the lawn (BOX 8.6. and calculation table 8.8.).

BOX 8.6. How to carry out a two-way non-parametric ANOVA test: specific hypotheses

GENERAL DETAILS	THIS EXAMPLE
1. Hypotheses to be tested **H_0:** The medians of the samples do not follow the predicted order. **H_1:** The medians of the samples do follow the predicted order.	**1. Hypotheses to be tested** **H_0:** The median density of *Bellis perennis* does not follow the predicted order. **H_1:** The effect of the fertilizer on the median density of *Bellis perennis* will be greater on the cricket pitch than on the lawn so that A – C > B – D.
2. To work out $z_{calculated}$ i. Each block is given a weighting (λ) that reflects the order identified in your specific hypotheses. ii. For each block calculate λ^2, λR, λn, $\lambda^2 n$. (Remember some of the values may be positive and some negative. You need to ensure you handle these signs correctly. For example if $\lambda = -1$, then $\lambda^2 = 1$.)	**2. To work out $z_{calculated}$** i. There are four blocks (A–D) each with four observations. In this example we assign a weighting that reflects the proposed effects. A – C > B – D λ +1 –1 +1 –1 To use these weightings the equation first needs to be reorganized. $A - C - (B - D) > 0$ Therefore the weightings become: A C B D λ +1 –1 –(+1 – 1) A and C stay as they were +1 and –1 respectively. By reorganizing the equation the weighting for B and D change sign so they become –1 and +1 respectively. ii. Table 8.8.

BOX 8.6. Continued

GENERAL DETAILS	THIS EXAMPLE
iii. Sum these across blocks, to calculate $\Sigma n = N$; $\Sigma(\lambda R)$; $\Sigma(\lambda n)$; $\Sigma(\lambda^2 n)$.	iii. $N = 4+4+4+4 = 16$ $\Sigma(\lambda R) = 27.5 - 57 - 10 + 41.5 = 2.0$ $\Sigma(\lambda n) = 4 - 4 - 4 + 4 = 0$ $\Sigma(\lambda^2 n) = 4+4+4+4 = 16$
iv. Calculate the terms L, E, and V. $L = \Sigma(\lambda R)$ $E = \frac{(N+1)(\Sigma \lambda n)}{2}$ $V = (N+1) \times \frac{\left(N(\Sigma \lambda^2 n) - (\Sigma \lambda n)^2\right)}{12}$ v. Calculate the test statistic z $z_{calculated} = \frac{(L-E)}{\sqrt{V}}$	iv. $L = 2.0$ $E = \frac{(16+1)(0)}{2} = 0$ $V = (16+1) \times \frac{[(16 \times 16) - 0]}{12} = 362.66666$ v. $z_{calculated} = \frac{(L-E)}{\sqrt{V}} = \frac{2.0 - 0}{\sqrt{362.66666}} = 0.10502$
3. To find $z_{critical}$ This is a one-tailed test. z is independent of sample size. Therefore you do not need to know either the sample size or degrees of freedom.	**3. To find $z_{critical}$** For a one-tailed test at $p = 0.05$, $z = 1.645$
4. The rule If $z_{calculated}$ is more than $z_{critical}$ then you may reject the null hypothesis (H_0).	**4. The rule** $z_{calculated}$ (0.105) is less than $z_{critical}$ (1.645), therefore you may not reject the null hypothesis (H_0).
5. What does this mean in real terms?	**5. What does this mean in real terms?** The effect of the fertilizer on the median density of *Bellis perennis* is not greater ($p = 0.05$) on the cricket pitch than on the lawn.

8.7. Scheirer–Ray–Hare test

The ANOVAs we have described in tests 8.3–8.6. can easily be worked out using a calculator as long as you do not have many observations. If you had a larger data set you would probably prefer to carry out an analysis using computer software. Unfortunately, at present this software does not include the tests outlined in 8.4.–8.6. An alternative method may then be helpful. This is an extension of the Kruskal–Wallis test and can be used to test general hypotheses only. It is based on the parametric ANOVA; however, the rank values are used in place of the actual observations (Scheirer *et al.*, 1976).

8.7.1. To use this test you:

1. Wish to test for differences in population medians.
2. Have two treatment variables each with at least two categories.
3. The design is orthogonal.
4. Have non-parametric data that can be ranked.

To illustrate this test we will use the data from Example 8.3. These data meet the criteria for this test in that there are two treatment variables (location and fertilizer treatment) and each has two categories. The design is orthogonal and the observations are numbers of daisies, which is a non-parametric measure that can be ranked.

8.7.2. The calculation

The ranks have already been assigned and it is these that are used in the analysis (Table 8.9.). The three pairs of hypotheses being tested are those given in BOXES 8.5.A, 8.5.B and 8.5.C.

Table 8.9. Ranking values assigned to the density of daisies from two locations treated or not treated with fertilizer for the Scheirer–Ray–Hare test

Ranking values for Example 8.3. The effect of fertilizer treatment on the density of *Bellis perennis* (daisy) at two locations on the University of Worcester campus, 2003

	Cricket pitch	Lawn
Fertilizer	5	14.5
	6.5	12
	6.5	14.5
	9.5	16
No fertilizer	1	8
	2.5	9.5
	2.5	12
	4	12

Table 8.10. ANOVA table for Scheirer–Ray–Hare analysis of data from Example 8.3. The density of *Bellis perennis* (daisy) on the University of Worcester campus, 2003

Source of variation	SS	Degrees of freedom	s^2	$H_{calculated}$	$H_{critical}$
Variable 1 (Location)	*232.5625*	*1*		232.5625/22.4 = 10.382	$p = 0.01$ $H_{critical} = 6.64$
Variable 2 (Fertilizer treatment)	*68.0625*	*1*		68.0625/22.4 = 3.0385	$p = 0.05$ $H_{critical} = 3.84$
Interaction	*0.25*	*1*		0.25/22.4 = 0.01116	$p = 0.05$ $H_{critical} = 3.84$
Within samples	*35.125*	*12*			
Total	*336.0*	*15*	336/15 = 22.4		

A two-way parametric ANOVA as described in BOX 7.8. is carried out on these data, generating the values shown in italics in the ANOVA table (Table 8.10.). For more computation details see the Online Resource Centre.

At this point the analysis diverges from that described for a two-way parametric ANOVA in Chapter 7. First, an s^2_{total} is calculated by adding all the SS values together and dividing the total by the total degrees of freedom (Table 8.10.). An H value is calculated instead of an F value as SS/s^2_{total}. The critical values for H are found in a chi-squared table for the degrees of freedom associated with the SS (Table 8.10.). The rule is the same as that given for the parametric two-way ANOVA (7.7., BOX 7.6.) in that if $H_{calculated}$ is greater than $H_{critical}$ then you may reject the null hypothesis. In our example it can be seen that there is a significant difference ($H = 10.38$, $p = 0.01$) between the median density of daisies in the two locations, but there is no significant difference for the fertilizer treatment or the interaction. The outcome is therefore the same as that for the non-parametric ANOVA described in BOXES 8.5.A–8.5.C., the advantage being that this calculation can be carried out using computer software.

8.8. Experimental design and non-parametric tests of hypotheses

In this chapter we have covered the non-parametric equivalents of the parametric tests covered in Chapter 7. Therefore, the experimental designs suitable for using these tests have equivalents in Chapter 7. The difference is largely in that the observations will be non-parametric measurements and some of the criteria for the use of these tests differ, e.g. sample sizes.

8.8.1. One variable

In designs that test one variable there are only non-parametric equivalents to the matched and unmatched tests and not to the Latin square.

i. Unmatched data

The Mann–Whitney U test may be suitable, if all other criteria are met, for an experiment with the design shown in BOX 7.4. with one variable and two categories. The sample size is determined by the criteria in 8.1.1. Each observation in a sample is a **replicate**, although not usually referred to as such. One of the two samples could be a control and this test will allow you to make a comparison between the variation in a control sample and that of a sample exposed to a particular treatment.

The one-way non-parametric ANOVA may also be appropriate for this design particularly where more than two categories are present. The advantage here is that specific and general hypothesis may be tested if the criteria are met (8.3.1. and 8.4.1.). These tests are the non-parametric equivalent of the parametric one-way ANOVA followed by a Tukey's test.

ii. Matched data

The Wilcoxon's matched pairs test is the non-parametric equivalent of the matched t and z tests (Table 7.22.). The hypotheses being tested relate to one variable and two samples where the observations are not independent but are made on the same item. One set of observations could be a **control** and the other as a result of a treatment on the same item, e.g. 'before' and 'after', but need not always be so. The number of samples is determined by the criteria given in 8.2.1.

8.8.2. Two variables

As we indicated in Chapter 7 with two variables you may need to consider several more elements about an experimental design including whether the design is orthogonal, if the variables are fixed or random, and therefore if the ANOVA is Model I, Model II, or a mixed model. With the non-parametric tests we have included only orthogonal designs can be tested, and ideally where equal numbers of observations (replicates) are included in each category. This design is illustrated in Table 7.23. There is no non-parametric test that equates to the nested design (Table 7.24.) and this needs to be considered when designing experiments.

The two-way analyses examine three pairs of hypotheses, one of which relates to the possible interaction between the treatment variables, in addition to which specific hypotheses may be devised and tested. In the non-parametric ANOVA these specific hypotheses are more flexible than the parametric equivalent test (Tukey's), which is restricted to a comparison between pairs of mean values.

Two-way designs with replicates are organized into blocks, as are the two-way parametric equivalents. This can allow for a certain degree of randomization and replication within the experiment and so help to minimize the impact of **non-treatment** variables.

8.8.3. Three variables

The Sheirer–Ray–Hare test is the only non-parametric test that we have included that may be extended to more complex designs, such as three variables. The comments relating to three-way parametric ANOVAs remain pertinent to this non-parametric test.

Q3 **We are designing an experiment with one treatment variable and three categories, one of which is a control. Which test should we consider using and can we make a comparison that will allow us to analyse the difference between the control and other treatments?**

A3 With this design if the data are non-parametric we should consider using a non-parametric one-way ANOVA and test both general and specific hypotheses (8.8.1.). This will allow us to ask specific questions about the control and other treatments. If the data are parametric (7.12.1.) we should consider the parametric one-way ANOVA, which, if significant, could be followed by a Tukey's test.

Summary of Chapter 8

- We consider several non-parametric tests in this chapter. The Mann–Whitney U test for unmatched data (8.1.) and the Wilcoxon's matched pairs test (8.2.) for matched data may both be used to analyse data from experiments with one treatment variable, with two categories. The one-way non-parametric ANOVA (Kruskal–Wallis test, 8.3.) is particularly appropriate for a design with one treatment variable but three or more categories, and the two-way non-parametric ANOVA (8.3.–8.6.) and Scheirer–Ray–Hare tests (8.7.) may be appropriate for designs with two or more variables.
- These non-parametric tests are distribution free and therefore the criteria for their use are often easier to meet than for the parametric tests (Chapter 7). Most of the tests are ranking tests: central steps are the determination of ranks and sum of ranks as outlined in Chapter 3 (BOX 3.3.).
- In Chapter 7 we consider the parametric equivalents to the tests covered in this chapter. If your data are parametric (BOX 3.2.) or can be normalized by transforming them (3.9.) you should refer to Chapter 7.

- The non-parametric ANOVAs allow you to test both general and specific hypotheses (4.1., 8.3.–8.6.). This provides you with considerably more explicit information about the effects of your treatment variables and is a design that should be used whenever possible.
- In general these non-parametric tests can be applied to more restricted experimental designs compared with the parametric equivalents considered in Chapter 7, especially when an experiment sets out to investigate the effect of two or more variables.
- The Online Resource Centre includes interactive exercises that test your understanding of this chapter with other topics, particularly those considered in Chapters 2 and 7.

9

Research, the law, and you

Research and the law are intertwined in numerous ways and at all levels from the individual to the institution in which the research is carried out. It is a huge area, and our aim in this chapter cannot be to cover all relevant legislation. Instead we provide an introduction to the areas that appear to be most relevant to undergraduates and graduates. Again, from necessity we have restricted the details to law in England and Wales. On the whole these two countries have a very similar body of law that impinges on research and we can therefore cover both countries at the same time. In other parts of the UK, such as Scotland and Northern Ireland, there are sufficient differences to make it impossible to clearly explain the variation and the similarity within one chapter. However, this chapter will still be relevant to students studying outside England and Wales since the majority of the principles underlying most of the legislation in England and Wales are enshrined in very similar laws in Scotland and Northern Ireland. The Internet in general is an invaluable and usually reliable resource in relation to the law for any student requiring further details. We provide links to relevant websites for Scotland and Northern Ireland, as well as England and Wales, in our Online Resource Centre.

Clearly, as an undergraduate or graduate student your institution carries some responsibility for you and what you do. As a result your institution will make available to you guidelines and information that it is your responsibility to read and respond to. Some of the legal issues we cover in this chapter may therefore be addressed by your institution centrally and through staff such as your supervisor and technical staff. You have responsibilities both to your institution and under the law to comply with legislation whether it is channelled through your institution or by some other means. For example, your institution will have health and safety guidelines in place that put into practice the health and safety legislation. You are required to comply with these guidelines and be aware of the relevant legislation. However, if you work with wild plants and animals your institution is less likely to have a central policy; nonetheless you are responsible for being aware of the relevant legislation and complying with it.

Unlike other chapters there are not many exercises here. Instead we have suggested a number of discussion topics at the end which will allow you to relate real undergraduate project proposals to the areas of law we cover in this chapter. It will take you about 1 hour to consider these discussion topics in detail.

9.1. About the law

In England and Wales the body of law is developed from two sources: international law and national law. We use the term 'law' to mean any legislation or regulation derived from an Act of Parliament or other ratification of international agreements which is potentially binding on the individual and where failure to comply may lead to sanctions.

9.1.1. International law

Many nations voluntarily negotiate agreements at an international level. These agreements, however, are not legally binding until ratified by that country.

An example of this is the Council of Europe where the European Convention on Human Rights and Fundamental Freedoms was developed. This was later ratified within the UK as the Human Rights Act 1998. Another example is the Convention on International Trade in Endangered Species of Wild Flora and Fauna (CITES). Originally agreed in 1975, this was ratified in the UK in 1976 as the Endangered Species (Import and Export) Act.

In England and Wales other international law derives from the European Community (EC). Agreements usually take on one of four formats: Regulations, Directives, Recommendations, and Opinions. Regulations take immediate effect as law in all member states and may subsequently also be included in national law. A Directive is binding but it is not law in a particular country until the national authorities pass an appropriate Act. Recommendations and opinions do not have a binding effect on nation states. An example of EC law is the Habitats Directive 1992, which became incorporated into legislation primarily in the Conservation (Natural Habitats etc.) Regulations 1994.

9.1.2. National law

In England and Wales the body of national law comes from three sources: common law, statutes, and local law.

i. Common law

The legal system in England and Wales is based on common law derived from legal decisions made during specific court cases, i.e. we have a doctrine of legal precedent. These cases make rules that must be followed subsequently by all courts of similar or lower standing in the court hierarchy. Common law has been particularly important in shaping the law relating to trespass.

ii. Statute law

These are laws enshrined in an Act of Parliament, e.g. the Health and Safety Act 1974. The Acts of Parliament are divided into chapters, sections, and parts. Detailed information is usually included in Schedules at the end of the Act. For example, the lists of species covered by specific parts of the Wildlife and Countryside Act 1981 are included in the Schedules.

Law introduced through an Act of Parliament may empower other authorities to make further provisions or amendments. The provisions are often known as Orders or Regulations. Principal among those empowered to make secondary legislation are ministers in a particular government office, e.g. the Home Secretary and the Welsh Assembly. The law may also empower government departments to licence certain otherwise prohibited activities and in developing the conditions for the licence further statutory controls are introduced. For example, to obtain a licence under the terms of the Animals (Scientific Procedures) Act 1986 an applicant must adhere to the code of practice produced by the Home Office for the housing and care of animals used in scientific procedures.

iii. Local law

Parliament has identified certain national bodies as being able to make laws. These bodies include the Crown, local authorities, and public corporations, such as the Environment Agency, Forestry Commission, Countryside Agency, English Nature, and the Countryside Council for Wales. These bodies may establish laws within certain remits prescribed by the Statutory Instruments Act 1946 and Local Government Act 1972 and often in response to other specific statutes such as the Wildlife and Countryside Act 1981. Where a public body develops these laws they are usually known as codes of conduct, and where local authorities do so they are known as bye-laws.

iv. Quasi-legislation

Apart from the source of law outlined above there are many individuals, organizations, etc., which also develop policies, codes of conduct, etc. These are voluntary codes and do not have the force of law. This can be confusing, especially where such terms as 'code of conduct' are used. You may therefore need to confirm whether a code of conduct is enforceable under the law or not. Failure to follow some recommendations and codes of practice may be used as evidence that a law has been broken even though these guidelines may not themselves be part of statute law.

v. Enforcement

The agency responsible for enforcement of legislation varies enormously and is outlined in the original legislation. The police are responsible for enforcing most bye-laws and most statute laws, but there are many other agencies with powers of enforcement. For example, Customs and Excise staff have a major role in enforcing laws relating to imports and exports whereas the Animals (Scientific Procedures) Act is enforced through an Inspectorate which reports to the Animals Procedures Committee and to the Home Secretary.

9.2. Health and safety

You must work in such a way that your safety and that of others is paramount. To do this you need to identify the hazards of each step of your research protocol and then incorporate practices that minimize the risk to health. At the end of this process you need to make a judgement as to whether the risk to your health and others is sufficiently small that the research may proceed. There are a number of laws that cover health and safety, including the Health and Safety at Work Act 1974 updated by the Management of Health and Safety at Work Regulations 1992 and a number of detailed regulations including the Control of Substances Hazardous to Health Regulations 1988 (as amended) (COSHH); Chemicals (Hazard Information and Packaging for Supply) Regulations 2002 (CHIP); Environmental Protection Acts 1990, 1995, and the Special Waste Regulations 1996.

Health and safety is something that is usually considered in laboratory work where chemicals or microorganisms are being handled. However, health and safety must be considered in all types of research from carrying out a questionnaire survey to sampling soil in the field. The method for considering health and safety is a risk assessment. This can be divided into seven steps:

1. Hazard identification and rating. What hazards may be present if you carry out your research as you have designed it? In what way may these hazards affect your own or other people's health?
2. Activity. How might you become exposed to the hazard and what is the possible scale of exposure?

3. Probability of harm. Given the nature of the hazard, the possible effect on health and the possible scale of exposure what is the probability of harm?
4. Minimize the risk. Decide what precautions are needed to control or reduce this risk.
5. Risk evaluation. Is this an acceptable level of risk?
6. Emergencies. Be aware of, or prepare procedures to deal with accidents and emergencies.
7. Action. Ensure these precautions are in place and followed.

9.2.1. Hazard identification and rating

What hazards may be present if you carry out your research as you have designed it? In what way may these hazards affect your own or other people's health? In most student research the hazards will fall into three broad categories:

i. The environment

Your working environment may contain a number of hazards. This is true of both laboratory work and working in the field. In the laboratory you should consider physical features including temperature extremes such as working in a cold room or in a greenhouse in the summer. There are often many electrical sources, equipment with moving mechanical parts, or dust in the atmosphere from, for example, grinding bones, sorting soil, allergen studies.

In the field you almost certainly will encounter particularly hot or cold days. The environment may have rough terrain and may be subject to flash flooding. Additional hazards can come from living things in the environment that are not strictly part of your research. For example, you may need to consider threats from strangers; bites which may cause skin damage and introduce infections; poisonous or allergenic plants or fungi, etc.

You must also be aware of activities that may be carried out in the environment of your research that you are not involved in but may be a source of hazards. For example, areas that have been sprayed with pesticides, chemical spills, the use of microbes, the presence of a UV source, etc. Chemical hazards found in the environment are considered further in 9.2.1.ii.

One example where the environment may have hazards that need to be considered is seen in Example 5.2. The genetics of flower colour in *Allium schoenoprasum*. In this experiment the investigator is working in a greenhouse. Common hazards of a greenhouse include broken glass which may cause cuts, electrical sources in contact with water which may result in electrocution and death, soil-borne pathogens which can bring about serious illnesses, and allergens from fungi and plants which may cause asthma. Most universities and other similar institutions carry out general risk assessments for areas such as laboratories. You may therefore be able to make use of these when carrying out your own risk assessment.

ii. Chemicals

Obtaining, storing, using, and the safe disposal of chemicals is covered primarily by the Dangerous Substances and Explosive Atmospheres Regulations 2002 and by the Control of Substances Hazardous to Health Regulations 2002 (COSHH). The first of these two Acts covers all flammable substances. COSHH covers 'hazardous substances' that are to be used during work, substances generated during work, naturally occurring substances, and biological agents such as bacteria and other micro-organisms. The term 'hazardous substance' is taken to mean a substance or mixtures of substances classified as dangerous to health under the Chemicals (Hazard Information and Packaging for Supply) Regulations 2002 (CHIP). These regulations do not cover all substances, however, and specific regulations cover the safe handling of gases such as helium and the use of pesticides, medicines, cosmetics, lead, and radioactive substances. Under the terms of the Health and Safety at Work Act 1974 these substances must also be considered in hazard identification.

To identify the hazard relating to any substance that you will use in your work you can refer to the label on the container in which the substance is supplied. Further details can easily be obtained by searching the web using the chemical name or by referring to suppliers' catalogues or through Health and Safety Executive (HSE) publications such as the *Approved Supply List*. When carrying out research you may not always know how substances you are using will react and therefore may not be able to predict what substances will be produced during your work. If you are not sure for any particular protocol you should refer this matter to your supervisor.

The potential effects of a chemical hazard are described using a number of specific terms. For example 'caustic' is the chemical burning that occurs on exposed surfaces such as skin, eyes, or internal organs if they become exposed to the substance. The hazard is therefore potentially damaging if you eat/drink it or allow your skin or eyes to come into contact with it (9.2.2.i.). Other common descriptive terms are flammable, explosive, toxic, harmful, mutagen, carcinogen, corrosive or strongly oxidizing, irritants, radioactive, harmful to the environment. Links to definitions of these 'risk phrases' are included on many chemical and biochemical hazards websites and are included in suppliers' catalogues.

Q1 **Use one of the sources outlined above to identify the hazard(s) associated with concentrated ethanoic acid (glacial acetic acid).**

A1 Corrosive, very harmful if swallowed.

R10 Flammable
R20 Harmful by inhalation
R21 Harmful in contact with skin
R22 Harmful if swallowed
R35 Causes severe burns

The R (risk) codes are used as a shorthand when providing some safety information. They are recognized standard descriptions of different hazards. R10 always means flammable, etc. (There are also S (safety) codes which indicate the safety precautions that should be followed.)

iii. Biological agents

When biological agents are discussed in health and safety, most thought is given to microorganisms. For example, in the field you may encounter *Leptospirosis* sp., which is a bacterium that can be found in water contaminated with the urine of infected animals. You also need to be aware of organisms that may present a potential hazard, including the common intestinal parasites of dogs and cats (*Toxocara* spp. (ascarid) and *Ancylostoma* spp. (hookworms)), which can be distributed in the environment from animal faeces.

Microorganisms (bacteria and fungi) present a different challenge from most chemical and environmental hazards. If you are working with microbes in the laboratory you may not know which of the possible thousands of species you are incubating and the potential virulence and toxicity of a species may vary between strains. It is therefore sensible to consider all microorganisms as potential pathogens unless the potential for harm is known to be higher or possibly higher. Information about pathogenic microorganisms can be obtained from a number of sources such as the Advisory Committee on Dangerous Pathogens (ACDP). In general the hazards associated with microorganisms are: infection (e.g. botulism, MRSA); the production of toxic substances by the microorganism (e.g. *Aspergillus flavus* and *Aspergillus parasiticus* are both known to produce aflatoxin) and the potential allergenicity of live or dead airborne microorganisms leading to the onset of asthma or dermatitis.

iv. Rating the hazard

Having identified all the hazards that you may encounter in your work, you need to rate them. The institution you are affiliated with should give guidance on this. In general, however, a major hazard would be one that may cause death or serious injury; a serious hazard may cause injuries or illness resulting in short-term disability; and a slight hazard may cause less significant injuries or illnesses.

In Example 7.1., the investigator was studying *Littorina* species on the mid to low areas of a rocky shore. What hazards might she encounter, and how would you rate them?

The hazards and ratings include: being swept out to sea and drowning (major), falling over and breaking a bone (serious), or other injury such as a cut (serious/slight) or sprain (slight), effects of temperature (hypothermia, sunburn, etc.) (major–slight), water-borne pathogens causing an infection (major–slight) or toxic reaction (major–slight), being attacked by other people or their dogs (major–slight).

9.2.2. Activity

How might you become exposed to the hazard, and what is the possible scale of exposure? The way in which you encounter, use, and/or generate hazardous substances will determine the probable extent to which you may become exposed to the hazard. Some substances you handle may be in a powdered form and so you may risk exposure through inhalation. Some substances can penetrate the skin and others damage the skin.

i. Hazards arising from the activity

Activities may in their own right be potentially hazardous and these need to be recognized as part of your evaluation of health and safety. Examples include where you or others need to move equipment, to sit or stand in awkward or cramped conditions, or make repetitive movements.

Hazards in some activities will be the result of a particular substance being exposed to a particular treatment and the potential hazard will differ from either the hazard related to the substance or the hazard relating to the activity alone. For example, agar in its crystalline powder form is believed to present a negligible risk to health. If you then wish to make agar plates the protocol requires the agar to be placed in a container (usually a glass beaker or bottle) with distilled water and heated to 100 °C. The agar needs to be poured when it is still hot, at about 80 °C. Splashes from the hot liquid agar could cause serious scalds, which will be hard to treat because the agar sticks to the skin. Therefore, the hazard as a result of the activity is both different from the substance and activity alone and more serious.

ii. Scale of exposure

Having considered in what way you may become exposed to the hazard, you then need to grade this possible exposure. You should be given guidance as to how to grade the likelihood of exposure. However, in general you would assign the hazard to a 'high' category where it is certain harm will occur, 'medium' when harm will often occur, and 'low' where harm will seldom occur. For example, concentrated ethanoic acid is caustic (Q1/A1) and would be rated as a major–serious hazard. You may be provided with 1 ml of a very dilute solution of this (e.g. 5% w/v) of which you will pipette 1 μl into a small tube. As you are using small volumes of a solution with a low concentration the scale of exposure would receive a 'low' rating. (In fact this dilution of ethanoic acid is commonly found in kitchens as vinegar!)

iii. Storage and disposal

With some potential sources of hazards, such as chemicals and microorganisms, you need to consider more than just your use of them in an investigation. These substances need to be stored and safely disposed of. Your consideration of hazards and risks needs to extend to this full lifetime of such substances. This is of particular importance, as it is during these times that other people are likely to become exposed to the potential hazard. Usually technical staff can advise you about safe storage and disposal, and regulations such as the Special Waste Regulations 1996 and the Dangerous Substances and Explosive Atmospheres Regulations 2002 provide statutory guidance.

9.2.3. **Probability of harm**

Given the nature of the hazard, the possible effect on health and the possible scale of exposure what is the probability of harm?

i. Probability of harm

This, the risk assessment step, is not easy and is usually subjective and qualitative. Most institutions provide guidelines to help you. For supplied substances the HSE has developed a generic web-based guide to risk assessment called *COSHH Essentials.*

In a risk assessment you need to make a judgement about each hazard you have identified using the information:

- The hazard and the nature of its possible effect on health (major, serious, slight).
- The possible scale of exposure as a result of the way in which the hazard will be encountered/used (high, medium, low).

The probability of harm is the first of these combined with the second. Therefore, in 9.2.2.iii. the major caustic hazard of ethanoic acid will only be present in a form where the risk of exposure is low. In this example we would grade the overall probability of harm as low.

ii. Variation between individuals

Human beings all differ and some of this variation can increase or decrease risks from any one hazard. Points that often need to be considered include height, handedness, sensitivity to allergens, suppressed immune system, the side effects from medication (such as antihistamines causing drowsiness), and pregnancy. For example, you may not be pregnant but you may need to alert those who are to a hazard that is of more significance to them than to yourself.

iii. Occupational exposure limits

Some substances have occupational exposure limits. As an undergraduate you are unlikely to be using such substances although as a graduate this is possible, e.g. radioactive substances. There are two types of these occupational exposure limits: the maximum exposure limit (MEL) and the occupational exposure standard (OES). Further information about occupational exposure limits is available from the HSE.

9.2.4. **Minimize the risk**

Decide what precautions are needed to control or reduce this risk. It is impossible to cover in this section all possible types of hazard with suggestions as to how you may reduce the risk from each hazard. We have concentrated on the most common, but for each of the hazards you have listed you must consider how to prevent the hazard or reduce the risk (i–ii) and then re-evaluate the probability of harm (iii).

i. Prevention

No risk to health is really acceptable; therefore, ideally you should prevent exposure to the hazard. Prevention can be achieved by changing the process or activity (e.g. using a different field site) and/or by using a safer alternative (e.g. using a pelleted form of a substance rather than a powder).

ii. Reduction

If prevention is not reasonably practicable you must adequately control exposure and therefore you need to put in place control measures suitable for the activity and consistent with your risk assessment.

a. Wear personal protection You may wear and/or provide personal protection appropriate to the nature of the hazard. In the biosciences this protective clothing most often consists of laboratory coats, general safety glasses or face masks, UV light shields, and gloves known to exclude the particular hazard you may be exposed to (thermal or chemical). In the field additional items providing personal protection include strong boots, waders, waterproof clothing, suncream, and insect repellent. It should be noted, however, that the suncream and insect repellent are substances and as such their use should also be considered under COSHH.

Additional medical protection may be appropriate, for example vaccinations (e.g. tetanus or hepatitis), medication to control allergies, and medical/biological monitoring. If you are ever in doubt or feel unwell you should always seek medical assistance and draw to the attention of the medical staff the nature of your research and the hazards you have identified.

b. Reduce quantities You may reduce the hazard by ensuring, for example, that the quantity of substances you are storing, using, or will need to dispose of are limited. For example, you may keep most of your stock solution in safe storage and only take out the volume required at the time.

c. Control ventilation You may use controlled ventilation as in a fume hood or laminar flow hood to limit exposure to a hazard. This is common practice where volatile substances are used or produced and in microbiology.

d. Organize space The most common method used in risk reduction is to organize your space. Eating and drinking should be prohibited in all laboratories to prevent accidental ingestion of a hazardous substance. In the field the equivalent practice is to ensure you are able to wash your hands with clean water before you eat. Other examples of good practice are to keep flammables away from a Bunsen burner and to keep electrically powered equipment away from liquids. For some research a dedicated area to which access is restricted is appropriate (e.g. microbiology areas and areas where radioactive sources are used and stored). For some substances such as flammables and poisons special storage should be provided. If you are in doubt your supervisor or technical staff will be able to advise you.

e. Take a break This is especially important if you are doing a repetitive task or if you are using light sources such as a microscope.

f. Inform It is essential that you alert other people to potential hazards. With stocks of substances this is usually already done for you by the manufacturers. Solutions or mixtures that you prepare yourself must be similarly labelled with details of the contents including concentrations, the hazard(s), and the date. There are internationally recognizable symbols that are used to indicate certain hazards. Laboratories often carry stocks of sticky labels with the symbols for chemical hazards such as 'explosive' (Fig. 9.1.) or 'oxidizing' (Fig. 9.2.). These symbols can be stuck onto the containers holding your solutions or substances.

If you will be working in the field then providing information is important in a number of ways. For example, you may wish to leave markers protruding from the ground to indicate the location of permanent quadrats. If so, you should consider the need to alert other people to their presence so that they will not fall over them. When working in the field you should ensure that at least one responsible person knows exactly where you are working and when you are expected back. It is also advisable to take a phone so that you can keep in touch.

g. Company It is general practice in most institutions where students may be working independently to require or at least to recommend that the

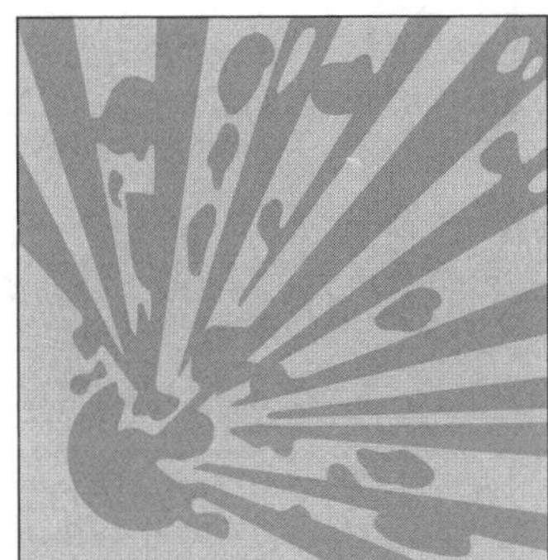

Fig. 9.1. Standard symbol for explosive hazard.

Fig. 9.2. Standard symbol for oxidizing hazard.

researcher always has someone else with them or near by. In laboratories this can most easily be achieved by agreeing a schedule of work that is compatible with another student also working in that laboratory. For those working in the field (both urban and rural areas) this may not be so easily arranged but should be achieved if at all possible.

h. Be informed Part of the process in risk assessment is planning for emergencies. This level of planning is a form of risk reduction. We consider this in 9.2.6.

iii. Re-evaluate the probability of harm

Having put into place control measures to reduce the risk you can now revaluate both the hazard and the nature of its possible effects on health (major, serious, slight) and the possible scale of exposure as a result of the way in which the hazard will be encountered/used (high, medium, low). Most control measures work by addressing the second of these. You will now have two measures of the probability of harm, one where there are no control measures (9.2.3.) and one where the control measures are in place and effective (9.2.4.)

9.2.5. **Risk evaluation**

Is this an acceptable level of risk? Using these two sets of risk evaluations you are now in a position to determine if the level of risk is acceptable. We cannot provide detailed guidance on this as there are so many differing potential hazards. You must therefore seek specific guidance from your supervisor. However, in general undergraduate work will normally be allowed to continue if the risk, with the controls in place, is low and occasionally moderate and where the controls are not likely to fail and where if they did the risk was no more than moderate. Graduates may be allowed to accept a higher risk, but in our experience a severe risk of injury to health is not usually deemed acceptable for any student.

9.2.6. Emergencies

There are two types of emergencies that you need to consider before you may finally complete this part of your research preparation. The first point you need to consider is what to do if one or more of your measures introduced to control the hazards fail. The second is what to do if there is a fire or other external emergency that means you must leave your research immediately.

i. Failure of control measures

You must consider here both in what ways these protective measures may fail and what needs to be done in response. For example, if a laminar flow hood fails you are more likely to become contaminated with the micro-organisms you are handling. You may therefore need to carry out biological monitoring or be aware of symptoms of any illnesses that may develop as a result. You may also wish to satisfy yourself that the laminar flow hood is appropriately maintained. A second example of such an emergency would be if you spilt a hazardous substance. You will need to know who needs to be informed and how; how quickly does anyone else need to be informed; who can tidy the spill and how; does the area need to be evacuated; if so who is responsible for this. In the field this preparation can include ensuring that you have emergency phone numbers, additional food, water and clothing, basic first aid, and to have worked out an emergency escape route.

ii. External emergencies

In preparation for such events as a fire alarm sounding you should discuss with your supervisor and technical staff what needs to be done to ensure the hazard(s) in your work continue to be contained and other hazards do not arise. There may be a need, for example, to turn off a Bunsen burner or safely store hazardous substances. As part of their emergency procedures your institution will have guidelines as to who should carry out such actions: it may not be you.

iii. Information

If your control measures fail or you experience an external emergency then the risk assessment that you carried out will inform yourself and those around you as to what action is required. Therefore, it is essential that you keep it with you. For example, if you splash yourself with something, you may need urgently to be able to communicate details about that substance and its hazard to others. Therefore, you should keep a copy of your risk assessment including your emergency plans in your field or laboratory notebook and keep the notebook with you.

9.2.7. **Action**

Having carried out your risk assessment and determined that the risk is acceptable you can then carry out the research. However, you will need to:

- ensure that the precautions to control the hazard(s) are in place, are followed and that they are effective
- ensure that your emergency planning information is with you when you are working and others in the vicinity are aware of this
- contact medical assistance immediately if you start to feel unwell, and take your risk assessment details with you.

9.3. **Access and sampling**

In this section we consider the law that particularly relates to the studying or sampling of wildlife. The legislation relating to access and to handling plants and animals is extensive. We cover some of the key areas most often encountered by undergraduates, but have restricted ourselves largely to land-based work.

9.3.1. **Access**

The laws relating to access are primarily derived from national law. The statutory regulations are included in a number of Acts, such as the Environmental Protection Act 1990, the Countryside and Rights of Way Act 2000, and the Water Industry Act 1991.

In essence in England and Wales you may become a trespasser:

- if you enter land without the permission of the landowner
- if you go outside the area which you are allowed on because you have a statutory right or you have permission from the landowner
- if you use land in a way in which you have not been authorized to do so

Trespass is not usually a criminal offence in that you cannot be prosecuted but you can be sued. This may result in being required to pay compensation. In addition the landowner can use reasonable force to eject a trespasser and obtain an injunction if necessary to prevent you from returning. Trespass can be a criminal offence on Ministry of Defence land, land adjacent to railways, and some nature reserves. In addition to trespass you may also commit offences of criminal damage if you put up signs, dig holes, etc., without the permission of the landowner.

i. Public right of way

There are some areas such as registered footpaths where you have a public right of way. This, however, only means that you may use the path to travel, usually on foot, from point A to point B. You may therefore be considered to be trespassing if you carry out other activities such as taking samples without the permission of the landowner.

ii. Open country

There are some parts of England and Wales that are considered to be 'open country'. There are two main pieces of legislation relating to access in open country: the National Parks and Access to the Countryside Act 1949 and the more recent Countryside and Rights of Way Act 2000 (CRoW). Both of these provide for the possibility of access to 'open country' such as mountains, moors, heath, and down. The way in which these rights are managed varies slightly in that within the first Act details about access are largely set out in an agreement or order made in conjunction with the owner. In the more recent Act access is determined more by the terms of the Act than by local agreements. Whichever Act covers a particular piece of open ground normally there will be a right of access on foot for open-air recreation, such as walking, bird-watching, picnicking, running, and climbing, but not usually for driving a vehicle on the land, using boats, hunting, fishing, collecting anything from the area including rocks or plants, camping, or lighting fires. There are variations and exceptions to both Acts; for example, some areas within the designated 'open country' regions in the Countryside and Rights of Way Act 2000 will not be subject to the new access rights. These include buildings and livestock pens; land ploughed or drilled during the previous 12 months to grow crops or planted with trees; quarries and other active mineral workings; land used as a golf course or race course; and land where military bye-laws apply.

iii Waterways

There are a number of different types of waterways: tidal and non-tidal rivers, and streams, canals, natural and man-made lakes, and reservoirs.

Ownership and responsibility for the management of these areas varies. However, none of these areas has a general right of way unless they are included within the regions of 'open country' identified in the Countryside and Rights of Way Act 2000. However, many bodies responsible for waterways allow some rights of navigation and access. This permissible access and associated activities will be laid out under agreements, codes of conduct, or bye-laws. Key bodies that may be consulted in relation to waterways are the Environment Agency and British Waterways.

iv. Foreshore

The foreshore between high- and low-water mark is the property of the Crown except where it has been sold or leased usually to a local authority. Again access is customarily permitted usually subject to bye-laws. The Countryside and Rights of Way Act 2000 gives the Secretary of State powers to propose the extension to the right of access to coastal land, but no significant changes have occurred to date. Most beaches above the high-water mark are owned by the local authority and access is subject to bye-laws.

v. Commons, and village and town greens

Commons, and village and town greens are not necessarily places with a right of public access. For most of this type of land consent is usually given by the landowner to allow access and some activities. This permission is outlined either in an agreement or bye-laws under the terms of statutory law such as the Commons Act 1876 and 1899 and the Village and Town Greens Act 1965. Local authorities have in the last 50 years or so clarified the ownership of and public rights of way on commons, etc. Therefore, if you need to contact an owner, the council may be able to help you. Village and town greens may have increased public rights of access if local people have openly used the land for recreation for at least 20 years without the permission of the owner. These increased rights still only relate to access and recreational activities.

vi. Other areas

There are some areas where you may feel you can roam at will, e.g. National Trust properties and country parks. Most of the latter are owned by the local authority. Acceptable activities in these areas will also be outlined in for example, a code of conduct or bye-laws. Similarly in woodlands and forests you do not have a general right of access. There may be public rights of way such as footpaths passing through the woodland, or local agreements giving some access. Again, these local agreements usually relate to walking and a few prescribed activities such as bird-watching.

9.3.2. Theft

Taking things from the environment, such as stones, wood, earth, etc., is a form of theft unless you have been authorized to do so by the landowner. Therefore, if you wish to remove leaf litter or soil, etc., you should ensure you have discussed this first with the landowner. Many species are protected in this regard by additional legislation (9.4.3.), but wild plants and fish not covered by the additional legislation are the property of someone (e.g. Theft Act 1968). Therefore, it is illegal, for example, to uproot any wild plant for commercial purposes without authorization from the land owner.

9.3.3. Plants, animals, and other organisms

Currently the key legislation in England and Wales that provides statutory protection for organisms is the Wildlife and Countryside Act 1981, the Wildlife and Countryside Act (Amendment) 1985, the Conservation (Natural Habitats, &c.) Regulations 1994, and the Countryside and Rights of Way Act 2000. Some species are subject to additional legislation, for example the Protection of Badgers Act 1992. This wildlife protection falls into two categories: protection of specific species and the protection of habitats. In this section we consider the protection of specific species under three subheadings: birds, animals, and plants. The Schedules in the Wildlife and Countryside Act 1981, which list the birds, animals, and plants covered by the legislation, are reviewed every 5 years. A review is imminent. Therefore, for current lists of Scheduled species, we refer you to the Joint Nature Conservation Committee website: Wildlife and Countryside Act 1981.

i. Wild birds

a. General protection Within England and Wales all (approximately 500) species of wild birds are covered by the general legislation in the Wildlife and Countryside Act 1981 (1981 Act) and the Countryside and Rights of Way Act 2000. These acts make it an offence to:

- kill, injure, or take any wild bird
- take, damage, or destroy the nest of any wild bird while that nest is in use or being built
- take or destroy an egg of any wild bird or stop it from hatching
- possess any live or dead wild bird or any part of a wild bird
- possess any egg or part of any egg from a wild bird

There are exceptions to this in that it is not an offence to kill a bird if it has been mortally injured or to take a wild bird to treat it if it has been hurt, but only where the bird has been injured in a way that does not contravene the Act. You may also be exempt if the outcome is 'the incidental result of a lawful operation and could not have been avoided'.

b. Enhanced protection Some species have enhanced protection under Schedule 1 of the 1981 Act. For the species listed in this Schedule the penalties for the offences described above are higher and it is an offence to:

- disturb any wild bird included in Schedule 1 while it is building a nest or is in, on, or near a nest containing eggs or young
- disturb dependent young of such a bird.

Some of the birds listed in Schedule 1 are given this additional protection for the whole year whilst for some it only covers the spring and summer (February–August). The exceptions that relate to the general protection of wild birds also apply to this enhanced protection.

There are other regulations relating to game birds, 'pests', and swans. Game birds are covered by Schedule 2 part 1 of the Wildlife and Countryside Act 1981. This provides regulations concerning the killing of game birds. Some birds such as pigeons may be 'pests' and Schedule 2 part 2 of the 1981 Act provides regulations concerning the control of these birds. Wild swans belong to the Crown and whilst they are subject to the 1981 act are also Crown property unless they have been tamed and are on private waters.

ii. Wild animals

The Wildlife and Countryside Act 1981 prohibits certain methods of killing and taking wild animals. Further protection, however, is only given to Scheduled species. The species in Schedule 5 at present includes certain mammals, reptiles, amphibians, fish, butterflies, moths, beetles, hemipteran bugs, crickets, dragonflies, spiders, crustaceans, sea-mats, molluscs, annelid worms, sea anemones, and their allies.

Under the terms of Section 9 of the 1981 Act for the animals listed in Schedule 5 it can be an offence to:

- intentionally kill, injure, or take such an animal
- possess or control a live or dead animal, part or derivative
- damage or destruct or obstruct access to any structure or place used by the scheduled animal for shelter or protection
- disturb the animal occupying such a structure or place
- sell, offer for sale, possess, or transport for the purpose of sale such an animal (live or dead, part or derivative)
- advertise for buying or selling such things.

You may be exempted from these laws for the same reasons given for wild birds: for example, if you take a Scheduled animal to tend it if it has been injured due to a reason that does not contravene the Act.

a. Badgers Badgers are not included in the 1981 Act but are protected under different legislation including the Badgers Act 1973, the Wildlife and Countryside (Amendment) Act 1985, the Badgers (Further Protection) Act 1991, and consolidated in the Protection of Badgers Act 1992. This combined legislation provides the same type of protection given to the animals on Schedule 5 of the 1981 Act but with additional offences relating to the use of dogs, cruelty, and enhanced protection relating to reckless disturbance and causing damage to a badger sett.

b. Deer Deer are also not included in the Schedule 5 of the 1981 Act but in the Deer Act 1963 and Deer Act 1980 there are laws which specify which species may be killed, when and how.

c. Fish Some fish species are included in Schedule 5 of the 1981 Act. In addition to this, common law treats fish in non-tidal waters as property. Therefore, you will need the landowner's consent before carrying out any work relating to freshwater fish. In addition, the Salmon and Freshwater Fisheries Act 1975 prohibits killing freshwater fish by certain methods, including firearms and spears, or using lights. In UK tidal waters and seas anyone can fish but you are subject to some statutes and bye-laws.

d. Bats Bats are protected by a number of different statutes including the Wildlife and Countryside Act 1981, the Conservation (Natural Habitats, etc.) Regulations 1994, the Wild Mammals (Protection) Act 1996, and the Countryside and Rights of Way Act 2000. Under this legislation all bats are listed as 'European protected species' and it is an offence for any person to:

- deliberately capture, kill, injure, or take a bat
- possess or control a live or dead bat, any part of a bat, or anything derived from a bat
- damage, destroy, or obstruct access to any place that a bat uses for shelter or protection
- deliberately disturb a bat
- sell, offer, or expose for sale, or possess or transport for the purpose of sale, any live or dead bat, any part of a bat, or anything derived from a bat
- set and use articles capable of catching, injuring, or killing a bat (for example a trap or poison), or knowingly cause or permit such an action. This includes sticky traps intended for animals other than bats
- make a false statement in order to obtain a licence for bat work
- possess articles capable of being used to commit an offence, or to attempt to commit an offence.

Again there are exceptions to this legislation which are similar to those described in relation to wild birds.

iii. Wild plants

In most UK legislation the working definition of the term 'plants' is wide and includes algae, lichens, and fungi, as well as true plants such as mosses, liverworts, and vascular plants. Protection is provided at three levels: general, enhanced national, and international protection.

a. General protection All wild plants are covered by legislation relating to property (9.3.2.). In addition all wild plants are given general protection in the Wildlife and Countryside Act 1981, which makes it illegal to uproot any plant for any reason without the permission of the landowner, but you may legally pick plant material for the plants not covered under the enhanced protection of, for example, Schedule 8 of the Wildlife and Countryside Act 1981.

b. Enhanced protection In the Wildlife and Countryside Act 1981 some plants are included in Schedule 8. For these plants it is an offence to:

- intentionally pick, uproot, or in any way destroy such a plant
- sell, offer for sale, possess, or transport for the purposes of sale any such plant (live, dead, or derivative)
- advertise for buying or selling such things.

This enhanced protection, therefore, includes collecting seeds or spores of any of the species listed on Schedule 8 and there is no exception for the owner of the land on which these plants are found. Again there is an exemption made under the terms of the 1981 Act which is where plants are damaged or destroyed as an incidental result of a lawful operation and the damage could not reasonably have been avoided.

In addition to the enhanced protection given to some plant species in the Wildlife and Countryside Act 1981 the local authority can place tree preservation orders, for example, to protect ancient trees and/or to preserve a certain ambience in urban areas. If a tree is included in a tree preservation order then it is an offence to cut down, uproot, or wilfully damage or destroy the tree covered by the order without the consent of the planning authority.

9.3.4. Protection in special areas

Within England and Wales species are also protected by being within certain designated areas. This protection comes primarily from the Wildlife and Countryside Act (1981 and as amended), the Council Directive 92/43/EEC on the Conservation of Natural Habitats and of Wild Fauna and Flora (EC Habitats Directive), which became UK legislation through the Conservation (Natural Habitats &c.) Regulations 1994 (as amended). The areas we consider in this chapter are those protected because of their biological importance; however, some areas are protected because they are of archaeological or geological value, e.g. limestone pavements.

Areas within England and Wales may be given one of several different types of designation and the level of protection varies depending on the status assigned to the area. We briefly review the current types of designation and the protection offered to the species within these areas that is in addition to that outlined in 9.3.3.

i. Habitats and Natura 2000

Under the terms of the Habitats Directive and EC Birds Directive, Member States are required to put forward a number of national sites for consideration as Special Areas of Conservation (SACs) or Special Protection Areas (SPAs). These will form a network of protected areas known as Natura 2000, which aims to protect certain rare or endangered species and habitats. At present these candidate sites are being identified within the UK by organizations such as the Joint Nature Conservancy Council (JNCC). Once a site is designated the regulations that might reduce the impact of future use and disturbance to the habitat will be reviewed and if necessary improved.

The amount of legislative protection for habitats is increasing, as seen in the establishment of Natura 2000 sites. In particular wetlands and woodlands are subject to additional specific legislation. Wetlands of International Importance especially waterfowl habitat are covered by the Ramsar Convention (1971), which was ratified in the UK in 1976. Many of the UK's Nautra 2000 sites are also Ramsar sites. The Forestry Act 1967 originally focused on forests as purely economic enterprises. This has changed with the Wildlife and Countryside (Amendment) Act 1985 where the terms outlined the need to find a balance between commercial forestry, amenity use and conservation.

ii. Nature reserves

These are areas that have a special conservation interest either for a species or habitat or an unusual geological feature.

a. Statutory local and national nature reserves The National Parks and Access to the Countryside Act 1949 allows English Nature and the Countryside Council for Wales to establish a nature reserve. These can be in private ownership or owned by the council. If the JNCC thinks a nature reserve is very important then it may be designated as a national nature reserve. Bye-laws are then established for each nature reserve which may, for example, control access and in other ways protect the species on the nature reserve. A few marine reserves have been established in this way.

b. Non-statutory nature reserves These are reserves which are run by non-statutory bodies such as the Woodland Trust, Wildlife Trusts, and the Royal Society for the Protection of Birds. Some of these non-statutory nature reserves are run by County Trusts which come together under the title 'the Royal Society of Wild life Trusts'. None of these management bodies can make bye-laws and they are therefore dependent on voluntary codes of conduct and national legislation to provide protection to the species within the nature reserve.

iii. National Parks, Areas of Outstanding Natural Beauty (AONB), and Sites of Special Scientific Interest (SSSI)

Land within these areas remains in the same, usually private, ownership as they were prior to designation. However, owners of the land and authorities such as the planning authority then have a statutory responsibility for these areas and may be limited as to what they are able to do.

iv. Ministry of Defence land, National Trust land

Some land other than statutory nature reserves, for example, land owned by the Duchy of Cornwall, the Windsor Estate, and the Malvern Hills, Worcestershire, may also be covered by bye-laws, etc., which may control access and provide additional protection for species within the area.

9.3.5. **Movement, import, and export**

There are several national statutory controls and EC Regulations that control the movement of animals, plants, and other organisms. Controlling potential pests and diseases requires restrictions on the movement of materials around the globe and suitable containment if work is carried out in the UK. For example, the Plant Health (Great Britain) Order 1993 prohibits the import, movement, and keeping of certain plants, plant pests, and other material, including soil. Some species are listed in the Wildlife and Countryside Act 1981 Part II of Schedule 9 which prohibits any person from releasing and allowing to escape into the wild certain plants such as the *Fallopia japonica* (Japanese knotweed). The Act also prohibits anyone from releasing or allowing to escape any wild animal that is not ordinarily resident or a common visitor to the UK. Some 'alien' animals have become established in the wild but it is not considered to be desirable to add to their number by allowing further escapes/introductions. These are included in Part I of Schedule 9 of the 1981 Act.

At an international level, the most substantive regulation in relation to rare and endangered species is the Convention on International Trade in Endangered Species (CITES) 1975 where 160 Member States have ratified controls regulating the movement of more than 2,500 animals and 25,000 plants. There are three levels of protection. Those species listed in Appendix I of the convention are those that may be threatened with extinction and where international trade is only allowed in exceptional circumstances and where the prohibition extends to dead or live individuals or derivatives such as ivory and furs. The trade in species in Appendix II is monitored through a licensing system and in the final category (Appendix III) are those species not threatened on a global level but protected under national legislation within Member States.

In addition to international and national legislation carriers within the UK, such as the Post Office, have regulations concerning what may be carried and the method and appropriate packaging required. These regulations extend to microorganisms and other biomaterials such as blood or tissue samples.

9.3.6. Permits and licences

If you wish to work on a species or habitat that is protected by legislation or wish to transport a protected species, etc., it may be possible to obtain a licence or permit that allows you to carry out otherwise proscribed activities on protected species or in protected areas. The bodies responsible for granting these permits or licences are usually listed within the Act of Parliament. Most commonly these are the Department for the Environment Food and Rural Affairs (Defra), English Nature, and the Countryside Council for Wales (CCW). For example, taking photographs of bats is considered to be unlawful disturbance and you therefore need a licence from English Nature or CCW to do so. More unusual arrangements exist for some species; for example, sturgeon and whales in British waters may only be taken under licence from the Crown.

9.4. Animal welfare

Much of the relevant legislation for working with animals in the wild has been considered in 9.3. However, for animal work that involves working with an animal in an institution such as a university or working with a domestic or captive animal there is additional legislation. At present there are two main Acts which are most pertinent to this: the Protection of Animals Act 1911 and the Animals (Scientific Procedures) Act 1986 (as amended). If you are working with agricultural animals there is considerably more legislation that is likely to be relevant, including the Welfare of Animals (Slaughter or Killing) Regulation 1995 and the Welfare of Farmed Animals Regulations (England 2000, Wales 2001). For more information about working with agricultural animals and for more details in general about animal welfare we refer you to Radford (2001).

9.4.1. Protection of Animals Act 1911

This Act is the current core of much animal protection legislation. The essence of the law is that it is an offence to be cruel to any captive animal which in this context includes any 'wild' bird, fish, and reptile confined or in captivity as well as domestic animals. The offence of 'cruelty' can be applied to a wide variety of circumstances and events and this has been the strength of this law. The notion of what is cruel has changed over time in response to greater scientific understanding and changes in social attitudes. Therefore, what may have been acceptable practice when the bill was first passed may now be considered to contravene the Act.

The definition of cruelty varies but relates primarily to the idea of causing unnecessary suffering. Cruelty in the Act is defined in terms of types of conduct including to:

- cruelly beat, kick, ill-treat, override, overload, torture, infuriate, or terrify any animal; or cause, procure or being the owner, permit any animal to be so used
- wantonly or unreasonably do or omit to do any act, causing unnecessary suffering to any animal; or cause, procure or being the owner permit any such act
- to convey or carry any animal in such a manner or position as to cause it any unnecessary suffering; or cause, procure, or, being the owner, permit any animal to be so conveyed or carried
- wilfully, without a reasonable cause or excuse, administer any poisonous or injurious drug or substance to any animal; or cause, procure, or, being the owner, permit such administration or wilfully, without any reasonable cause or excuse, cause any such substance to be taken by any animal
- subject any animal to any operation which is performed without due care or humanity; or cause, procure or being the owner, permit any animal to be subjected to such an operation
- without reasonable cause or excuse, abandon an animal in circumstances likely to cause it any unnecessary suffering; or cause, procure, or being the owner permit it to be abandoned.

Further legislation has introduced amendments and extensions to this Act; for example, the Welfare of Animals (Transport) Order 1997 relates to carrying animals in the course of trade and the Protection of Animals (Anaesthetics) Act 1954 adds legislation referring to anaesthetization of mammals during an operation.

9.4.2. Animals (Scientific Procedures) Act 1986 and Amendment, 1998

The Animals (Scientific Procedures) Act 1986 regulates scientific procedures which may cause pain, suffering, distress, or lasting harm to protected animals. Unlike the definition of animals in 9.3. and 9.4.1. in this instance a protected animal includes fetal, larval, and embryonic forms (within certain limits) of any living vertebrate (excluding humans) and the invertebrate *Octopus vulgaris*.

There are a number of procedures covered under these regulations including:

- any experimental or other scientific procedure applied to a protected animal which may have the effect of causing it pain, suffering, distress, or lasting harm
- anything done for the purpose of, or liable to result in, the birth or hatching of a protected animal if this may have the effect of causing it pain, suffering, distress, or lasting harm
- the administration of an anaesthetic or analgesic to a protected animal, or decerebration or any other such procedure, for the purposes of any experimental or other scientific procedure
- the humane killing by certain methods of a protected animal when this is for experimental or other scientific purposes and carried out in designated establishment
- the removal of blood or tissues from a live protected animal.

Some procedures are not included in the Act, such as those carried out as part of the normal veterinary, agricultural, or animal husbandry practices; ringing or other methods of tagging animals to allow their identification provided that this causes no more than momentary pain and distress; the humane killing of a protected animal by a method listed in Schedule 1 of the Act and the administration of materials to animals as part of a medicinal test in accordance with the Medicines Act 1968. However, in these circumstances the Protection of Animals Act 1911 still applies.

Under the terms of this Act a system of licensing enables such work to be carried out when the benefits that the work is likely to bring (to humans, other animals, or the environment) outweighs the pain or distress that the animals may experience. Other criteria that have to be met before a licence is granted are that there are no alternatives, the procedure uses the minimum number of animals, involves animals with the lowest degree of neurophysiological sensitivity, and causes the least pain, suffering, distress, or lasting harm. In addition, within the terms of the Act certain types of animal must also be obtained from designated breeding or supplying establishments.

The Act includes the arrangements for a three-level licensing system where those carrying out procedures must hold a personal licence, which will not be granted unless they are qualified and suitable. The programme of work must also be authorized in a project licence and the work must also normally take place at a designated-user establishment. However, in specific circumstances (such as field trials) work can be carried out elsewhere with the Home Secretary's authority. Clearly, there will be considerable variation between institutions in terms of which type of licence if any is already held. Therefore, if you are intending to work with such animals you will need to obtain advice about the existing licences for your institution from your supervisor.

9.4.3. Animal welfare

At present animal welfare is considered as a small element in a number of specific Acts, for example the Animal Health Act 1981, Zoo Licensing Act 1981, Welfare of Animals at Market Order 1990, and the Welfare of Farmed Animals Regulations England 2000, Wales 2001. Further consideration is given to animal welfare through the regulations relating to licensing all those who work with or handle animals. For example, points relating to animal welfare including quality of care, accommodation, environmental conditions, etc., come into the terms of granting a licence in the Animals (Scientific Procedure) Act (as amended) 1998.

There is currently no real legal definition of animal welfare. Best practise is outlined in guidelines produced by the Farm Animal Welfare Council (FAWC) and has been described as the five freedoms. Animals should have:

- freedom from thirst, hunger, and nutritional deprivation
- freedom from discomfort
- freedom from pain, injury, and disease
- freedom to express normal behaviour
- freedom from fear and distress.

Since there is currently no overarching legislation on this subject you may not be required to adhere to such good practice but can aspire to it.

9.5. Working with humans

After the atrocities of World War II the Nuremberg Code (1947) was developed and is the first internationally recognized code of human ethics. This code has informed policies relating to work on humans since this date and has been restated and extended in other international agreements such as the World Medical Association's Helsinki agreements of 1964 and 2000. The Nuremberg Code and subsequent agreements recognize the need during research to minimize harm; to find a balance between the risks to the individual and the benefits to the individual and to society; to recognize the importance of obtaining fully informed and voluntary consent from humans participating in research; and to ensure that the research has real validity.

Within England and Wales there is a considerable body of legislation that also impinges on studies involving humans, including the Obscene Publications Act 1964, the Sex Discrimination Act 1975, the Race Relations Act 1976 and the Race Relations (Amendment) Act 2000, the Disability Discrimination Act 1995, the Computer Misuse Act 1990, the Data Protection Act 1998, the Human Rights Act 1998 (as amended), the Freedom of Information Act 2000, health and safety legislation (9.2.), and the Special Educational Needs and Disability Act 2001. Apart from this legislation there are also extensive guidelines that come from professional bodies and codes of practice within your institution. It is impossible for us to cover in detail all this legislation and the guidelines, and therefore in this section we look instead at the general approach taken in research involving humans that is both good science and reflects appropriate practice. Different legislation covers the use of human tissues and we do not consider this here. If you are working in a laboratory using human tissues there should already be procedures in place which you will need to follow.

Most institutions have some form of ethics scrutiny which considers research involving human volunteers, so you are unlikely to be left making decisions about ethics in isolation. This is essential in research for several reasons. First, researchers themselves, especially if they are inexperienced, may not foresee circumstances where their subjects may be harmed. Second, the researchers have their own ethical and moral principles and these will influence the design of a project. Researchers may also have a vested interest in carrying out this particular research project. For most students this will relate to their wish to complete and be successful on a degree or higher degree programme but can also relate to career opportunities and obtaining further funding. Therefore, it is advisable to have an independent scrutiny system in place for all human-based studies. There can be some variation in terms of practice between different disciplines, so a subject-specialist should be a key part of such a review.

9.5.1. Harm, risks, and benefits

There are two basic principles behind an ethical basis for research: these are non-maleficence and beneficence. Non-maleficence is the principle of 'doing no harm' as a consequence of the research. This can be applied both to direct action such as exposing volunteers to a substance and more widely, for example, in the collection and use of confidential information. Therefore, as part of your research planning, you must seek to minimize any harm or potential risks to your volunteers. This may not always be easy as it can be difficult to know what another individual will be sensitive about. For example, someone may be sensitive about their weight, height, status, etc. This notion of 'do no harm' also covers those who may be affected by the results such as particular groups in society, other researchers, and the institution to which you are affiliated.

Beneficence is the requirement to 'do good'. Again, this is applied broadly across all aspects of working with humans. Both terms (non-maleficence and beneficence) are essential, although at first glance they appear to be two sides of the same coin. However, to 'do good' does not necessarily mean that we do no harm. For example, some studies using questionnaires may produce information that can be used to improve some aspect of human experience but there may be a potential for unknown and unquantifiable harm in terms of psychological risks to those you collect the information from especially where questions are personal. Collecting confidential information may also carry the risk of this information becoming accessible to inappropriate individuals. In medical science novel treatments or the use of a placebo may be 'harmful' to the patient but the understanding gained about a novel treatment may be beneficial to new patients. Clearly, a balance needs to be struck between the potential for doing harm and the benefits, and this balance needs to be in favour of the volunteer not the research.

For example, in the study of the effectiveness of tea-tree oil (Example 2.2.) the student took hand swabs from staff at a GP's practice. Clearly, if it became known that certain staff had 'dirty' hands then there may have been adverse consequences for those individuals. Therefore, the student carrying out the research assigned numbers to each sample. In this way neither she nor the managers of the practice could identify which sample belonged to which person. When the results were known it was clear that some individuals carried much higher bacterial loads on their hands than others and may therefore have posed an increased risk to patients. The practice managers were able to act on this information in a general way whilst the confidentiality of the volunteers was maintained.

9.5.2. **Consent**

Another strand in an ethical approach to research is autonomy. Autonomy is being able to make choices and decisions for oneself and by oneself. In practice this means both providing appropriate information about the research to allow volunteers to make an informed decision when giving or withholding their consent and exerting no pressure on a volunteer to take part in the research programme.

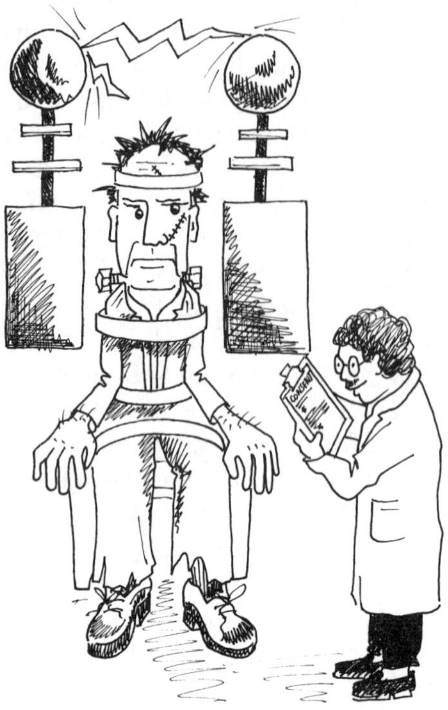

The information you provide to a potential participant should make it clear what you are asking the volunteer to do, why you wish to carry out this research, what you hope to gain from it, and what the person is being asked to do. There may be some risks, for example if you are examining human physiological responses. These need to be made explicit in the information you provide. You may need to exclude certain people from your research either on the grounds of health and safety or due to your experimental design. If you have exclusion criteria these need to be outlined and the reason for them explained. The name of the person who will carry out the work or is responsible for the project, and the location, may need to be included in the information you provide. Where external funding has been obtained it is also ethical to advise any potential volunteers as to who is funding the work. Some volunteers may not wish to take part, for example, if your work is funded by a tobacco or pharmaceutical company, etc. Information about what you are going to do with the data must also be included; for example, how will you ensure confidentiality, to whom will the results be made available, and in what format. Finally, you need to make it clear that the volunteer does not need to take part and may stop/leave at any time.

By asking your volunteers to sign a consent form or in some other way indicate their consent you are not passing your responsibilities for the well-being of your volunteers to them. You remain responsible for their health and safety and for the ethical conduct of your research.

Information should be provided in a written format that can be taken away and referred to by the volunteer at a later date if they wish. The information should be explained in language which the informant can easily understand. For most work involving humans you should obtain written consent. If you do have written consent forms you must consider where you keep this information as it is subject to the Data Protection Act 1998 and is central to your ability to preserve confidentiality.

To obtain honest informed consent you cannot directly or indirectly put pressure on (coerce) someone to take part in your research. Most coercion is negative; for example, there may be some sort of penalty for not taking part or not completing the research. But coercion can be quite subtle. For example, a department may expect all students to take part in each others' projects, or you may ask your family or friends to take part, or you may only provide the information about the project just before you intend to carry it out and so allow no time for the potential volunteer to reflect on the information. Coercion can be positive, for example, if you provide a reward. This is not only unethical but can lead to bias in sampling and you may not obtain the representative sample you need for your research (1.6.)

There are some circumstances where informed consent may not be obtained. These fall into two areas: either you are working with young people (under 16 years of age) or vulnerable people (for example mentally incapacitated adults or elderly people), or the nature of your investigation is such that you do not want to provide too much information in advance as it may influence the outcome. We consider the first of these in 9.5.3. For the latter you should first make certain that there is no alternative approach, that the benefits justify not asking for informed consent, that you will not mislead participants, and then if you cannot provide information before the study this should be done afterwards and the volunteers given the opportunity to withdraw retrospectively.

9.5.3. Special cases: children and vulnerable people

For young or vulnerable people, such as elderly people and adults with learning difficulties, there are additional codes of practice and legal requirements in place to protect them even if the work is indirect, for example, observing behaviour at playtime from outside the school premises. If a child is under 16 years old then you will normally be expected to obtain informed consent from parents, carers, or guardians. If you are working with children or vulnerable adults in a school or other similar setting then you should also obtain the consent of the school from an appropriate person such as the head teacher or chair of governors. In addition, when working with young or vulnerable people you will be required to obtain clearance from the Criminal Records Bureau (CRB). This disclosure scheme is designed to enhance public safety by providing criminal history information on individuals who wish to work with or in the course of their work may come into contact with vulnerable adults or children. This requirement for obtaining a disclosure includes undergraduates and graduates carrying out projects. The institution you are affiliated with should have a well-established procedure for obtaining a CRB disclosure.

Ethically there are difficulties working with vulnerable people and with whether it is acceptable to ask someone else to provide informed consent. It is, therefore, even more important to be sure that all elements of your project are ethical. In some areas of research such as within the Health Service this difficulty is recognized and it is therefore considered to be unethical to conduct research on children where there is no direct benefit to the individual child. If it is feasible, young and vulnerable people should also be asked for their consent having been given information in an appropriate format.

9.5.4. Equality

In Chapter 1 (1.6.) we discussed the notion of a representative sample. This is very pertinent when studying humans where it can involve additional planning and preparation to ensure that all potential volunteers can take part in a research project. In addition, some legislation (e.g. Race Relations Act 1976) makes it an offence to exclude certain groups of people through either practices or attitudes. The Human Rights Act 1998 prohibits discrimination on any grounds, such as gender, race, colour, language, religion, political or other opinion, national or social origin, association with a national minority, property, birth, or other status. The Disability Discrimination Act 1995 requires you to make reasonable adjustments to enable a disabled person to take part in your research. Making a 'reasonable adjustment' can involve the provision of materials that are more accessible by individuals with dyslexia or with restricted sight, providing an environment more suitable to some one with a physical disability, etc. We touch on some of these points in 2.3.4. Most higher education establishments have student support service staff who will be able to provide more specific advice on this matter.

Other rights enshrined within the Human Rights Act 1995 that need to be considered when designing and carrying out research to ensure equality in relation to both inclusion and communication of results are that individuals have a right to freedom of expression; a right to freedom of thought, conscience, and religion; a right to respect for private and family life; and a right to be treated with justice and fairness.

9.5.5. Anonymity, confidentiality, information storage, and dissemination

When working with humans you may wish to ensure that each volunteer's contribution is either anonymous or confidential. The rationale for doing so is that you may reduce potential harm to the volunteers, and participants may be more willing to give full cooperation if they will not be identified personally. To ensure anonymity no one, including the researcher, must know which data derive from which individual. This can be achieved, but it is difficult to ensure total anonymity where written informed consent is obtained or where personal information or visual records are required. Although anonymity in human research is usually the ideal, more often the research is confidential. Here only one researcher can identify the individuals within the research and when reporting the researcher endeavours to ensure anonymity. To protect confidentiality or anonymity data collection, storage, and dissemination has to be handled with care and consideration. Legislation that impinges on these activities includes the Human Rights Act 1995 and the Data Protection Act 1998. Under the terms of these Acts you must consider what data you collect and how, how the data will be stored, who will be able to access the data and through what media, and how long the data is to be stored. The overriding principles are those outlined in earlier sections including an individual's right to privacy, to not be misrepresented, and to not be harmed.

Confidentiality can be broken inadvertently either by storing information where names are linked to data and this stored data is accessed by a third person or by communicating information in reports, etc., with enough detail for certain individuals to be identified. For example, if you have a relatively small cohort, known to a third person, and you report the data from the only individual over 50 years of age, it will be easy to identify this individual. In undergraduate and graduate work you may therefore need to remove certain information from reports, such as a location, age, or gender, to ensure that individuals cannot be identified. For example, when the undergraduate in Example 2.2. reported on her findings relating to bacterial loads on the hands of staff in a GP's surgery she did not include information about the geographical location of the GP's surgery.

There may be some occasions when you may need to reveal a person's identity. This should not be done unless you have received the volunteer's written permission. You may also need to extend this to members of the public if they are included in photographs, video, or film. There is considerable sensitivity at present about recording and storing images of children. If your work requires this you must discuss it with your supervisor first and ensure that the information you provide to those giving informed consent makes it clear that this is what you intend doing.

9.6 Discussion topics

Consider the following scenarios. What issues arise from these research proposals in relation to the law?

Q3 An undergraduate keeps stickleback at home. She wishes to use them in her honours year project and investigate their behaviour when presented with different food sources in a number of different environments.

Q4 An undergraduate wishes to compare the effectiveness of two teaching media (book versus computer) in a primary school. The class will be divided and half will cover a topic from the national science curriculum using a book resource, while the other half will use a computer-based resource. The children will be tested on the topic before and after.

Q5 An undergraduate wishes to test the effect of music on cats in a cattery. Individual cats are observed during times when music is playing and when it is not. Several types of music are tested.

Q6 A student wishes to follow the colonization of a new pond in a primary school in Worcestershire where the great crested newt is relatively common.

Q7 To investigate the bactericidal properties of *Allium* species an undergraduate grew a laboratory strain of *E. coli* on a solid agar media and added small pieces of bulbs from a number of different *Allium* species to the plates.

Q8 A student wishes to compare the diet of elderly patients in hospital with their normal diet at home.

The answers to these discussion topics are given at the end of this chapter.

Summary of Chapter 9

- The aim of this chapter is to provide a necessarily brief overview of the nature of law and regulation in the UK (9.1.).
- The chapter provides an introduction to the law in relation to four key areas; health and safety (9.2.), access and working with wildlife (9.3.), working with animals in captivity (9.4.), and working with humans (9.5.).
- The information in this chapter should help you to carry out a risk assessment and to carry out your research safely (9.2.).
- You are encouraged to be aware of your responsibilities in relation to research involving wild species and be able to extend and update your knowledge of this body of law (9.3.).
- You are encouraged to be aware of your responsibilities in relation to research involving captive animals and to be able to extend and update your knowledge of this body of law (9.4.).
- You are introduced to some of the legislation relating to working with humans. We consider ethical approaches to research and the ideas of non-maleficence and beneficence. These are discussed in the contexts of consent, children and vulnerable people, equality, and data storage and handling (9.5.).
- A number of examples from undergraduate honours project proposals are included in the chapter for you to discuss and to challenge your understanding of the topics covered in this chapter (Q3–7).
- The Online Resource Centre includes interactive exercises that test your understanding of this chapter with other topics, particularly those considered in Chapters 2–8.

Although every effort has been made to ensure that the information in this chapter is accurate, it should not be taken as a definitive statement of the law, nor can responsibility be accepted for any errors or omissions.

Answers to Q3–8

A3 The two main areas relevant to this proposal are first, that the student will need to complete a health and safety assessment. Second, the work will be subject to the laws outlined in the Protection of Animals Act 1911 and the Animals (Scientific Procedures) Act 1986 (and amendment 1998).

A4 All the topics covered in 9.5. are pertinent. The student would have to obtain a CRB disclosure. Decisions about consent would need to be made. To reduce harm to the children the student would have to consider how to protect confidentiality or anonymity, how to ensure that no child's learning was affected, etc. To protect the school from harm the student would have to consider how the school's anonymity is protected. The student would have to ensure that the 'harm' to the children is outweighed by the benefit. The student should also carry out a risk assessment. Risks might include picking up infections, head lice, etc.

A5 A health and safety assessment would need to be carried out by the student. The student would need to comply with the requirements under the licence given to the cattery. Permission would need to be sought from the cats' owners. The student would be subject to the Protection of Animals Act 1911.

A6 This project has three elements. First, the student will need to carry out a health and safety assessment particularly in relation to water-borne pathogens. Second, the student will need permission from the school to ensure (s)he is not trespassing. She will need a CRB disclosure as (s)he is working in an area where children may be present. The great crested newt is a European Protected Species. The student may continue to sample from the pond until such time as (s)he has confirmed that great crested newts have colonized the pond. Should this happen (s)he will need to apply for a licence under the terms of the Wildlife and Countryside Act 1981 from English Nature or the Countryside Council for Wales.

A7 One *Allium* is currently listed on Schedule 8 for the Wildlife and Countryside Act and the student should not therefore be using this species in the research. In addition, the student must carry out a health and safety assessment.

A8 The student must consider the topics and review the legislation introduced in 9.5. such as obtaining consent, ethics, and data handling. The student would probably require CRB clearance. In addition, the student must complete a risk assessment for themselves both in relation to working in a hospital and in relation to visiting other people's homes (9.1.).

Reporting your research

There is little point in carrying out research unless your findings are communicated. The most common way to do this is to write a report or paper. Generally publications of practical research in science adhere to a specific format. This is the structure also followed by most bioscience departments most of the time for their science reports. In this section we include a guide to the general format common in scientific reports. The format outlined in this chapter should be used in conjunction with local guidelines, either those relating to the journal you wish to publish your article in or your course or department requirements.

When learning how to write a scientific report or improving your skills it is best if you have a journal article to look at. Try to have one from your subject area to hand, either as hardcopy or online, as we will use it to illustrate our points. We include some links to online journals in our

Online Resource Centre. It will take you about 4 hours to complete these exercises.

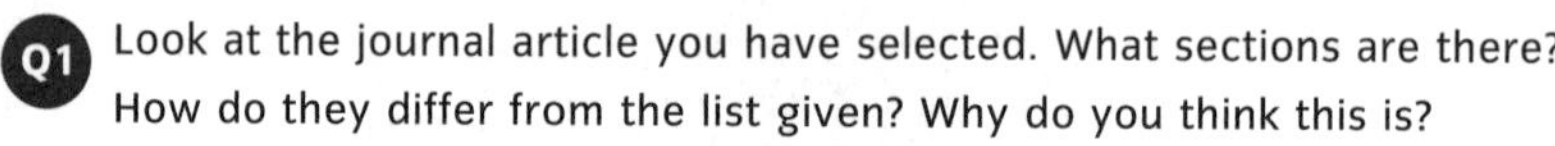

10.1. General format

In general there are nine sections to a full journal article: Title; Abstract; Keywords; Acknowledgements; Introduction; Method; Results; Discussion; and References. Of these the abstract, keywords, and acknowledgements are invariably not required for most undergraduate practical reports. When writing an honours project report or thesis you may be required to include an abstract, and most people chose to include acknowledgements. In addition, in unpublished reports such as honours projects and theses you may include an appendix for additional material such as the names of suppliers of chemicals, details about how stock solutions were made up, etc.

Q1 Look at the journal article you have selected. What sections are there? How do they differ from the list given? Why do you think this is?

Your whole report should usually be written in the third person, past tense. The 'third person' means that you are not referring to anyone in particular so you do not use 'I', 'you', 'we', etc. The 'past tense' means that you write about what was done and not what is being done or will be done. Have a look at the Introduction and Method in the article you are examining and note the tense that is used.

Q2 Rephrase the following sentences so that they are in the third person, past tense.

a. I added 5 mg of sodium chloride to water and stirred until it was dissolved.

b. In my reading I found that ice is less dense than water.

A2

a. 5 mg of sodium chloride was added to water and stirred until it was dissolved.

b. It is clear from the literature that ice is less dense than water.

Your report must be referenced correctly throughout. This is very important. Failure to do so is called plagiarism and usually carries severe penalties. We discuss plagiarism under 'cheating' (10.6. Common errors) and referencing systems are discussed in 10.10.

In larger studies you may have carried out more than one investigation. Some small elements may be redundant to the main thrust of your report. If this is the case then leave them out or include this detail in an appendix. If your research extends over several months or years you may tend to write about your work in the historical order in which it was carried out. This chronology may not be relevant to the actual outcome from the research and can be detrimental to the structure of your report or paper.

10.2. Title

This is the first thing a reader looks at and it should capture their attention. After all your hard work you want people to read about your investigation. The title should be concise and should relate to the aim and conclusion. If you are publishing in a journal your title may be part of the information made available for electronic searches so you need to consider who you want to read your article and include the terms they are likely to use in an internet search.

Common errors

Too long It is not necessary to write 'An investigation into . . . '. Editors of journals like to save ink and will not wish to include unnecessary words. Your title will be more forceful if you go right to the subject under investigation. In undergraduate reports the same is true. Your title will be much more dynamic if written in this way.

Too short You are not writing tabloid headlines. Your title does need to be complete and informative.

Q3 Read the title of the journal article you have. Does it do everything we recommend? If not, how can it be improved?

10.3. Abstract

An abstract is defined in the *Concise Oxford Dictionary* as a 'brief statement of content'. Most journal publishers do not want the 'brief statement' to be an extract copied from part of the report, so you have to write an abstract, not cut and paste it. This is also usually the case for undergraduate work where you are asked to write an abstract.

The format for an abstract varies between publications. In some journals the key findings are given as a list; in others the abstract is written as prose with sentences organized into paragraphs. Whichever format is required you need to include enough background information so that your aim, which you also include, can be understood. This is followed by a summary of your data, usually with numerical details, and your conclusion. There needs to be enough detail in the abstract for your conclusion to be understood. You should not include any new information in the abstract that is not elsewhere in your report.

Common errors

The most common error in undergraduate reports where an abstract is required is insufficient detail. This can include forgetting to include the aim or the conclusion. The conclusion also forms part of the discussion and is often incorrectly written (10.9.). Clearly these errors can then carry through to the abstract.

Do not look at the abstract of the journal article you are examining. Instead, read the paper and write your own abstract. How does it compare with the author's version? What are the differences? Which is the better version and why?

10.4. **Keywords**

Keywords are given as a short list of single words or linked words but not long phrases. For example, linked words might be 'resource protection' or 'agri-environment schemes'. The keywords you select from your paper will depend in part on who you are writing for and how the keywords will be used. Most often keywords are only required when publishing a paper and there will be guidelines concerning them in the journal's guide to authors. In this context the keywords are used by a reader to quickly see if the article is of interest to them and by indexing systems. Indexes of keywords are used by the publishers of your paper to draw attention to your paper. They are also used by bibliographic journals that provide a quick method of locating articles of interest across a wide number of journals. Increasingly the use of internet searching facilities means that keywords are becoming less important as often the whole paper is subjected to a search for a specific term.

In the same way as the title, the keywords are your way of flagging up your work for a particular audience. So you need to ensure that the terms they are likely to use when carrying out a literature search are included in your list. To achieve this you should consider using one or two broad terms and one or two specific terms. For example, if you had carried out an investigation into the levels of pesticide residue on organic and non-organic fruit you might include 'organic' as one broad term and the name of one of the pesticides as a specific term. Keywords may include terms not in the title.

How many keywords are there in the paper you are examining?
How many of these terms are broad and how many are specific?
Do all these terms appear in the title?

10.5. Acknowledgements

Many people feel that they would like to include acknowledgements at the completion of a significant piece of work. Those who may be acknowledged fall into three categories: the funding body, those who provided professional assistance, and those who provided personal support. It is usually a requirement as well as a courtesy to include an acknowledgement to any organization or person who has provided you with funding or support in kind (e.g. seed supplies, an item of equipment) that has enabled you to carry out the research you are now reporting. You may also have received professional assistance, such as being given permission to carry out your work in a particular location; help with technical aspects of the research; help in identification of species, etc. Again it is important that the input of these people is acknowledged. In journal articles it would be unusual to find personal acknowledgements, although in honours project reports and theses it is a common practice to thank friends and/or family for their support. This is personal to you, and most universities have regulations that allow the inclusion of these acknowledgements. It is not appropriate to make these acknowledgements too personal in either a positive or negative way, nor should your acknowledgements include tenuous supporters such as drinking friends and distant cousins!

10.6. Introduction

An Introduction is the part of a report that sets the scene and leads up to a statement of your aim.

Q6 Have a look at the format of the Introduction in your journal article. How many paragraphs are there? How does this compare with the discussion? Where is the aim? Are there any figures or tables? What information is contained in brackets throughout the Introduction?

This quick look at your journal article should have drawn your attention to several features of an Introduction: it is written as prose with fully referenced sentences and paragraphs; there are not usually any figures or tables and the aim is at the end.

Writing an Introduction is like telling a joke: you must not give the punchline first or no one will understand it. The punchline in an Introduction is the aim. It will be uppermost in your mind so it is tempting to place it first but if you do this your reader, when reading the aim, will not understand the reasons for the investigation or fully understand the aim itself. Writing the aim first is a format often used in schools but you should now use the format current in the scientific literature instead. The aim therefore is something you lead up to and should be in the last paragraph at the end of the Introduction.

Your Introduction should cover enough background information for a reader to understand the aim. If you are not sure how to achieve this, first look at your aim and note down critical words. Your Introduction should include details about all these words. Second, complete your literature review as notes. Be guided by the words you have identified in your aim and draw a mind map or flowchart which develops step by step the background to the research, linking all the parts together and ending in the statement of your aim. Use this flowchart to guide your writing.

Common errors

The Introduction reads like an essay An Introduction is not an essay. It performs a very specific and different function. Even if you are asked to include an extensive literature review, it should normally be written so that each paragraph takes you one step closer to being able to state your aim. One feature that tends to be associated with 'essay' introductions is their length. If in doubt, count how many paragraphs you have in the Introduction compared with the Discussion. If the Introduction is considerably longer than the Discussion then you have probably made this error.

The Introduction lacks detail This is a feature that has become more common in recent years with the advent of electronic literature searching facilities. Many publishers make abstracts from journal articles available online and there is a temptation to rely on these as the prime source of information when reading about a topic. Abstracts do not provide enough detail for anyone to be able to really appreciate the value or rigour of the research or the detailed arguments presented within the paper. A reliance on abstracts carries this lack of critical detail through to your report, leading to a weak Introduction (and Discussion).

The Introduction is aimless It is surprising how often students forget to include their aim in their Introduction. This leaves the reader in utmost confusion, usually for the remainder of their time reading the report. Given report formats, it is extremely difficult to fully comprehend what the aim of the research is unless you explicitly state it.

The Introduction is not focused on the actual piece of research that has been carried out This usually occurs because you are using sources of information that are looking at a slightly different topic and you can get swept along. Sometimes you may find a lot of information on a small specific part of your study and so you tend to note down everything you have found about that topic at the expense of the areas where information is harder to find. You can avoid this by using the flow diagram and planning your Introduction before you start writing. For example, one of our undergraduates investigated the opinions of students in relation to over-the-counter genetic testing. She might have written an introduction all about genetic testing and never touched on 'over-the-counter' or how people develop their opinions. All of these topics should be covered in the Introduction, but since information is easiest to find on genetic tests it would not be surprising if the student was led astray. (We are pleased to say that she wasn't!)

You use quotations In science we are rarely interested in how someone else has phrased a sentence; we are interested in the facts within the sentence. Therefore, unlike other subject areas you will not see quotations being used. When you are reading around the subject you should extract the facts in note form only. You then use these succinct notes to write your own Introduction. This makes for a much more readable piece of work. It also avoids inadvertent plagiarism (see Common errors – cheating). The exception to the use of quotes tends to be if you are defining a word, or are taking a figure or table from another source.

Look at your journal article. How many quotations are included in the Introduction?

Expectations It appears to be coming increasingly common for students to declare what they are expecting as an outcome from their investigation in their Introduction. (We use the term 'expectation' here as it might be used in general spoken English. Elsewhere we use the term in a statistical way (Chapter 5).) This is a really bad idea for two reasons. First, you should approach any investigation with an enquiring and open mind. If you start out with an expectation than you are likely to bias your design, your execution of the investigation, and your evaluation of the data. This can lead to either a flawed piece of research and/or completely incorrect conclusions. Second, if you include your expectation in your report your readers will know that you have not come to this with an enquiring and open mind and be prejudiced against your work as it is more likely to be biased.

More about the 'punchline too early' phenomenon Another error we come across is where detailed elements usually relating to the Method appear in the Introduction. In your Introduction you need to keep yourself aloof from what you actually did. The nearest you get to referring to what you did is the aim. The Introduction is for background information only.

Cheating (plagiarism) The computer age has brought easier access to sources of information through electronic searching and the 'cut and paste' facility. It has also tempted some undergraduates to cheat, either by downloading whole reports or essays or by cutting and pasting paragraphs from articles. Assessments, including reports of investigations, are supposed to be your own work, which means that they should be your own ideas and your own words. If the ideas or words are not your own you should indicate the source by referencing them. Failure to do this is called plagiarism. Not only will you not learn anything by cheating but you will also be subject to your Institution's or the journal's penalties. Scientists make their living by the facts they discover and the communication of these facts. This is why you must always acknowledge the sources of your information (10.10.) and this is why there are severe penalties if you do not.

If you are still tempted, please bear in mind that if you can find these articles so can others and it is surprisingly easy to spot the use of essay bank work and content that has been cut and pasted from another source.

10.7. Method

The Method outlines the procedure(s) you have undertaken during your investigation. Methods can vary in their content and you should check your local regulations or guide to authors to confirm the content. A Method can include a site description, the method, and details about the statistical analyses used.

10.7.1. **Site**

This section is included where the site is an important part of the research and its evaluation. You may either have carried out your research at a particular site or sampled from a particular site, and there is a possibility that the site has given your sample or study unique characteristics. The site description is written in such a way that your site can be relocated and the features of the site that impinge on your investigation need to be reported.

For example, an investigation was carried out into the percentage cover of five plant species growing on and/or off anthills in grassland. In this study the site is important because the surrounding habitat and its management will affect the flora found in the locality. Therefore, you need to include a description of the area and its geology, typical plant species, and details about the management, e.g. grazing, presence of rabbits, cutting regimen, spraying, etc.

The location of your site is given in terms of Ordnance Survey (OS) coordinates. These are found using an OS map of the area. The complete coordinates consist of two letters and six numbers, e.g. the University of Worcester is at SO835556. The letters are printed on the map itself and are large and pale blue, e.g. Worcester is in the 'SO' compartment. Some OS maps may include details from more than one compartment, each of which has the two blue letters printed within the relevant area on the map. The six numbers comprise two groups of three numbers. The first three numbers come from the *x*-axis (west to east) and the second set of three numbers refer to the *y*-axis (north to south) (Fig. 10.1.). Along the *x*-axis of the map there are a series of numbers. Locate your site in relation to these numbers first. You then subdivide the single grid square into 10, and this gives the third value for the *x*-axis. Repeat this for the *y*-axis. The full OS coordinates are the two letters the three *x*-axis numbers and the three *y*-axis numbers.

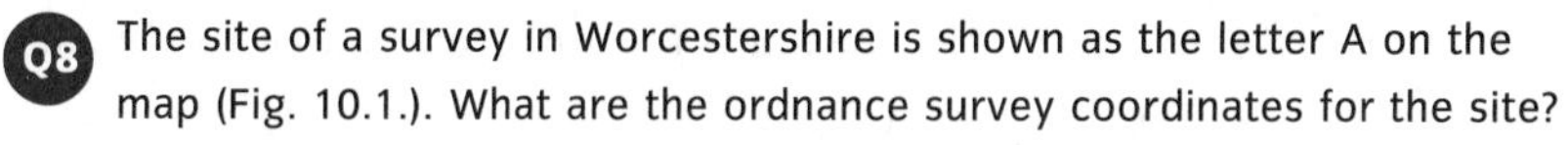
Q8 The site of a survey in Worcestershire is shown as the letter A on the map (Fig. 10.1.). What are the ordnance survey coordinates for the site?

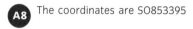

A8 The coordinates are SO853395

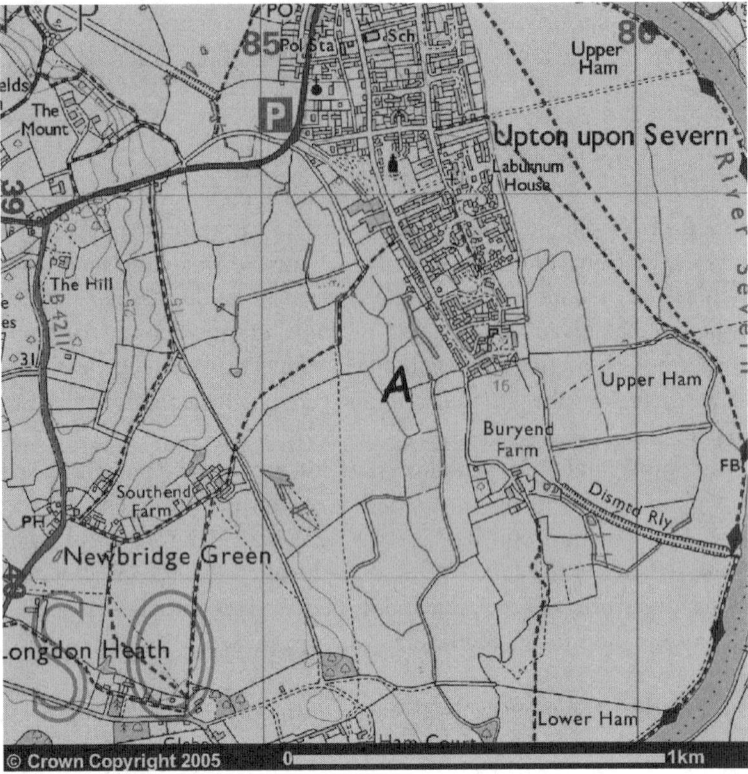

Fig. 10.1. Working out Ordnance Survey coordinates.

10.7.2. Method

A method describes the processes you carried out to investigate your aim. There needs to be enough detail included so that someone else can repeat your work. There is some variation in how much detail is required. For example, in undergraduate work you do not usually have to indicate the name of the company that supplied your chemicals but you may explain how you made up stock solutions which were then used to make other solutions. In journal articles the reverse tends to be the case. You should include information about the equipment used at each step in your protocol and if the equipment has critical characteristics you should name these and the company that supplied the equipment. Similarly some substances (e.g. compost) vary from one supplier to another and in these instances details about the supplier and the full trade name of the substance should be given.

As this information about materials is an essential part of a Method, this section can be titled Materials and Methods. However, the materials detail is integrated into the method and is not given as a separate section in the method.

Q9 Look at the Method in the article you have been examining and imagine yourself repeating the experiment. Is there enough detail included for you to do this? If not what more is needed?

The Method, as you will have noticed from examining the journal article (Q9) is written as paragraphs of prose usually in the third person, past tense. Often in school you are required to write a method as a list of steps and often with a separate section for materials. For undergraduate work this format is no longer appropriate.

Some investigations are time-sensitive. Most commonly this is where you have collected samples from the field or carried out work outside the laboratory. If the outcome from your Method is likely to be influenced by the seasons, you should include the date in your Method. Some investigations are influenced by the time of day, for example, the behaviour of animals. In this instance you need to report the time at which you made your observations. Time and the effect of seasonal changes can be non-treatment variables and should be considered when planning your investigation (2.5.).

10.7.3. **Statistical analyses**

In some Method sections where experiments have been carried out you will also see a brief outline of the statistical tests used to test the hypotheses. This needs to be no more than a brief statement of the test used and a reference indicating the resource used to help you learn how to carry out this test. For example, in Chapter 2 we discussed how the effect of wind speed on seed dispersal might be investigated and we explained which test would be appropriate (2.2.8.). When reporting this in a Method you would state that the distance travelled by the seed in relation to the wind speed was 'analysed using a Kruskal–Wallis test (Holmes *et al.*, 2006)'. For more details about referencing see 10.10.

Common errors

Too little or too much information Methods most often suffer from too little or too much information. Like Goldilocks you need to get it 'just right'. To check if you have included too little information, give your method to a friend and ask them to tell you how the investigation was carried out. This will quickly identify the gaps. Too much information can arise when you are using the 'materials and methods' format from school or you include unnecessary statements, e.g. 'the data were collected and summarized'. If you have developed a method during the research process and so in each use there are slight differences you do not have to report the whole method each time. Reporting the full method once and indicating the subsequent revisions will provide sufficient detail for your work to be understood.

It is difficult to find the correct balance when writing a Method but you can learn to do this by checking with your lecturers/supervisor and learning from the feedback you have received about earlier reports.

Symbols Since most reports are now word-processed, errors relating to the use of symbols have become common. Most versions of Microsoft Word have the following facilities that enable you to write symbols correctly:

- *Subscripts* (e.g. H_0) and *Superscripts* (e.g. m^2) are generally located under Format–Font
- *Symbols* can usually be found either in Insert–Symbol (e.g. μ, σ, Σ) or as part of Equation Editor at Insert–Object–Microsoft Equation (e.g.$\Sigma, \overline{x}$)

When using symbols be consistent in their use in terms of the font you choose and other text characteristics such as italics.

Cheating If you are provided with a protocol for a Method this does not mean either that it is written in the format appropriate for a report that you necessarily have permission to report it verbatim in your report, nor that you necessarily followed it exactly without modification. Methods that you have been given need to be considered with as much care when you are preparing a report as a Method you have developed yourself. Failure to acknowledge the source constitutes plagiarism; failure to follow the appropriate format or provide accurate details indicates learner incompetence.

Read through the method in your article. Does it follow our guidelines? If not, in what way does it differ? Why do you think this is? Check your Department's or the journal's guide to authors. What do they say about a Method? How does it compare with the information in this chapter?

10.8. Results

The whole point of a Results section is that you draw to your reader's attention the main trends you have found in your data. You achieve this by first organizing your data in tables (10.8.1.) or figures (10.8.2.) and/or using summary statistics (3.3.–3.5.). By taking an overview of your data you are more likely to be able to identify the apparent trends in your data. Second, if you have hypotheses to test (Chapter 4), carry out the statistical analysis and draw conclusions from this (10.8.3.). Finally, look at the raw data to see if any specific further points can be made. This information is what needs to be communicated in your Results section. It is not easy to write a Results section: it requires practice. Like the Introduction it helps if you plan it out first.

Q11 Have a look at the Results section in your article. How many paragraphs of text are there? How many figures and how many tables?

A scientific report is written as words. This seems to be stating the obvious but it is amazing how many people forget to write any words in their Results section. Instead they include a series of tables and/or figures and leave the reader to work out what is going on for themselves. To write a Results section look at the information you have about the trends in your data and prepare a flowchart/mind map that presents all the points you want to make in a coherent way. Then write your Results section using the flowchart. Only then should you consider whether to use any figures or tables. The role of figures and/or tables in a Results section is a supporting one only. They allow you to include detail and to illustrate your trends. To review good practice in relation to figures and tables, see 10.8.1. and 10.8.2.

Common errors

'Therefore' It can often feel very awkward when writing a Results section to limit yourself to highlighting the trends seen in the data without referring to possible explanations for these results. If you find yourself including 'because' or 'therefore' this is probably what you are doing. The Discussion section is the place for all these explanations. The reason there is a need to have separate sections is that explanations for the trends in your data are rarely straightforward and conclusive. If you embarked on lengthy discussions in the middle of presenting your data the reader would lose any overall sense of your findings. It is better to keep all discussion separate from the presentation of your results.

Repetition Many students suffer from over-enthusiasm for tables and figures. The tendency, which again appears to come from early training in school, is to include a table of your data and then represent this with a figure. Such repetition shows that you have not thought through how best to support the points in your text. It is very rare indeed that a table and figure are needed to illustrate the same data. The exception to this would be a desire to demonstrate the overall trend using a figure but a need to then identify details in the data, which is best done using a table.

10.8.1. Tables

A long string of numbers does not help you to understand your raw data or help you to explain it to someone else (e.g. Example 2.4.). Therefore, it is common practice to organize the raw data in a summary table. There are a number of different summary tables depending on the type of data you have (3.1.). For discrete data (nominal or ordinal) the categories reflect those in which the observations naturally fall (e.g. Example 5.3.). For continuous data (interval) you may construct a frequency table where you select the categories within certain constraints. These divisions are often artificial in that they have no scientific significance or meaning. The categories have to clearly and fairly demonstrate the trend you wish to illustrate, abut but not overlap, and should normally all be the same size (e.g. Table 5.1.). You then record the number of observations that fall within each category.

As you evaluate your data you may also construct a calculation table; we have seen many examples of these in Chapters 5, 6, 7, and 8 (e.g. Tables 7.4. and 7.5.). In scientific papers or reports you may include a summary table and occasionally a calculation table. Which table will depend on the points you have made in your Results section. You do not, however, include the calculation itself.

When writing for a particular journal or preparing other reports there may be local requirements that you need to follow when producing a Table. For example, it is usual when reporting summary statistics to include the mean (3.7.), the standard error of the mean (3.7.), and n. In doing so you have provided sufficient detail for a reader if they wished to be able to calculate confidence intervals for your data.

A tables should always:

- *Have a title* This should clearly indicate what the data are. For most publishers the title should have enough detail so that the table can be taken out of context and still be understood. In most scientific journals the title for a table is placed above the table. The title may be followed by additional details about the contents of the table.

- *Be numbered* Numbering tables makes it much easier for you to explain which table you are referring to. Table numbers should be sequential.
- *Be labelled (rows and columns)* Most tables are organized into rows and columns. It is important to make it clear what information each row and each column contains.
- *Indicate the units of measurement* If a row or column includes data, you should indicate the units of measurement e.g. mm, °C, etc., in the column and/or row headings. Units should also be included in the table title (if they apply to the entire table) or in column or row headings. Having indicated the units in the title or column and row heading you do not then include units in the central body of the table.
- *Distinguish between zero and missing data* Sometimes investigations do not go according to plan and you have 'missing data'. You may then need to distinguish between a zero measurement and missing data (Table 10.1.). If you are going to analyse your data using computer software this is also very important. If you enter a 0 for missing data on a spreadsheet the software usually assumes this is a real value.

EXAMPLE 10.1. The response of tobacco explants to auxin

In an undergraduate investigation into the effect of auxin on the growth of tobacco in tissue culture the relative increase in diameter (mm) of leaf explants was recorded after 2 weeks. Some of the samples e.g. explant 1, auxin 1, showed no increase after 2 weeks. Two of the samples became contaminated and died. These missing data are indicated by a dash (–) (Table 10.1.).

Table 10.1. The relative increase in diameter of explants of tobacco after treatment with auxins

Explant	Relative increase in diameter (mm) after 2-weeks tissue culture	
	Auxin 1	Auxin 2
1	0	1
2	2	0
3	—	1
4	1	—

- *Be simple* If the table is complicated, break it down into several different tables.
- *Use appropriate classes in a frequency table* Where data are measured on a continuous scale 'classes' must be used. Two common errors relating to the use of frequency tables are choosing the wrong number of size classes and choosing size classes that overlap.

In Example 2.4. we described some research where the natural variation in growth of rye seedlings was examined and the researcher recorded heights (mm) after 3 weeks for 15 seedlings. In her work book the researcher recorded the following: 12.0, 12.0, 11.5, 18.0, 14.0, 11.0, 14.5, 11.5, 10.0, 10.0, 19.5, 19.0, 21.0, 15.5, 14.5.

This is continuous data and therefore best summarized in a frequency table with size classes of your choosing. We illustrate three possible sets of size classes for this data so that you can see which class sizes best illustrate the trend in the data.

It is also a common failing to choose size classes that overlap. If you were planning a frequency table for the rye data (Example 2.4.) you may have chosen classes 10.5–11.0 m, 11.0–11.5 mm, etc. With these classes there is an overlap. If you had a value of 11.0 mm, which class would you place it in? The classes in Table 10.4. abut but do not overlap and are therefore correct.

Q12 Examine the tables in your paper. Do they follow all these guidelines? Find a table you have prepared. Are there any improvements you could make?

10.8.2. Figures

If you believe that a figure best illustrates the points you are making in your Results section then you need to decide which is the correct figure to use. The five most common types of figures for data are a **scatter plot, pie chart, bar chart, line graph,** and **histogram.** For a review of more figure types see Crothers (1981) and Hawkins (2005). The type of figure you use depends on the point you are making, as some styles will illustrate

Table 10.2. Frequency distribution of leaf length in rye seedlings after 3 weeks growth

	Size classes											
Length of leaf (mm) of rye seedlings	10.0–10.4	10.5–10.9	11.0–11.4	11.5–11.9	12.0–12.4	12.5–12.9	13.0–13.4	13.5–13.9	14.0–14.4	14.5–14.9	15.0–15.4	15.5–15.9
Number of individuals	2	0	1	2	2	0	0	0	1	2	0	1
Length of leaf (mm) of rye seedlings	16.0–16.4	16.5–16.9	17.0–17.4	17.5–17.9	18.0–18.4	18.5–18.9	19.0–19.4	19.5–19.9	20.0–20.4	20.5–20.9	21.0–21.4	
Number of individuals	0	0	0	0	1	0	1	1	0	0	1	

Too many size classes. It is not possible to see any overall trend in this data.

Table 10.3. Frequency distribution of leaf length in rye seedlings (mm) after 3 weeks growth

	Size classes	
Length of leaf of rye seedling	10.0–19.9	20.0–29.9
Number of individuals	12	1

Too few size classes. This summary does not allow you to see any detail.

Table 10.4. Frequency distribution of leaf length in rye seedlings after 3 weeks growth

	The frequency distribution of leaf length in rye seedlings (mm)					
Length of leaf of rye seedling	10.0–11.9	12.0–13.9	14.0–15.9	16.0–17.9	18.0–19.9	20.0–21.9
Number of individuals	5	2	4	0	3	2

Just about right. This summary table allows you to see some detail (compared with Table 10.3.) whilst allowing the trend to be evident (which it was not in Table 10.1.).

the points you are interested in more clearly than others. Choosing the correct format for a figure also depends on the type of data (3.1.) you have. Table 10.5. gives a guide to choosing the correct figure for certain types of data.

Table 10.5. Guide to the use of figures

Type of data	Pie chart	Bar chart	Line graph	Histogram	Scatter plot
Nominal	✓	✓			
Ordinal	✓	✓			
Interval, one **variable**. Few points within the variable are observed. The variable is usually under the control of the investigator.			✓		
Interval, one variable. Many points within the variable are observed. The variable is not usually under the control of the investigator.				✓	
Interval, two variables					✓

EXAMPLE 10.2. The behaviour of captive orang-utans

An undergraduate examined the behaviour of orang-utans at a zoo in the UK. She categorized the various behaviours, e.g. grooming, playing, solitary, eating, and showing aggression, and recorded the time spent on each.

If the data from this study were to be reported and the key trends illustrated using a figure these data would best be presented as a pie chart (Fig. 10.2.) or bar chart (Fig. 10.3.) because there is one variable (behaviour), the data collected are nominal, each type of behaviour is a **discrete** category independent of the others, and the types of behaviour cannot be arranged in any particular order.

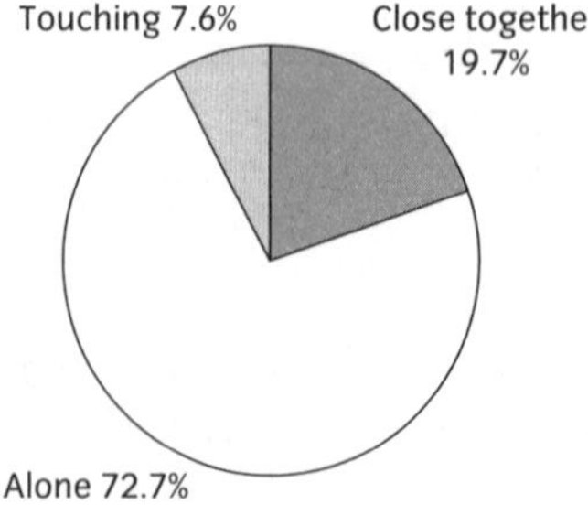

Fig. 10.2. Relative number of occasions (%) that particular behaviours were shown by one female orang-utan during a 4-hour observation period.

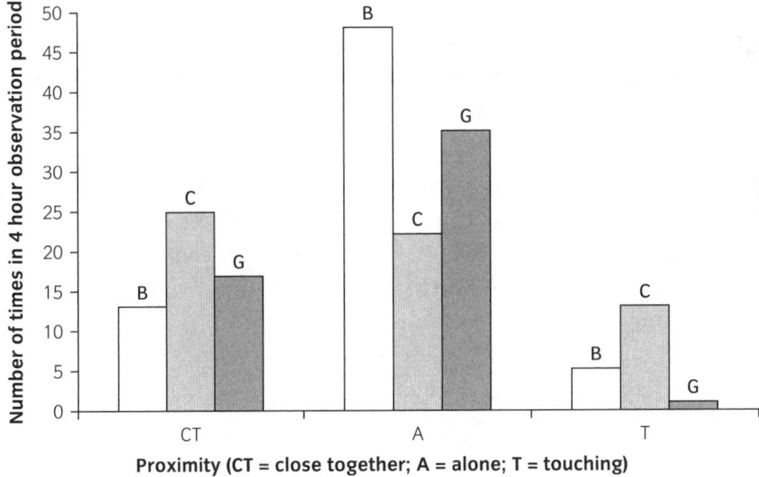

Fig. 10.3. Number of times particular behaviours were shown by three orang-utans (B, C, and G) during a 4-hour observation period. CT, close together; A, alone; T, touching.

EXAMPLE 10.3. The effect of petrol on the growth of *Lolium perenne*

An undergraduate investigated the effect of petrol on the relative increase in leaf length of a grass (*Lolium perenne*) over a 6-week period. Three concentrations were examined.

In this example, there is one variable (concentration of petrol). Concentration is a **continuous** scale of measurement. However, only a few (three) discrete points along this scale have been examined. The variable (concentration of petrol) was determined by the investigator. Therefore, the results are best shown as a line graph (Fig. 10.4).

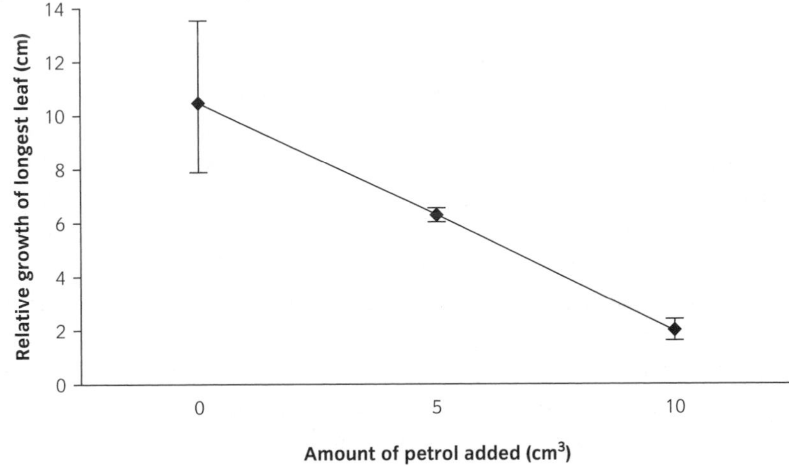

Fig. 10.4. Relative growth of longest leaf (cm) of *Lolium perenne* exposed to petrol in the soil.

Q13 Represent the data relating to the growth of rye seedlings (Example 2.4.) by an appropriate figure. You may wish to repeat this exercise in Excel or SPSS. How does the output compare?

A13 An interval scale of measurement (length of leaves (mm)) has been used but the points along this scale of measurement have not been determined by the investigator and are scattered. To impose some order on the representation of this data a histogram with classes (Table 10.4.) is more useful than a line graph (Fig. 10.5.).

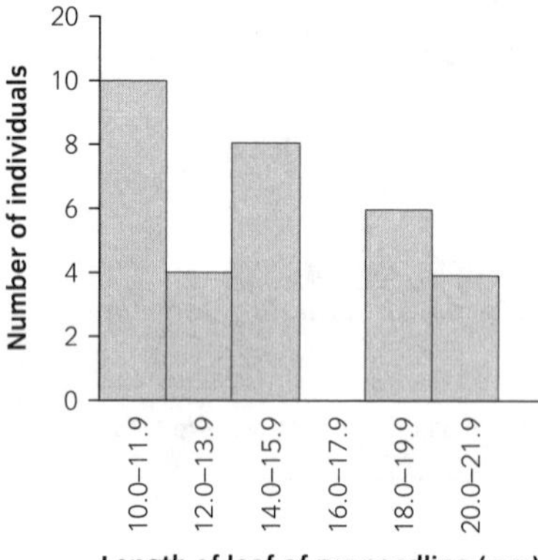

Fig. 10.5. Frequency distibution of leaf length in rye seedlings after 3 weeks growth.

If you are using a figure to illustrate your data it is common practice to include confidence limits (3.7.). These are added as vertical lines about the point or bar on your figure (e.g. Fig. 10.4.). For regression analyses confidence limits for the regression coefficient may also be calculated (Sokal & Rohlf, 1981).

For some studies, e.g. microbiology, molecular biology, and ecology, you may have photographs that illustrate your results. These used to be called plates but more recently have tended to be included in with the figures. If they are called plates then the numbering system should be sequential for all plates included.

More errors in figures are appearing with the use of computer software, often, although not always, through user inexperience. If you are going to use a software package it is worth roughing out the figure beforehand to ensure that you are quite clear in your own mind what is required. As with tables you are likely to have local regulations about how you present figures in your reports, papers for publication, etc. The main aim is always to make figures explicit and appropriate for the data. Therefore, a figure should:

- *Include a title* Most journals require figure titles to contain enough detail so that they can be taken out of context and still be understood. You do not need to include such things as 'a figure, to show . . .'. It is clear that this is a figure, so you do not need to include this in your title. In most scientific reports the standard practice is to place the title below the figure. You may also add brief additional information about the figure.

- *Be numbered* This enables you to explain clearly which figure you are referring to at any one time. The numbering should be sequential.
- *Have clearly labelled axes, stating the units in use* If you fail to label your axes then your figure will convey no useful information at all.
- *Provide a key if necessary* You may be using your figure to allow easy comparison between several **treatments.** Figure 10.3 shows the behaviours of several orang-utans. Each column is shaded in a different manner and a key given so that you can identify which individual is which. Where shading or colour are used to identify particular **samples** or treatments this coding should be consistent for all figures. So if sample 1 is shaded blue in the first figure it should be shaded blue in all subsequent figures.
- *Use the bar chart or histogram correctly* Both bar charts and histograms are drawn as bars on a figure. It is therefore tempting to think that they are the same, but bar charts have a gap between each bar (Fig. 10.3.) whereas histograms do not (e.g. Fig. 3.4.). The gaps indicate, on a bar chart, that the data are measured on a discrete scale (e.g. blood groups). If the data are nominal then the order of bars on a bar chart are imposed by the reporter and do not reflect any meaningful order. A histogram with no gaps shows that the data are measured on a continuous scale (e.g. height (mm)) and the order of the bars is derived from the data and is meaningful.
- *Make it clear if you have dependent and independent variables* For some data it is possible to identify a **dependent** and an **independent** variable. These terms are used by different authors in different ways (glossary). When we use these terms, we mean that the independent variable is taken without sampling error, is usually under the investigator's control and that there is probably an association between the two variables.

  In Chapter 6 we considered two examples: Example 6.2. (Heavy metal contamination of soil under electricity pylons) and Example 6.3. (Lower arm (cm) and lower leg (cm) length of a small cohort of female undergraduates). In both these investigations two variables were being considered. In Example 6.2. soil samples are taken at regular points from the electricity pylon. This treatment variable (distance from pylon) is under the investigator's control and is the independent variable. The level of lead contamination is not under the investigator's control and is said to be the dependent variable. In Example 6.3. neither parameter is being manipulated within the experiment and neither could sensibly be said to be dependent on, or independent of, the other.

Where you have a dependent and an independent variable, the dependent variable is plotted on the *y*-axis and the independent variable is plotted on the *x*-axis (Fig. 6.12.). Where there is no clear case for the variables being classed as dependent and independent then either variable may be plotted on the *x*- or *y*-axis (e.g. Fig. 6.14.).

Q14 Cows in calf were fed known amounts of supplementary corn. The calves were weighed when born. The investigator wanted to test if there was any association between the amount of corn the cows were fed and the calves' birth weight. Is there a dependent and an independent variable? If so, which is which?

A14 Yes. The amount of corn given to the cows is under the investigator's control and is the independent variable. The weight of the calves at birth is the responding dependent variable.

- *Use trend lines correctly* Lines may be fitted to data in two ways: either by joining the points on a graph together as a trend line or as the result of a statistical analysis, e.g. regression (Chapter 6). Trend lines are very problematic. Before using one you should ask yourself whether it is sensible to draw a line between two points on your graph and read off an intervening value. It is usually appropriate to draw trend lines on line and scatter plots but not on histograms and bar charts.

  In Example 10.4. only three concentrations of petrol contamination of soil were investigated. The data are measured on a continuous scale (amount of petrol added (ml)) and the data are plotted as a line graph. A trend line that links each point on this line graph makes sense: you could read off an intervening point and it would be meaningful (Fig. 10.4.). If you have continuous data which are organized into classes and plotted as a histogram it is not appropriate to link these classes together with a trend line. The imposition of classes has effectively taken away the continuous nature of the scale when visualized as a figure (Fig. 3.4.).

In Example 10.3. the various behaviours of an orang-utan have been recorded. When these data are plotted on a bar chart (Fig. 10.3.) it is quite clear that any line joining the tops of these bars would be meaningless. This is particularly so since the data are nominal and the categories (in this case different behaviours) have no inherent order.

Beware: computer software is very keen to draw in trend lines. Do not assume this is correct. Where it is not correct, do not leave the trend line in your figure.

- *Distinguish between a trend line and a regression line* If a line is drawn, you must indicate clearly if this is a trend line or one derived from statistical analysis. Some statistical tests (e.g. regression analysis, 6.4.) enable you to work out a mathematical equation that describes the line that best fits your data (Figs. 6.14 and 6.15.). If you have a regression equation you should add this to your figure; if you do not then you must make it clear that the line is drawn in only to indicate the possible trend.
- *Have a restricted line* If you are drawing a line on your figure do not extend it beyond the range of your data. You do not know what happens outside this range and you could be suggesting a trend that is quite wrong. If you need to know what is happening outside the data set you have you need to repeat your experiment with more observations in the area you are interested in. Chapter 6 (Q3) illustrates this graphically.

Q15 Examine the figures in your paper. Do they fulfil all the guidance on good practice? If not how might they be improved? If there are no figures included in the paper, do you think it would have been useful to include some? Why? To avoid overlap with data already represented in a table would any tables need modification? In what way?

10.8.3. Reporting statistics

When reporting the outcomes from your analysis, there is very little detail that is required but what is needed is critical to making sense of your analysis. In each BOX in Chapters 5–8 where we have illustrated worked examples for a number of statistical tests we have included one way of reporting our results. However, there are several conventions that may be used. These conventions vary depending on the statistical test you used, the p value, and the local requirements of the journal or your Department. The following, therefore, needs to be used in conjunction with these other guides.

i. When $p = 0.05$

Throughout Chapters 5–8 we have indicated the usual format for reporting the results from statistical analyses. This reporting requires the inclusion of details from your hypothesis, the calculated value of your test statistic, an indication as to whether you reject or do not reject the null **hypothesis,** and the level of probability (p) at which this decision has been made. In each BOX in Chapters 5–8 we have written this information formally so that you become used to the inclusion of these various elements. For example, in BOX 5.2. the outcome from the analysis is reported as 'There is no significant difference ($\chi^2_{calculated} = 1.92,\ p = 0.05$) between the observed lengths of ladybirds compared with that expected if the data are normally distributed'. This formal phrasing includes all the information required. However, in most scientific papers a looser phrasing is used.

ii. When $p < 0.05$

We explained in Chapter 4 how the p value you initially use in hypothesis testing is $p = 0.05$ but that if you may reject the null hypothesis at $p = 0.05$ you should go further and identify the smallest p value at which the null hypothesis is rejected. For example, in BOX 5.1. the study examined the distribution of holly leaf miners on a single tree using a chi-squared goodness of fit test. It was found that $\chi^2_{calculated}$ (155.47) is greater than $\chi^2_{critical}$ (5.99) at $p = 0.05$ and therefore we reject the null hypothesis. Looking again at the chi-squared table of critical values you can see that in fact at $p = 0.001$ the $\chi^2_{critical}$ is 13.82. Therefore, we can reject the null hypothesis at this higher level of significance and have a greater degree of certainty that the decision we have made is correct. It is important to convey this increased confidence and this can be done using one of two conventions.

You may emphasize how significant the difference is between your samples by using particular phrases. These are a widely adopted standard and should be adhered to.

p value at which null hypothesis is rejected	Phrasing
0.05	Significant difference
0.01	Highly significant difference
0.001	Very highly significant difference

In our example from BOX 5.1. we would therefore write: 'There is a very highly significant difference ($\chi^2_{\text{calculated}}$ 155.47, $p<0.001$) between the numbers of holly leaf miners found at three different levels on the tree.'

As an alternative to using these particular phrases a system of asterisks (*) can be used to indicate the level of significance at which you rejected your null hypothesis. If you are using a table that includes the data from your investigation, then it is common practice to incorporate the * symbol within the table. The number of asterisks used is again a standardized convention that you should follow.

p	Symbol
>0.05	NS (not significant)
≤ 0.05	*
≤ 0.01	**
≤ 0.001	***

These symbols tend to be used when reporting the outcome from an ANOVA and the ANOVA table is included in your Results section (e.g. Table 7.10.).

iii. Regression analysis

If you have carried out a regression analysis (Chapter 6) the data are usually represented as a scatter plot and the regression equation and level of significance on the regression line of the figure is added (e.g. Figs 6.13., 6.14., and 6.15.). In 10.8.2. we discuss good practice in relation to lines on figures including regression lines.

iv. Tukey's test

The outcome from a multiple comparisons test such as the Tukey's test is often reported in a visual way, where lines are used to link non-significant values. Examples of this can be seen in Chapter 7 (7.6.2.). A similar method is the use of superscripts (7.6.2.).

v. Confidence limits

An estimate of the probability that the population mean lies within a particular range around our sample estimate is known as a confidence limit (3.7.). This statistic is usually represented in a figure as a vertical line (Fig. 10.4.). Confidence limits can be added to most figure types as long as you have appropriate data to summarize.

Where data are summarized in a table you may indicate the **variation** around a **mean** value using either the standard deviation from the data (s) or the standard error of the mean (SE). The first of these provides a valuable indication of the variation in the data in the same units as the original observations. In this case you would report $\bar{x} \pm s$. If you use the mean and standard error of the mean this allows a reader to calculate confidence limits for the mean value. In this case you would be reporting $\bar{x} \pm \text{SE}$. Variation around a mean is more usually indicated using SE. Either way you must make it clear in table headings which terms you are using.

Q16 Look in the journal article you have been reading. How have the statistics been reported?

10.9. Discussion

In the Results section you have *reported* your results, the key trends identified in your data and any outcomes from testing hypotheses. In your Discussion you *evaluate* these findings. This is achieved by starting with the points you have identified in your data and one by one providing explanation and comment about each point. Each part of this discussion would usually draw on other reported scientific literature and here we remind you not to rely on abstract sources only (10.6. Common errors). You need to have an overall plan for your discussion so a mind map or flow chart devised before you start writing will help to ensure that you address all the trends identified in your results in a logical coherent way.

As part of the evaluation you need to be open about any assumptions you have made or any possible bias arising from your design (2.2.10.). You also need to demonstrate your awareness of faults and limitations (2.1.3.) within your investigation. There may be some weaknesses in the report as a result of practical requirements; however, careful planning before carrying out the research should ensure that these are minimal.

Recommendations may also form part of your Discussion section. For example, you may have recommendations arising from the development of a site management plan, or you may have ideas about how to take your research forward or strengthen it. This information needs to be integrated into your overall discussion and should not be tacked on at the end so that it appears to be an afterthought.

At the end of your Discussion you should include a conclusion. Some report formats place the conclusion in its own section; however, in most journal articles the conclusion is an integral and final part of the Discussion. Conclusions are one of the key parts of your report as this is what the reader will 'take away' with them and will remind them of the contents of your paper or report. It is worth thinking carefully about the conclusion and making sure that it only refers to what you found in *your* study and is supported by your data analysis. A conclusion should not include any new material: sometimes people wrongly embark on another discussion in the middle of a conclusion. Often there is a desire to make a great flourish at the end of a report, especially if you are writing a lengthy report for an honours project or thesis. This flourish needs to be avoided as it tends to lead you to overemphasize your findings and make them out to be more than they are.

Read the Discussion in the journal article you have been examining. Note how the Discussion is constructed, how the information from other studies is linked to the findings from this study, and the location and construction of the conclusion. How might you improve your own Discussion section? Read your Institution's or Department's guidance. Does it differ? How?

Common errors

Essays A Discussion can suffer from the same problems that the Introduction section is prey to (10.6.). Your Discussion should be structured and progress in a logical manner, leading to the conclusion. A common error is that a Discussion can become essay- like and lose its focus, which should be your results. This can be evident if you do not explicitly link information from other sources to a debate about your results. If you have free-standing facts then you have probably made this error. Planning your Discussion before you start writing it and making sure you cover each point identified in your Results section in a balanced systematic manner should produce a balanced discussion.

No results For many students having just written their Results section there is a sense that the results have been dealt with and the Discussion then plunges directly and exclusively into a consideration of other people's work with little or no reference to your own results. To avoid this, make a list of all the points arising from your Results section and tick them off as these are covered in your Discussion.

10.10. References

Correctly referencing your work is essential (10.6. Cheating). There are two parts to referencing a report or paper: you must acknowledge your sources of information in the body of the report and then include full details of these sources at the end. There are several systems for referencing. In biosciences the most widely used is the Harvard system and this is what is covered here. You should check with the journal's guide to authors or your Department's regulations to confirm that they are using the Harvard system. Within the Harvard system you will find some slight variation in the use of punctuation, etc., and again should check which version is in use by referring to your local regulations. The Harvard system does not cover referencing electronic sources and here we draw on best practice from a number of different higher education institutions.

The essence of the Harvard system is to identify in your text the source of the information using the author and date of publication as a 'tag'. This 'tag' allows you to cross-reference to the reference list at the end of your paper or report where the full details of the source of information are given. Clearly, there is a wide range of sources that may be used, from journal articles to videos and television. In Tables 10.6 and 10.7. we summarize the Harvard system and current best practice in referencing.

10.10.1. Referencing in the text

You must acknowledge the sources of information at the point at which you incorporate them into your report. If one sentence contains information from one source and the next sentence contains information from another source, you need to reference at the end of each sentence. If all the facts in a paragraph have been derived from a single source then the reference can be placed at the end of the paragraph. The reference in this context is the author and date of publication, or where this is not possible some other unique tag that allows unambiguous cross-referencing to the full information in the Reference section.

Table 10.6. Summary of formats used in one version of the Harvard referencing system in the main body of text in a report

Context	Referencing in the text	Example
You refer to the author in the sentence.	Only the date is added in brackets after the author's name.	In an excellent study, Herbert (2005) identified...
You refer to the facts in the sentence.	The author (or other, see below) and date of publication are given in brackets.	A rare moth was recorded in Shropshire during a study of *A. moschatellina* (Holmes, 2005).
You are using a table or figure taken from another source.*	You should make it clear that this is taken from another source, and include the author (or other) of that work and date of publication.	Fig. 11.1. The variation in lead concentration in a soil sample from Hartlebury Common (taken from Capper, 2003).
You are using more than one paper by the same author from the same year.	To distinguish between several papers add a, b, c, d, etc., after the date.	(Weaver, 2005a; Weaver, 2005b; etc.)
You use information which is shown to have come from somewhere else but you have not read this original source.	You give details of both in the text.	(Sokal & Rohlf, 1981, as cited in Holmes *et al.*, 2006)
You need to indicate particular parts of a document.	The additional information (usually a page number) should be given after the year, but within the brackets.	(Ruxton & Colegrave, 2003, p. 64).
There are two authors for a journal article, book, or conference paper.	You include both author's surnames and the date of the publication. The author's names are given in the same order as that used in the source.	(Dine & Mortimore, 1999)
There are more than two authors for a journal article, book or conference paper.	Only include the first author's surname and then add et al. and the date of publication. et al. is Latin and used to be placed in italics. This is a less common practice now than formerly.	(Alma et al., 2004) or (Alma *et al.*, 2004)
Contributor to a book.	Include the surname(s) of the contributor and date of publication.	(James, 2001)
The work is anonymous or the author not given in the source.	Anonymous may be used in place of the author's surname, or use details of the source.	(Anonymous, 2004) (Defra, 2004) (*The Independent*, 2004)
You have personally been provided with unpublished information.	Indicate that this is a personal communication, who from, and include the date you received the information.	(J. Huffer, personal communication, 7.4.97)
e-journals	Author's surname and date of publication.	(Davis, 2005)
Web material other than e-journals	Author's surname (or other) and date of publication (if known) or date accessed.	(The Bat Society, accessed 21.3.05)
Television programme, film	Give the title and year of publication or date screened.	(The boy with the incredible brain, 23.5.05)
CD ROM and DVDs	Use the author's surname (or other) and year of publication.	(Joseph, 1994).

*If you are preparing a paper for publication (rather than a student essay or internal report) you need to obtain permission from the author and copyright holder to reproduce previously published material.

Table 10.7. Format of information in a reference list following one version of the Harvard system

Context	Format in Reference list	Example
Article in a journal, one author	Author's SURNAME, and INITIALS., (the year of publication). The title of the article. *The title of the journal* **Volume** and (part number) where known: page numbers of article.	HOLMES, D.S. (2005). Sexual reproduction in British populations of *Adoxa moschatellina L. Watsonia* **25**(3):265–273.
Article in a journal with two authors	As above but list both authors. Use 'and' or '&' between the two authors.	DINE, D. & MORTIMORE, D. (1999). Nutrition and school meals. *Schools Today* **35**(5): 21–24.
Article in a journal with more than two authors	As above, listing all authors names and use 'and' or '&' before the last author.	ALMA, P., PIP, D. & HANNAH, M. (2004). The population genetics of tepal colour in *Allium schoenoprasum. Journal of Ecology and Genetics* **10**: 54–63.
An authored book	Author(s) SURNAME and INITIALS., (the year of publication). *The title of the book*. Edition (if not a first edition). Publisher, Place of publication.	RUXTON, G.D. & COLEGRAVE, N. (2005). *Experimental design for the life sciences*, 2nd ed. Oxford University Press, Oxford.
Contribution in a book	Contributors SURNAME, and INITIALS., (year of publication). Title of contribution. INITIALS and SURNAME of the author(s) or editor(s) of the book, if the latter include ed. or eds. *The title of the book*. The publisher and place of publication, page number(s) of contribution.	JAMES, R. (2001). Bactericide properties of herbs. D. CHRISTOPHER, ed. *Herbs Today*. Worcester University Press, Worcester, p. 35.
Conference paper	Contributors SURNAME, and INITIALS., (year of publication). Title of contribution. INITIALS and SURNAME of editor of conference proceedings *Title of conference proceedings* and date and place of conference. Publisher and place of publication, page numbers of contribution.	JOSEPH, J. (1992). Art and the biosciences. C. JOSEPH, ed. *American Society of Bioscience meeting* 24.3.94. New Orleans. Houston Press, Houston, p. 13.
Thesis	Author's SURNAME, and INITIALS., (the year of publication). *Title of thesis*. Award (PhD, MSc etc.). Name of Institution to which work submitted.	HOLMES, D.S. (1986). *Selection and population dynamics in Allium schoenoprasum*. D. Phil., University of York.
Publication from a corporate body	NAME of body, (year of publication). *Title of publication*. Publisher, place of publication, report number (if any).	BOTANICAL SOCIETY OF THE BRITISH ISLES (1999). *Code of conduct for the conservation and enjoyment of wild plants*. Botanical Society of the British Isles, London.
Newspaper article	Authors SURNAME, and INITIALS., (if known) or TITLE OF NEWSPAPER (year of publication). Title of article. *Title of newspaper*. Day and month, page number(s) and column number.	VERKAIK, R. (2005). Police investigate retaliation attacks. *The Independent*. 9 July, p. 21.

Table 10.7. continued

Context	Format in Reference list	Example
Personal communications	Do not include in the reference list.	—
e-journals	Authors SURNAME and INITIALS., (year of publication). Title of article. *Title of e-journal.* **Volume** or part if known. Publisher. url [date accessed].	DAVIS, G. (2005). How to keep active during your retirement. *Lifelong Learning* **40**. Gill Press. **www.gillpress.com** [24.5.05]
Web material other than e-journals	Authors or editors SURNAME if known and INITIALS., (year of publication, last updated), url [date accessed].	Bat Conservation Trust. (last updated 2.7.05), **www.bats.org.uk** [6.7.05].
Television programme, film	*Title*, (year of production), Type of material (video, TV etc), Directors SURNAME, and INTIALS., Production details – place, organization. [date screened or seen].	*The boy with the incredible brain* (2005). TV documentary. Oxford Scientific, C4. [23.5.05].
CD ROM and DVDs	Authors SURNAME, and INTITALS., (year of publication), *Title*, {type of medium}, (Edition). Publisher and place of publication. Any identifier number. [Date accessed]	JOSEPH, A. (1992). *The wild child.* {CD ROM}. Houston Press, Houston. [14.3.92].

There are many possible scenarios for referencing. For example, the type of source may be a journal article or a video or television programme, there may be one or many authors, you may be using a table or figure taken directly from a source or facts taken from the source but presented in your own words. It is neither feasible nor helpful to cover all possible combinations of these but the guidance in Table 10.6. is constructed so that you may use one or more rows to produce the appropriate reference tag.

Examine the journal article. Find ten examples of referencing in the Introduction and/or Discussion. How do the format and positioning of these various references differ?

10.10.2. Reference list

You may, at the end of a report or paper, be asked to prepare a bibliography or a reference list. A bibliography is usually a record of all sources of information you have used, whereas a reference list is those sources of information you have included directly in your report (i.e. you have referred to them). You will usually be asked for one or the other and need to ensure that you include the appropriate list of sources. In most cases you will be asked to compile a Reference list and it is to this that the Harvard referencing system really applies, although the principles can also be used when constructing a Bibliography.

There are two aims for a Reference list. First, the tag from the text (10.10.1.) must link uniquely to one reference, and second, that sufficient detail is provided in the reference list for the reader to be able to obtain the reference themselves. All the references in a Reference list need to be arranged in a consistent manner to enable the reader to be able to find the full reference details quickly. The reference list can be organized in a number of ways but most often is in alphabetical order using the author's surname (or other) first. If you need to you can also use the second author's name and the date of publication for ordering the references. The earliest paper by the same author(s) will come first at that point in the list. Where you have also used a, b, c, etc., in the tag then these references are ordered by these letters.

Q19 Examine the reference list in the journal article you have been scrutinizing. Are the references organised alphabetically by author? Are there any papers with very similar authorship? What criteria have to been used to order these papers?

As we outlined in 10.10.1. there can be many types of information and therefore many possible types of references. Again, we give a guide which by combining one or more rows will provide you with the requisite detail for dealing with most types of sources of information you will encounter (Table 10.7.).

Q20 The Harvard system is used extensively in bioscience publications, although there is considerable variation between journal styles especially in relation to the type of fonts, case, and punctuation used. Examine your journal article and note how the system differs from that described in Tables 10.6. and 10.7. Find your local regulations relating to referencing. How do these differ from the ones we have outlined.

Common errors

- *Not recording the sources of information* One of the most common errors in relation to referencing is a failure to record the sources of information at the time when you are taking notes. Clearly, making good this omission takes a lot of unnecessary work and can be easily avoided.

- *Referencing in the wrong place* If you are discussing your ideas and your results you need to check that any referencing in that sentence is in the right place. Often referencing is misplaced within a sentence and appears to imply that your own ideas or your own results are in fact those of another author. For example, an undergraduate wrote 'It is clear that the organic content of the soil in this study (Holmes *et al.*, 1999) should be higher'. This implies that Holmes *et al.* carried out the study and not the student. What the undergraduate should have written was 'It is clear from other studies (Holmes *et al.*, 1999) that the organic content of the soil in this study should be higher'. Referencing in the wrong place can also lead to the reverse implication, i.e. that you carried out work that was in fact the result of someone else's efforts. Although this will have occurred in error, it is plagiarism (see cheating).
- *Not a bibliography* You are usually asked to include a Reference list. Therefore you should not include sources of information that have not been referred to in the text. Your Reference list should be complete, and all reference tags in the text should link to full details in the Reference list.
- *Only include the sources you have read* Books are a frequently used source of information. These are invariably derived from other (primary) sources which are referenced in the book. If you have not read these primary sources you need to make this clear in your referencing (Table 10.6.). This is particularly important as the authors of the book may have misunderstood the original report and misrepresented the findings. If you have not read the original source you will not know this, but the error will appear to be yours and not the author's of the book, unless you make it clear that you have read the book and not the primary source.
- *Cheating* A failure to reference correctly both in the main body of the text and by not providing a Reference list when asked to is plagiarism. This is considered to be a form of cheating (10.6. Common errors). Cheating in this way is increasing with the ease of accessing electronic forms of information. Make sure you avoid this by referencing correctly throughout all written work.

10.11. Appendix

An Appendix is not part of a published paper, but may form part of undergraduate and graduate reports, including theses. There is considerable variation in what may be included in an Appendix, which reflects not only differences in practice between departments but also variation in research areas. The golden rule when considering whether to use an Appendix is never to include anything that a reader needs to refer to in order for them to understand your report. Therefore, you should not include critical tables or figures in the Appendix of your report.

For many reports it would be appropriate to include evidence supporting your compliance with the law (Chapter 9), such as your risk assessment, ethics approval, confirmation of permission for working in an area, etc. If you have needed to use a consent form an unsigned copy may be incorporated into the Appendix. However, you should not include anything that might compromise the confidentiality of any volunteers, such as signed consent forms or completed questionnaires, if these could be used to identify individuals.

The Appendix can also be useful as a repository for raw data but only if this is not required in the Methods or Results section. Including this type of information in an Appendix can be useful both for the Institution and for yourself; however, before you commit yourself to including pages of numbers check with your supervisor. The data on which most of the worked examples in this book are based come from undergraduate projects and for many of these the raw data were included in the Appendix.

If you have been carrying out an experiment that required the use of stock solutions (e.g. 0.5 M hydrochloric acid) you may wish to include details on how the stock solution was made. Similarly, it is usually more appropriate to include details about how well-known solutions or compounds (e.g. Feulgen stain) were made in the Appendix rather than clutter your Method with these details. Details about suppliers can also be an important piece of information that will allow someone to repeat your experiment. These can also be included in the Appendix. If you are in doubt as to what information your Institution or Department prefers in an Appendix you should discuss your draft Method and Appendix with your supervisor.

10.12. Your approach to writing a report

When writing a scientific report you need to consider how to use words, figures, tables and the scientific report format to best convey the findings from your research. The tendency is to write the report in the format order, i.e. Title, Introduction, Method, Results, Discussion and conclusion, References, and Abstract. In fact a much more efficient approach and one that will enable you to avoid some of the problems we have outlined in this chapter (e.g. essay-like Introductions) is to write your draft report as follows: Method (and site if relevant), Results, Discussion, Introduction and then on reviewing the draft add the conclusions, Abstract, and finally the Title. From our experience this order will require the least number of revisions, give you greater mastery of the overall report and lead to a better, more balanced report at the end.

The reason for this order is that the Method is usually the most straightforward section and is a matter of reporting from (hopefully) a clear set of notes. Completing one section such as the Method gives you a sense of satisfaction and progress which can give you a much-needed lift. The Results section is the key to your report as it is here that you present the information you will go on to discuss. The Results section usually takes the longest time to construct because all the data need to examined, summarized, and analysed before you begin to consider communication, and the communication itself can involve figures and tables that can take some time to select and properly construct. Therefore, we suggest that the Results is the second section to tackle. With your results in mind it is then usually easiest to move on to the Discussion. One advantage in writing the Discussion before the Introduction is that by drafting the Discussion you will have a clearer idea as to how you need to approach the Introduction to avoid unnecessary repetition. The Introduction is then the last main section to write in draft. By keeping this until towards the end of the writing process you will have a much clearer idea of what background material is needed in the Introduction to properly ground the rest of the report and allow you to introduce the points you have developed throughout your report.

Writing, even using a structured format, is a creative process and you may find yourself uncomfortable with tackling one section and more comfortable about tackling another. If you feel strongly about which section you wish to work on then you will probably be more efficient if you go with the flow.

Q21 Make a list of the areas where your report writing could be improved. Identify three points from those on your list that you will tackle when writing your next report.

Summary of Chapter 10

- The aim of this chapter is to encourage you to review and improve your writing skills in relation to writing scientific reports in the format usually used in science journals.
- We examine each section of a scientific report, look at its construction, and discuss approaches to identifying and addressing common errors.
- This evaluation is achieved by examining published research papers.
- The Online Resource Centre includes interactive exercises that test your understanding of this chapter with other topics considered earlier in this book.

Appendix a. How to choose a research project

Many undergraduates are expected to carry out a piece of independent research during their final year of a degree course. For some this may be literature-based. In some institutions the topics are listed and you indicate a preference. Some students, however, are asked to choose and design their own research project. This checklist is designed to help in particular the last group of students but some points are also relevant for the students choosing topics from a prescribed list.

Your honours project in your final year can be used to develop skills or contacts that will be valuable for obtaining employment, or the topic may be chosen because it particularly interests you. Above all it must be a project you are enthusiastic about since of all your degree work it is this research project that demonstrates your abilities as a scientist in your chosen field and as a graduate to be independent in your learning.

Most institutions do not require you, as an undergraduate, to carry out original work. You will, however, be expected to show some initiative. This means that, although you must not copy other people's work, it can usually be adapted. For example, you might apply the same method to a different location or species. However, you must check your local regulations to confirm this.

The checklist that follows is a series of prompts that will help you come to a better understanding of yourself and your interests and therefore help you to focus in on a topic of research that will most suit you.

A.1. How to choose a research project

Here is a checklist to help you choose a topic for your independent research.

1. Do you already have an idea?
   YES (go to14) NO (go to 2)
2. What do you wish to do when you complete your programme of study?
   You can often use your honours research project to strengthen links with future potential employers, develop skills relevant to your planned career, or provide you with school experience.
3. When would you like to carry out the work?
   Some research is seasonal and it may not be practical for you to carry out research at a particular time of year e.g. fieldwork, observations of reproductive behaviour in vertebrates or invertebrates, a survey of student opinion.
4. Where would you like to carry out the work?
   You may prefer to work from home; you have contacts abroad, etc.
5. Do you have any potentially useful outside contacts?
   Undergraduates may be able to set up projects with schools, special needs groups, support groups, businesses, or industry. Speak to your supervisor/personal tutor about this as there will be institutional regulations that relate to such work.

6. Do you prefer fieldwork or laboratory work?
   Most people have a preference for one or other of these and this can guide you when choosing a project.
7. Which part of your course have you most enjoyed?
8. Has there been a particular subject in your course that you would like to investigate further?
9. Often lecturers and PhD students have small research projects they would like to see carried out. Ask.
10. Organizations such as English Nature and the Wildlife Trusts invariably have a list of projects they would like undergraduates to carry out. Get in touch with them or ask a member of staff who is likely to have contacts with the organization that interests you.
11. Go to the library and look through a journal in your area of interest, e.g. *Journal of Biological Education, Journal of Ecology, Heredity*, etc. See what other research has been done. Can you extend or repeat part of this work?
12. Look at projects that has been carried out by students in the past. Research reports such as these are often held in the Institution's library.
13. Now do you have an idea?
    YES (go to 14) NO (see your supervisor/personal tutor)
14. Check that your institution has the equipment you will need and that you can be supervised by someone who works in a similar area. Contact the lecturer whose research area is most similar to the topic you are interested in. Ask them: Is this topic suitable?

A.2. Common problems

From our experience we are aware that students' choices of project often suffer from problems. The common ones we most often encounter are listed below. Check your ideas to make sure your project does not suffer from any of these.

A.2.1. The topic is too broad

Often students are enthusiastic and find it really hard to narrow down their topic and to remain focused. One approach is to use a flowchart or mind map to show how different sections interlink. Then select from this a central narrow area. You can always expand later if you have time.

A.2.2. The research topic is vague

Some of the proposals we have seen from our students are very vague, such as 'something on fish', etc! Vague choices usually reflect a lack of thought and planning. They can also occur when no thought is given to the justification for the research. It may well be possible to examine something, but do you have a scientific reason for doing so? To resolve this make sure you write an aim and objectives at the outset as this will help to guide your planning. Draw a flowchart of the investigation and the thinking behind it as this can also help you to see if your choice makes sense.

A.2.3. Time management

For nearly all researchers from undergraduates in their final year to research scientists there are time constraints that can have an impact on your research. These fall into two groups:

i. Not appreciating the seasonality of some work

We have referred to this in the checklist above, in that there are often time constraints on the work you may wish to carry out. For example, if you wished to examine the association between asthma and pollen, the work would have to be carried out when the pollen was present.

ii. Not thinking about how long it will take to complete the work

When carrying out research in a new area this is always a point to remember. For example, in Chapter 2 (Example 2.2.) we described an undergraduate project in which the student investigated the relative effectiveness of tea-tree oil and triclosan as anti-bactericides in a GP's practice. To collect the swabs from the volunteers and to inoculate the plates took 13 hours. The student had carried out a pilot study and so was aware that this part of the procedure would take a long time and she was able to make suitable arrangements before she started.

When planning your work it is beneficial to draw up a detailed timetable and if you are not familiar with the techniques have a trial run to enable you to judge how long it will take to complete each part of the investigation. If your work is to be reported then ensure you also leave enough time for panic, computer downtime, drawing figures, printing, and binding. A general rule of thumb for all time management decisions is to work out carefully how long you think it should take and then double the time you have allowed.

A.2.4. Independent learners

Undergraduate and graduate research projects are usually intended to allow you to show your abilities as an independent researcher. Therefore, they are invariably less formally structured and you will not be chased up by staff. Attending regular meetings with your supervisor and drawing-up and sticking to a timetable are essential.

Although research is meant to demonstrate your own abilities, this does not mean you should work in a vacuum. Students who do this nearly always perform badly. This is not what is intended by the words 'independent learners'. Although your peers are not necessarily having the same problems that you are, use their support but also use supervisory support and that of other established researchers.

A.2.5. Trials and tribulations

Research is almost always unpredictable and things go wrong. If you are getting upset see your supervisor and if they do not seem to be sympathetic go to another lecturer or a counsellor. Try to keep things in perspective.

A.2.6. Non-significant results

Most practical work carried out by undergraduates is designed to illustrate differences between treatments; therefore, you get into the habit of expecting significant differences when testing hypotheses. As a result a non-significant result is often automatically assumed to be a mistake and dismissed as such. You must be confident that your results are a true reflection of the real underlying biology. Be critical of your work but do not undervalue the importance of non-significant results.

A.2.7. Cheating

There are many temptations in research to cheat. Perhaps you have an **outlier** and would like to ignore it to simplify your analysis. You may have missing data because an organism died or a treatment did not work. Failure to acknowledge all your real data or creating data is cheating and will carry very significant penalties both as an undergraduate and as a professional scientist. You will not be penalized for being honest.

Appendix b. Which statistical test should I choose?

This book is about the process of designing and reporting research in the biosciences as outlined below. As an integral part of this process you will often use statistics. You must make the decisions concerning your statistics before you carry out your research as the statistical tests determine features such as sample size and the number of samples. Since statistics is an integral part of research design we touch on choosing statistical tests in virtually every chapter in this book. Here, we therefore briefly draw these elements together.

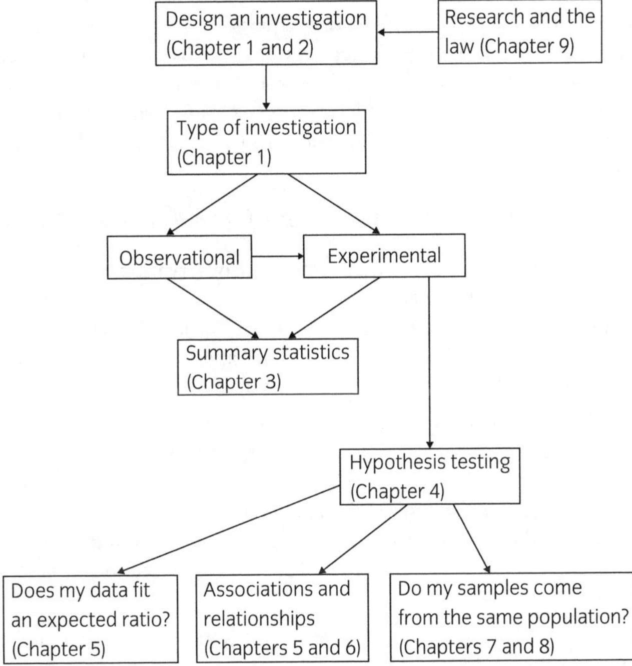

B.1. What type of investigation am I designing?

B.1.1. **Observational:** you are not starting out with a question, you are going to see what is there (go to B.3.).

B.1.2. **Experimental:** you are starting out with a question (**hypothesis**) (go to B.2.).

B.2. Which type of hypotheses am I testing?

There are three types of hypotheses which you need to choose between. If you are not sure which type of hypotheses you will be testing, read the information in B2.1.–B.2.3. before deciding. For more information about hypotheses and hypothesis testing, read Chapter 4.

1. Do the data match an **expected** ratio?
   or
2. Is there an association between two or more **variables**?
   or
3. Do samples come from the same or different **populations**?

B..2.1. Do the data match an expected ratio?

The term 'expectation' can be used in two ways, only one of which is correct in this context. You may have reason to expect that your observations will follow a particular ratio or mathematical relationship defined by biological or mathematical principles (an *a priori* expectation). For example, you may expect your data to be normally distributed and so fit a Gaussian equation or you may expect your data to fit a 1:1 gender ratio. This is the correct way to use the term 'expected'. The alternative is the notion of you having a personal hunch about the likely outcome. This is not an *a priori* prediction and plays no part in science at this level of your education. When we use the term 'expectation' it is the first of these meanings we are using.

There are three types of investigations where you may have an *a priori* expectation:

- You can reasonably argue that all samples or **observations** should have the same value.
  For example, if you sampled the numbers of beetles falling into different coloured pitfall traps you might expect that there should be the same number of beetles collected in each trap.
- You will be carrying out a genetic cross or sample in a population and have reason to expect a particular segregation ratio, population genetic ratio or sex ratio, etc.
  For example, you may use the Hardy–Weinberg theorem to predict the allele frequencies in the F1 generation of a population of *Drosophila melanogaster* exposed to a known selection pressure.
- You wish to confirm that your data have a particular distribution such as a normal or a Poisson distribution (3.2.2.).

How to choose the correct test

Your choice is determined by the number of variables and number of categories for each variable. This table directs you to the most likely statistical test. If you are not sure then go to the specific sections indicated and look at the examples to see if they are similar to the work you are planning. Each statistical test has specific criteria that need to be met. These are given for each test in the sections shown.

You have one variable and you have an *a priori* reason for expecting certain outcomes from your investigation. The variable has more than two categories.	Chi-squared goodness of fit test (5.1.) or G goodness of fit test (5.5.1)
You have one variable and you have an *a priori* reason for expecting certain outcomes from your investigation. The variable has only two categories.	Chi-squared goodness of fit test with Yates's correction (5.4.1.) or G goodness of fit test (5.5.5.)
You have one variable with two or more categories. You have more than two samples in your data set and wish to know if the samples are similar or different from each other.	Chi-squared test for heterogeneity (5.2.)
You do not have an *a priori* expectation. You have two variables. At least one of these variables has more than two categories. You wish to test for an association between the variables.	Chi-squared test for association (5.3.) *or* G test for association (5.5.2.)
You do not have an *a priori* expectation. You have two variables. Both variables have only two categories. You wish to test for an association between the variables.	Chi-squared test for association with Yates's correction (5.4.2.) *or* G test for association (5.5.2.)

One of the criteria for using a chi-squared test is that the expected values are greater than 5. Having calculated your expected values, if you find any that are less than 5 you should refer to 5.7. (The problem with small numbers).

B.2.2. Is there an association between two or more variables?

By association we mean where one variable is found to change in a similar manner to another variable. For example, is there a significant association between the hardness of shells of eggs laid by pullets and the amount of food supplement they have eaten? Tests for an association can allow you to model the association, make predictions and test the significance of the association.

How to choose the correct test

Your choice is determined primarily by whether you expect your data to be **parametric** (**interval**) or **non-parametric** (**interval**, ordinal, or **nominal**) and whether you wish to model the association or make predications as well as test the significance of the association. To tell it your data are parametric refer to BOX 3.2. This table directs you to the most likely statistical test. If you are not sure, go to the specific sections indicated and look at the examples to see if they are similar to the work you are planning. Each statistical test has specific criteria that need to be met. These specific criteria are given in the sections shown.

You have two variables. The data are counts arranged in discrete categories (nominal or ordinal). At least one of these variables has more than two categories. You wish to test for an association between the variables.	Chi-squared test for association (5.3.) *or* G test for association (5.5.2.)
You have two variables. The data are counts arranged in discrete categories (nominal or ordinal). Both variables have only two categories. The association does not need to be linear. You wish to test for an association between the variables.	Chi-squared test for association with Yates's correction (5.4.2.) *or* G test for association (5.5.2.)
You have two treatment variables. The data are **non-parametric*** and can be ranked (ordinal or interval). The association appears to be linear. You wish to test for an association.	Spearman's rank correlation (6.2.)
You have two treatment variables. The data are **parametric**. The association appears to be linear. You wish to test for an association.	Pearson's product moment correlation (6.3.)
You have two treatment variables, one of which is under the control of the investigator. You wish to test for an association, or you wish to model the association and/or predict *y* values for given *x* values within the range of your observations. The distribution appears to be linear.	Simple linear regression (6.5.)
You have two treatment variables and neither is under the control of the investigator. The two variables are measured on the same scale, e.g. mm. You wish to model the association and/or predict *y* values for given *x* values within the range of your observations. The distribution appears to be linear.	Principal axis regression (6.6.)
You have two treatment variables and neither is under the control of the investigator. The two variables are measured on different scales, e.g. grams and arbitrary units. You wish to model the association and/or predict *y* values for given *x* values within the range of your observations. The distribution appears to be linear.	Ranged principal axis regression (6.7.)
None of these tests seems to be right for your data. For example, you have three or more treatment variables; you have more than one *y* value for each *x* value.	Sokal & Rohlf, 1981; Legendre & Legendre, 1998; Grafen & Hails, 2002

*Parametric tests are more powerful; therefore, if you have parametric data you should always use the parametric test. If you have non-parametric data you should consider **transforming** it (3.9.).

B.2.3. Do samples come from the same or different populations?

These are the type of hypotheses that are most frequently investigated by undergraduates. For example, is there is a significant difference between the shell heights (mm) of periwinkles from the lower and mid shore at Aberystwyth? There are many tests that will test this type of hypotheses. These tests fall into parametric tests (Chapter 7) to be used when you have normally distributed data, and non-parametric tests (Chapter 8) when your data are not normally distributed or the distribution is not known. To tell if your data are normally distributed refer to BOX 3.2.

i. Parametric tests

These are largely selected on the basis of the experimental design – how many variables, how many categories in each variable, how many **replicates** in each category. This table directs you to the most likely statistical test. If you are not clear about the criteria refer to the specific section indicated and examine the examples to see if they are similar to the work you are planning. Each test has specific criteria that need to be met. These specific criteria are given in the sections shown.

You have one treatment **variable**. You are going to compare two samples. The data are **unmatched**.	*t* or *z* test for unmatched data (7.1. or 7.2.)
You have one treatment variable. You are going to compare two samples. The data are **matched**.	*t* or *z* test for matched data (7.3.)
You have one treatment variable. You are going to compare two or more samples. You wish to test general and specific hypotheses.	One-way parametric ANOVA and Tukey's test (7.5. and 7.6.)
You have two treatment variables. Each variable has at least two categories or classes and all categories from one variable are combined with all categories from the second variable. You wish to test general and specific hypotheses.	Two-way parametric ANOVA and Tukey's test (7.7. and 7.8.)
You have two treatment variables. Each variable has at least two categories. One variable is randomized or nested with regard to the second variable. You wish to test general hypotheses.	Two-way nested ANOVA (7.9.)
You have three treatment variables. Each variable has at least two categories and all categories from each variable are combined with all other categories from the other variables. You wish to test general and specific hypotheses.	Three-way parametric ANOVA (7.10.)
None of the above	Chapter 8 and Sokal & Rohlf (1981)

ii. Non-parametric tests

The choice of these tests is similar to choosing the parametric tests. Your selection depends on how many treatment variables you are planning to examine, how many categories in each variablem, and how many replicates in each category. This table directs you to the most likely statistical test. If you are not sure look at the specific section indicated for the test and examine the examples to see if they are similar to the work you are planning. Each test has specific criteria that need to be met. These specific criteria are given in the sections shown.

You have one treatment **variable**. You are going to compare two samples. The data are **unmatched**. You have 20 **observations** or fewer in each sample.	Mann–Whitney U test (8.1.)
You have one treatment variable. You are going to compare two samples. The data are unmatched. The data are measured on a continuous scale and you have more than 30 observations in each sample.	*z* test for unmatched data (7.1.)
You have one treatment **variable**. You are going to compare two samples. The data is **unmatched**. You have more than 20 **observations** in each sample.	Sokal & Rohlf (1981)
You have one treatment variable. You are going to compare two samples. The data are **matched**. You have less than 30 pairs of observations.	Wilcoxon's rank paired test (8.2.)
You have one treatment variable. You are going to compare two samples. The data are matched. You have more than 30 pairs of observations.	*z* test for matched data (Chapter 7 (7.2.))
You have one treatment variable. You are going to compare two or more samples. You wish to test **general and specific hypotheses**.	One-way ANOVA (Kruskal–Wallis test) (8.3. and 8.4.)
You have more than one treatment variable. You are going to compare two or more samples. You wish to test general and specific hypotheses. You will be using a calculator.	Two-way non-parametric ANOVA (8.5. and 8.6.)
You have more than one treatment variable. You are going to compare two or more samples. You wish to test general hypotheses. You want to use a computer.	Scheirer–Ray–Hare test (8.7.)

B.3. Are you going to report your results?

With most research you will wish (or be expected) to report your findings. We consider this in Chapter 10. When reporting your findings, you may use summary statistics and you may illustrate the trends in your data using a figure or table. These topics can be found in these sections:

Summary statistics
Central tendency
(***mean***, ***median***, ***mode***, ***skew*** and kurtosis) 3.4.

Variation
(***range***, *interquartile range*, *percentiles*, ***variance***, ***standard deviation***, *standard error of the mean*, *confidence limits*, *coefficient of variation*) 3.5.

Figures 10.8.2.

Tables 10.8.1.

Appendix c. Maths and statistics

In this book we come across a number of symbols and mathematical procedures. For those of you who are not familiar with these symbols or are not confident using them and the maths, we give further explanations and worked examples. There are questions at the end of each section to test your understanding. The answers are at the end of this appendix. If you work through all the questions it will take about 2 hours. In addition, in the Online Resource Centre we include the full step-by-step calculations for each of the worked examples that we have included in the book.

C.1. You and your calculator

In this appendix we encourage you to use some functions already integrated into your calculator. This requires you to own a statistical calculator. There are three types of calculators: ones that will carry out basic steps in maths such as addition and subtraction and no more; a statistical calculator; and a programmable and/or graphical calculator. The first of these does not offer you enough built-in functions to be useful and the last has considerably more than you will ever need and is unnecessarily difficult to master. For graphics and more complex statistics, use a computer package. Statistical calculators are not expensive; it will help if you note down the functions that we suggest you carry out using a calculator and take these with you when looking for a calculator to buy.

C.2. Add, subtract, divide, multiply, equals, more than, less than

In this book we use a number of symbols. These are a way of giving instructions to you, telling you what type of calculation to carry out.

You will be familiar with + (add), − (subtract), × (multiply), and ÷ (divide). Some of these processes can be written in other ways.

Multiply can be indicated as ab (i.e. $a \times b$) or if you have more complex sums you may use brackets $(1+2)(2+3)$. This tells you to multiply the result from the sum inside the first set of brackets by the result from the sum inside the second set of brackets. In this case this would be 3×5.

Division can be written as ÷, as in $4 \div 2$, but it is more usual to write 4/2. If the calculations are more complex division may be written for example, as $(6 \times 3)/(4 - 2)$.

The top part of an equation where you are asked to carry out a division is called the numerator. The lower part is called the denominator.

You will also be familiar with the symbol = (equals) as in $4 \div 2 = 2$, or $4/2 = 2$. There are a number of other symbols that can also be used here. For example, we use ≈, which means approximately equals. Two other symbols used in place of the equals sign are $<$ and $>$. The $<$ symbol means less than, and the $>$ symbol means more than. For example, $p < 0.05$ means that p is less than 0.05. If $p > 0.05$, then p is more than 0.05. These two symbols can be combined to set limits to a value. For example,

$0.05 > p > 0.01$ means that 0.05 is more than p which is more than 0.01, i.e. p is a value somewhere between 0.05 and 0.01. We also use these symbols when writing out specific hypotheses, e.g. A > B > C would mean that the median of sample A is greater than the median of sample B which is greater than the median of sample C.

Q1 Write in English your understanding of the following expressions:

a. 2/4

b. xy

c. a/b

d. $p < 0.001$

e. $12.2/2 \approx 6$

C.3. Brackets and absolute values

The next group of symbols are those that are placed round parts of a sum. These are collectively called brackets. The first set of brackets used is usually the round brackets ( ). If you need to surround round brackets with another set of brackets then you usually use square brackets [ ]. For example, $[(6 \times 2) - 2]/2$.

You carry out the calculation inside the brackets first. So for this example you first work out $6 \times 2 = 12$ and subtract 2, i.e. $12 - 2 = 10$, before carrying out the division, which will be $10/2 = 5$.

$[(5 - 1) \times (4 - 2)] - 3$ is an example of using two sets of brackets. Here, you work out the calculations for the inside brackets first, so $(5 - 1) = 4$ and $(4 - 2) = 2$. You now complete the sum inside the square brackets $[4 \times 2] = 8$ and finally complete the remaining steps in the sum, $8 - 3 = 5$.

Another symbol which can enclose parts of a calculation are the straight brackets | |. These are used to tell you that when the calculation within the brackets is completed the sign of the answer is positive, even if the calculation produces a negative number. These are called absolute values. For the example $|3 - 1| = 2$, the sign is already positive, so being an absolute value has made no difference. But for $|5 - 6| = 1$ the actual sum of $5 - 6$ would normally equal -1. Since this is an absolute value we ignore the negative sign, so the answer is $+1$.

Complete the following calculations:

a. $(10 - 2) / [(3 \times 2) - 2]$

b. $2(x - 1)$ where $x = 2$

c. $|(6 + 3) / -3|$

C.4. N, n, x, y, Σx, Σy, Σxy

To show you how the terms n, N, x and y can be used we have included a table from Chapter 7. In this study periwinkles from the lower and mid shore at Porthcawl have been collected and the shell heights recorded (Table 7.2.)

We have two columns of data. Either to simplify our explanations of calculations using this data or when explaining about choosing and constructing figures we need to

Table 7.2. Shell height (mm) in two groups of periwinkles from the mid and lower shore at Porthcawl, 2002

Shell height			
Periwinkles on the lower shore x		Periwinkles on the mid shore y	
5.5	4.0	3.3	6.7
8.4	5.0	6.3	5.7
5.0	6.2	6.1	4.2
5.0	5.0	8.0	6.3
5.6	7.7	13.5	6.0
4.8	6.0	5.3	7.2
8.4		6.7	

identify one column of data. For this we use the letters x and y. In this example the periwinkles on the lower shore are the x observations and the periwinkles on the mid shore are the y observations. There are 13 x observations which we would record as $n_x = 13$. There are also 13 y observations, so for this sample $n_y = 13$. In many calculations you need to work out $n - 1$, which for both these columns of data in this example would be $n - 1 = 13 - 1 = 12$. If we needed to know the total number of observations in our data this would be $N = n_x + n_y = 13 + 13 = 26$.

In some calculations (e.g. Chapter 7) you need to know the totals by adding all x observations together or all the y observations together. The symbol used to indicate this is Σ. In this example $\Sigma x = 5.5 + 8.4 + 5.0 + \ldots 5.0 + 7.7 + 6.0 = 76.6$. The same can be calculated for the Σy.

Another example of the use of x and y can be seen in data from Chapter 6. In this example, 13 pullets have been examined and the amount of feed each has eaten and the hardness of the egg each produced has been recorded (Table 6.1.). In this example there are 13 items (pullets) so for each column of data $n = 13$.

In this example instead of two separate samples as seen in Table 7.2., two observations have been recorded for each pullet: the amount of food supplement (g) and the hardness of shells. Although these are not independent samples, we still use x and y to indicate which column of data we are referring to. In this example the amount of food eaten is x and the hardness of egg shell is y. Unlike the previous example, one of these columns of data is said to be dependent and this is always called y; the other is said to be independent and this is always called x. We explain these terms in Chapter 6.

As we showed you, in relation to Table 7.2., you may need to calculate the totals for each column i.e. Σx and Σy. The idea of a sum (Σ) can be extended to the products xy. The symbol xy means that an x observation has been multiplied by its y observation $= xy$. We show this for the pullet example in Table 6.3. If all these 13 xy values are added together, this is $\Sigma xy = 661.0$. (In Table 6.3. the original data (Table 6.1.) has been reorganized into numerical order to simplify other parts of a calculation which we consider in Chapter 6.)

Table 6.1. The hardness of eggshells produced by 13 Maren pullets and their consumption of a food supplement (g)

Pullet	Amount of food supplement x	Hardness of shells y
1	19.5	7.1
2	11.2	3.4
3	14.0	4.5
4	15.1	5.1
5	9.5	2.1
6	7.0	1.2
7	9.8	2.1
8	11.6	3.4
9	17.5	6.1
10	11.2	3.0
11	8.2	1.7
12	12.4	3.4
13	14.2	4.2

a. What is the Σy for the data in Table 7.2.?

b. What is the Σx for the data in Table 6.1.?

c. What is the Σy for the data in Table 6.1.?

C.5. Powers (x^n) and roots ($\sqrt[n]{}$)

You will come across x and y with a superscript such as x^2 or y^2. This indicates that you must square the x or y observation by multiplying it by itself. For example if the x observation was 2, then $x^2 = 2^2 = 2 \times 2 = 4$. If a different superscript is used you multiply the number by itself as many times as the superscript indicates. So if $x = 3$, then $x^3 = 2 \times 2 \times 2 = 8$. This is described as x cubed. Although squaring most whole numbers is a straightforward process, which you can work out in your head, for more complex numbers use the x^2 button on your calculator.

This process can be reversed by taking the root of a value. In this book we only take a square root, and the symbol that denotes this is $\sqrt{}$. If you were to take the cube root the symbol would be $\sqrt[3]{}$. You may learn square roots from some numbers such as $\sqrt{4} = 2$, $\sqrt{9} = 3$, $\sqrt{81} = 9$. For other numbers you can use the $\sqrt{}$ button on your calculator.

Table 6.3. Calculating Σxy for the two variables hardness of shells and food consumption (g) in Maren pullets

Amount of food supplement x	Hardness of shells y	$x \times y$
7.0	1.2	8.4
8.2	1.7	13.94
9.5	2.1	19.95
9.8	2.1	20.58
11.2	3.0	33.6
11.2	3.4	38.08
11.6	3.4	39.44
12.4	3.4	42.16
14.0	4.5	63.0
14.2	4.2	59.64
15.1	5.1	77.01
17.5	6.1	106.75
19.5	7.1	138.46
		Σxy 661.0
n 13	n 13	

Taking squares and square roots is an important step in the calculation of a variance and standard deviation (e.g. BOX 3.1.) since the variance is the square of the standard deviation and therefore the standard deviation is the square root of the variance.

When calculating the variance or standard deviation two other important squared terms are used: $(\Sigma x)^2$ and Σx^2. These appear to be the same apart from the brackets but the brackets are especially important here. Remember, whenever you have brackets like this you first carry out whatever calculation is inside the brackets. Therefore $(\Sigma x)^2$ is the sum of all the x values (Σx), which is then squared. This compares with Σx^2, which means first square each x observation and then sum the squares.

These two terms can be calculated for the x observations in Table 7.2. We have already calculated $\Sigma x = 76.6$ so $(\Sigma x)^2 = (76.6)^2 = 76.6 \times 76.6 = 5867.56$. The term Σx^2 is calculated as $5.5^2 + 8.4^2 + 5.0^2 + \cdots + 5.0^2 + 7.7^2 + 6.0^2 = 475.5$. Clearly $(\Sigma x)^2$ and Σx^2 are very different calculations with very different results, so it is important to make sure you do not get them mixed up.

Q4

a. What is $(\Sigma y)^2$ for the data from Table 7.2.?

b. What is Σy^2 for the data from Table 7.2.?

c. What is $(\Sigma x)^2$ for the data from Table 6.1.?

d. What is Σx^2 for the data from Table 6.1.?

e. These calculations can also be used for the column of $x \times y$ values in Table 6.3. What is $(\Sigma xy)^2$?

f. What is $\Sigma(xy)^2$ for the data from Table 6.3.?

When you write y^2 this is sometimes referred to as y to the power 2. Here y is multiplied by itself so the calculation is $y \times y$: there are two y values in the calculation. This can be extended. For example y could be multiplied by itself three times, $y^3 = y \times y \times y$. If $y = 3$ then $y^3 = 2 \times 2 \times 2 = 8$. The power is 3. Powers may sometimes shown using symbols instead of numbers.

The binomial equation is written as:

$$y = \frac{n!}{x!(n-x)!} \times p^x \times q^{(n-x)}$$

Here there are two powers: p^x and $q^{(n-x)}$. If you encounter these you first need to identify what number the power symbol is; for example, what is x or what is $n - x$? Having found these values, and if you know the numerical value for p or q, you can work out the numerical value for p^x and $q^{(n-x)}$ using the powers button on your calculator. For example, if $p = 0{\cdot}5$ and $x = 4$ then $p^x = 0{\cdot}5^4 = 0.0625$. If $q = 6$, $n = 8$, and $x = 4$, then $q^{(n-x)} = 6^{(8-4)} = 6^4 = 1296$.

Powers can be negative, for example x^{-2}. In this case the sign is indicating that this calculation must be inverted (turned upside down). For example, if $x = 2$, and $x^2 = 2 \times 2$, then:

$$x^{-2} = \frac{1}{2 \times 2} = \frac{1}{4} = 0.25$$

You can use the power button (x^y) and the change sign button ($^+/_-$) on your calculator to work out positive and negative powers.

Q5 Use your calculator to work out the following powers:

a. 6^5

b. 6^{-5}

c. 0.5^2

d. 0.5^{-2}

C.6. **Factorials (*x*!)**

In the binomial equation (C5.) another symbol ! was used. Maths revolves around number patterns. These often occur so frequently that they are given a particular symbol. Factorials are one of these patterns that appear in this book. A factorial is the value that results from multiplying a whole number (integer) by all the integers less than itself. A factorial of $3! = 3 \times 2 \times 1 = 6$, a factorial of $4! = 4 \times 3 \times 2 \times 1 = 24$ and so on. In the binomial equation (3.5.1.) two factorials are included: $x!$ and $(n - x)!$. To work these factorials out you first need to know what x is, or what $n - x$ is. You could then work this sum out by hand but it is simpler to calculate factorials using the factorial button on your calculator.

Use a calculator to work out the following:

a. 10!

b. $x!$, where $x = 6$

c. $(n - x)!$, where $x = 6$ and $n = 13$

C.7. Constants: pi (π) and exponential (e)

Patterns in maths can also include constant numbers or constants. They are specific numbers that have been found to be needed in many calculations for the mathematical relationships to work. The two constants that appear in this book are pi (π) $\approx$ 3.14 and exponential (e) $\approx$ 2.72. π can usually be found on your calculator. e is invariably raised to a power as we see in the Gaussian equation (C10.). This button is indicated as e^x. To find the value of e itself press the e^x button, 1, equals.

a. What is π to five decimal places?

b. What is e to five decimal places?

c. What is e^2?

d. What is e^{-2}?

e. What is e^x when $x = 2(6 - 4)$?

C.8. Scales of measurement (%, log, ln, $\times 10^n$)

We use many scales of measurements. The one most familiar to you is our numbering system based on 10, where we have units (numbers 1–9), tens (numbers 10–99), hundreds (numbers 100–999), etc. Very large numbers and very small numbers can be very tedious to write out in full, so there is a shorthand, which we symbolize as $\times 10^n$, where n can be any number. As examples of using this shorthand 100 can be written 10^2 (i.e. 10×10), 1000 can be written as 10^3 (i.e. $10 \times 10 \times 10$), 200 this could be written as 2×10^2, 5000 could be written as 5×10^3, etc. The same idea can be used for very small numbers. We introduced the idea of negative powers in C6. Negative powers of 10 can be used as a shorthand when writing small numbers. For example, 0.1 can be written as 10^{-1} since this is telling us that the number is 1/10. So 0.01 is 10^{-2} or 1/100; 0.5 this could be written as 5×10^{-1}, i.e. $5 \times 1/10$, etc. Your calculator (and computer) will report large and small numbers in this way. If you used your calculator to answer Q5b. it would be reported as 1.28601×10^{-4}.

Apart from the base 10 system of numbering which we are very familiar with you will encounter other scales each indicated using a particular symbol. The four you need to be familiar with are percentage (%), log to the base 10 (log), natural log (ln), and inverse sine ($\sin^{-1}$). We cannot here explain how these scales are derived, but encourage you to find the relevant buttons on your calculator and feel confident using them.

On a scientific calculator, you will probably not have a % button. This is calculated from a proportion or ratio which is multiplied by 100. For example, 3/4 as a % = $(3/4) \times 100 = 0.75 \times 100 = 75\%$. Values may need to be converted to the $\log_{10}$ scale or converted from the $\log_{10}$ scale, taking an antilog. The first is achieved by pressing the $\log_{10}$ button on your calculator (sometimes labelled log). For example, $\log_{10}$

$2 = 0.30103$. To take the antilog for most calculators press shift $\log_{10} =$. This should return the log value back to the base 10 scale. The same process can be used for $\ln 2 = 0.69315$ and its antiln using shift $\ln =$. When using the sine functions, you again can take the sine (sin) or inverse sine ($\sin^{-1}$) usually by using the same button with or within the shift function first.

a. What is 5/20 as a percentage?

b. Rewrite 19080000 as a number $\times 10^n$.

c. What is $(3 - 6)/2$ as a percentage?

d. Rewrite 0.000000345 as a number $\times 10^n$.

e. What is log 4?

f. Take the antilog of 0.05.

g. What is ln 4?

h. Take the antiln of 0.05.

i. What is sine (sin) of 90°?

j. What is the inverse sine ($\sin^{-1}$) of 0.6?

C.9. Unpacking complex equations

The four basic procedures in maths are addition (+), subtraction (−), multiplication (×), and division (÷). These are straightforward in their own right but when there are several of these elements in one calculation it is not always clear in which order to carry out the procedures and this order can be critical to obtaining the correct outcome. To help convey which order should be followed, brackets are often placed around parts of an equation.

To help unpack complex equations you need to follow this order:

1. Brackets
2. Powers and roots
3. Division
4. Multiplication
5. Addition
6. Subtraction

We are going to unpack some of the equations frequently encountered in this book.

Example 1 (e.g. 3.6.1.)

$$\bar{x} = \frac{3 + 4}{2} = \frac{7}{2} = 3.5$$

In this example the first step is to add all the values in the numerator (the top of the fraction) together. Then divide the total by the denominator (lower part of the fraction).

Example 2 (e.g. 3.7.)

$$\text{SE} = \frac{10}{\sqrt{4}} = \frac{10}{2} = 5$$

Again there is a fraction, only this time the denominator has a square root sign ($\sqrt{}$) in front of it. You must first work out the square root of the denominator and then use this to divide the numerator.

Example 3 (e.g. BOX 5.3.)

$$\chi^2_{\text{calculated}} = \frac{(12-6)^2}{4} + \frac{(12-8)^2}{2} = \frac{6^2}{4} + \frac{4^2}{2} = \frac{36}{4} + \frac{16}{2} = 9 + 8 = 17$$

In the numerator there are three different procedures indicated by the symbols, a subtraction $-$, a square $(\ )^2$ and addition $+$. The subtraction is within brackets and must be carried out first. The value within the brackets can then be squared. This gives the numerator for each fraction. Each division is then carried out and finally the two numbers can be added together.

Example 4 (e.g. BOX 5.6.)

$$\chi^2_{\text{calculated}} = \frac{(|9-6.5|-0.5)^2}{2} + \frac{(|4-6.5|-0.5)^2}{2}$$

This example illustrates the use of another symbol $|\ |$. A number within these two vertical lines is called an absolute number. The sign of any number within these lines is ignored. The calculation is carried out following the same rules we have discussed above: brackets, then powers, then fractions, then addition. Absolute values also occur in t tests (BOX 7.3.).

$$\chi^2_{\text{calculated}} = \frac{(|9-6.5|-0.5)^2}{2} + \frac{(|4-6.5|-0.5)^2}{4}$$
$$= \frac{(|2.5|-0.5)^2}{2} + \frac{(|2.5|-0.5)^2}{4}$$

Ignore the sign within the $|\ |$

$$= \frac{(2.5-0.5)^2}{2} + \frac{(2.5-0.5)^2}{4} = \frac{(2.0)^2}{2} + \frac{(2.0)^2}{4} = \frac{4.0}{2} + \frac{4.0}{4}$$
$$= 2 + 1 = 3$$

Example 5 (e.g. 3.6.3.)

In some cases we may wish to calculate two values one at the top of the range and one at the bottom of the range. This is indicated by the symbol $\pm$. We use this in both the range for non-parametric data (3.6.2.) and confidence limits (3.6.3.).

For example:

$$\text{CL} = \bar{x} \pm (t \times \text{SE})$$

This symbol $\pm$ indicates that you need to work out the calculation twice once with an addition and once with a subtraction. If $\bar{x} = 15$, $t = 3$ and $\text{SE} = 2$ then:

$$\bar{x} \pm (t \times \text{SE}) = 15 \pm (3 \times 2) = 15 \pm 6 = \text{range is } 15 - 6 \text{ to } 15 + 6, \text{ or } 9 \text{ to } 21.$$

Example 6 (e.g. BOX 3.1. and BOX 7.6.)

A very frequently used equation is the one written here. It is used in parametric ANOVAs (Chapter 7) and is part of the calculation of a variance.

$$\text{SS}_{\text{sample}} = \Sigma x^2 - \frac{(\Sigma x)^2}{n}$$

We have already looked at the terms $\Sigma(x^2)$ and $(\Sigma x)^2$ in C5. To complete this calculation the fraction $(\Sigma x)^2/n$ is calculated first and then the subtraction. We worked out Σx^2, $(\Sigma x)^2$ and n for the x observations in Table 7.2. in C4. and C5., where $n = 13$, $\Sigma x^2 = 475.0$, and $(\Sigma x)^2 = 5867.56$. So for this example:

$$SS = 475.0 - (5867.56/13) = 475.0 - 451.35077 = 23.64923$$

Example 7 (e.g. 3.5.3.)
In several places in the book we have given some complex equations. These can be very off-putting, but when they are unpacked you will see that they are manageable. The Gaussian equation (3.5.4.) has been unpacked for you in 5.1.3. using the data from Example 3.2. Examine the Gaussian equation and decide in what order you would carry out the calculation. How does this compare with the steps in 5.1.3.?

Gaussian equation:

$$y = \frac{1}{\sqrt{(2\pi s^2)}} e^{-h}$$

where $h = \dfrac{(x - \bar{x})^2}{2s^2}$

Answers to Q3–8

a. Two divided by four

b. x multiplied by y

c. a divided by b

d. p is less than 0.001

e. 12.2 divided by 2 approximately equals 6.

a. First work out the values inside the round brackets. So $(10 - 2) = 8$ and $(3 \times 2) = 6$. Then work out the sum within the [ ] so $[6 - 2] = 4$. Then carry out the division of $8/4 = 2$.

b. First work out the sum within the round brackets where $(x - 1) = 2 - 1 = 1$. Then multiply this by $2 = 2 \times 1 = 2$.

c. First work out the sum within the round brackets $(6 + 3) = 9$. Then carry out the division within the straight brackets $|9/-3|$. When a positive number is divided by a negative number the answer will be negative, so $9/-3 = -3$. The straight brackets tell you this is to be an absolute value so you can ignore the sign. The answer to this sum is $+3$.

A3

a. $\Sigma y = 3.3 + 6.3 + 6.1 + \cdots + 6.3 + 6.0 + 7.2 = 85.3$ mm

b. $\Sigma x = 19.5 + 11.2 + 14.0 + \cdots + 8.2 + 12.4 + 14.2 = 161.2$ g

c. $\Sigma y = 7.1 + 3.4 + 4.5 + \cdots + 1.7 + 3.4 + 4.2 = 47.3$ units

---

A4

a. $(\Sigma y)^2 = 85.3^2 = 85.3 \times 85.3 = 7276.09$

b. $\Sigma y^2 = 3.3^2 + 6.3^2 + 6.1^2 + \cdots + 6.3^2 + 6.0^2 + 7.2^2 = 629.57$

c. $(\Sigma x)^2 = 161.2^2 = 161.2 \times 161.2 = 25985.44$

d. $\Sigma x^2 = 19.5^2 + 11.2^2 + 14.0^2 + \cdots + 8.2^2 + 12.4^2 + 14.2^2 = 2153.88$

e. $(\Sigma xy)^2 = 661.0^2 = 661.0 \times 661.0 = 436921.0$

f. $\Sigma(xy)^2 = 8.4^2 + 13.94^2 + 19.95^2 + \cdots + 77.01^2 + 106.75^2 + 138.46^2 = 51021.652$

a. 7776

b. 0.0001286

c. 0.25

d. 4.0

a. 3628800

b. 720

c. $(13 - 6)! = 7! = 5040$

a. 3.14159

b. 2.71828

c. 7.38906

d. 0.13534

e. $x = 2 \times 2 = 4$, so $e^4 = 54.59815$

a. $0.25 \times 100 = 25\%$

b. 1.908×10^7

c. $-3/2 = -1.5 \times 100 = -150\%$

d. 3.45×10^{-7}

e. 0.60206

f. 1.12202

g. 1.38629

h. 1.05127

i. 1.00 (note: set calculator to degrees not radians)

j. 36.87° (note: set calculator to degrees not radians)

Appendix d. Tables of critical values for statistical tests

D1. Critical values for the chi-squared test (χ^2) between $p = 0.05$ and $p = 0.001$, where v are the degrees of freedom, and p is the probability

D2. Critical values for the Spearman's rank correlation (r_s) between $p = 0.10$ and $p = 0.01$ for a two-tailed test, where n is the number of pairs of observations

D3. Critical values for the t test (and modified t tests) between $p = 0.10$ and $p = 0.001$ for a two-tailed test, where v are the degrees of freedom, and p is the probability

D4. Critical values for the F test that is carried out to confirm that the variances are homogeneous **before a z or t test**, where $p = 0.05$ and where v_1 is the degrees of freedom for the larger variance (numerator) and v_2 is the degrees of freedom for the smaller variance (denominator)

D5. Critical values for a two-tailed z test

D6. Critical values for an F_{max} test that is carried out to confirm that the variances are homogeneous **before a parametric analysis of variance**, where $p = 0.05$, a is the number of samples or treatments and v is the degrees of freedom

D7. Critical values for an F test for a **parametric ANOVA**, where $p = 0.05$, v_1 is the degrees of freedom for numerator and v_2 is the degrees of freedom for the denominator

D8. Critical values for an F test for a **parametric ANOVA**, where $p = 0.01$, v_1 is the degrees of freedom for numerator and v_2 is the degrees of freedom for the denominator

D9. q values for a Tukey's test at $p = 0.05$ and $p = 0.01$, where a is the number of samples and v is the degrees of freedom for the s^2_{within} from the ANOVA calculation table

D10. Critical values for the Mann–Whitney U test (U) at $p = 0.05$ for a two-tailed test, where n_1 is the number of observations in sample 1 and n_2 is the number of observations in sample 2

D11. Critical values for the Wilcoxon's matched pairs test (T) between $p = 0.1$ and $p = 0.002$ for a two-tailed test, where N is the number of pairs of observations used to provide ranks for the calculation (i.e. not those where $d = 0$)

Table D1. Critical values for the chi-squared test (χ^2) between $p = 0.05$ and $p = 0.001$, where ν is the degrees of freedom and p is the probability.

ν	p		
	0.05	0.01	0.001
1	3.84	6.64	10.83
2	5.99	9.21	13.82
3	7.81	11.34	16.27
4	9.49	13.28	18.47
5	11.07	15.09	20.51
6	12.59	16.81	22.46
7	14.07	18.47	24.32
8	15.51	20.09	26.12
9	16.92	21.67	27.88
10	18.31	23.21	29.59
11	19.67	24.72	31.26
12	21.03	26.22	32.91
13	22.36	27.69	34.53
14	23.68	29.14	36.12
15	25.00	30.58	37.70

ν	p		
	0.05	0.01	0.001
16	26.30	32.00	39.25
17	27.59	33.41	40.79
18	28.87	34.80	42.31
19	30.14	36.19	43.82
20	31.41	37.57	45.31
21	32.67	38.93	46.80
22	33.92	40.29	48.27
23	35.17	41.64	49.73
24	36.41	42.98	51.18
25	37.65	44.31	52.62
26	38.88	45.64	54.05
27	40.11	46.96	55.48
28	41.34	48.28	56.89
29	42.56	49.59	58.30
30	43.77	50.89	59.70

Table D2. Critical values for the Spearman's rank correlation (r_s) between $p = 0.10$ and $p = 0.01$ for a two-tailed test, where n is the number of pairs of observations and p is the probability. *(To find the critical values for a one-tailed test divide the p value in half. For example the critical values for a two-tailed test when $p = 0.1$ will be the critical values for $p = 0.05$ for a one-tailed test)*

n	p			
	0.10	0.05	0.02	0.01
5	0.900	—	—	—
6	0.829	0.886	0.943	—
7	0.714	0.786	0.893	—
8	0.643	0.738	0.833	0.881
9	0.600	0.683	0.783	0.833
10	0.564	0.648	0.745	0.794
11	0.523	0.623	0.736	0.818
12	0.497	0.591	0.703	0.780
13	0.475	0.566	0.673	0.745
14	0.457	0.545	0.646	0.716
15	0.441	0.525	0.623	0.689
16	0.425	0.507	0.601	0.666
17	0.412	0.490	0.582	0.645
18	0.399	0.476	0.564	0.625
19	0.388	0.462	0.549	0.608
20	0.377	0.450	0.534	0.591
21	0.368	0.438	0.521	0.576
22	0.359	0.428	0.508	0.562
23	0.351	0.418	0.496	0.549
24	0.343	0.409	0.485	0.537
25	0.336	0.400	0.475	0.526
26	0.329	0.392	0.465	0.515
27	0.323	0.385	0.456	0.505
28	0.317	0.377	0.448	0.496
29	0.311	0.370	0.440	0.487
30	0.305	0.364	0.432	0.478

Table D3. Critical values for the *t* test (and modified *t* tests) between $p = 0.10$ and $p = 0.001$ for a two-tailed test, where v is the degrees of freedom, and p is the probability.
(To find the critical values for a one-tailed test divide the p value in half. For example, the critical values for a two-tailed test when $p = 0.1$ will be the critical values for $p = 0.05$ for a one-tailed test).

v	p					
	0.100	0.050	0.025	0.010	0.005	0.001
1	6.314	12.706	25.452	63.657	127.320	636.620
2	2.920	4.303	6.205	9.925	14.089	31.598
3	2.353	3.182	4.176	5.841	7.453	12.941
4	2.132	2.776	3.495	4.604	5.598	8.610
5	2.015	2.571	3.163	4.032	4.773	6.859
6	1.943	2.447	2.969	3.707	4.317	5.959
7	1.895	2.365	2.841	3.499	4.029	5.405
8	1.860	2.306	2.752	3.355	3.832	5.041
9	1.833	2.262	2.685	3.250	3.690	4.781
10	1.812	2.228	2.634	3.169	3.581	4.587
11	1.796	2.201	2.593	3.106	3.497	4.437
12	1.782	2.179	2.560	3.055	3.428	4.318
13	1.771	2.160	2.533	3.012	3.372	4.221
14	1.761	2.145	2.510	2.977	3.326	4.140
15	1.753	2.131	2.490	2.947	3.286	4.073
16	1.746	2.120	2.473	2.921	3.252	4.015
17	1.740	2.110	2.458	2.898	3.222	3.965
18	1.734	2.101	2.445	2.878	3.197	3.922
19	1.729	2.093	2.433	2.861	3.174	3.883
20	1.725	2.086	2.423	2.845	3.153	3.850
21	1.721	2.080	2.414	2.831	3.135	3.819
22	1.717	2.074	2.406	2.819	3.119	3.792
23	1.714	2.069	2.398	2.807	3.104	3.767
24	1.711	2.064	2.391	2.797	3.090	3.745
25	1.708	2.060	2.385	2.787	3.078	3.725
26	1.706	2.056	2.379	2.779	3.067	3.707
27	1.703	2.052	2.373	2.771	3.056	3.690
28	1.701	2.048	2.368	2.763	3.047	3.674
29	1.699	2.045	2.364	2.756	3.038	3.659
30	1.697	2.042	2.360	2.750	3.030	3.646
50	1.676	2.008	2.310	2.678	2.937	3.496
100	1.661	1.982	2.276	2.625	2.871	3.390
∞	1.645	1.960	2.241	2.576	2.807	3.290

Table D4. Critical values for the F test that is carried out to confirm that the variances are homogeneous before a *z* or *t* test, where $p = 0.05$ and v_1 is the degrees of freedom for the larger variance (numerator) and v_2 is the degrees of freedom for the smaller variance (denominator).

v_2	v_1																		
	1	2	3	4	5	6	7	8	9	10	12	15	20	24	30	40	60	120	∞
1	647.8	799.5	864.2	899.6	921.8	937.1	948.2	956.7	963.3	968.6	976.7	984.9	993.1	997.2	1001	1006	1010	1014	1018
2	38.51	39.00	39.17	39.25	39.30	39.33	39.36	39.57	39.39	39.40	39.41	39.43	39.45	39.46	39.46	39.47	39.48	39.49	39.50
3	17.44	16.04	15.44	15.10	14.88	14.73	14.62	14.54	14.47	14.42	14.34	14.25	14.17	14.12	14.08	14.04	13.99	13.95	13.90
4	12.22	10.65	9.98	9.60	9.36	9.20	9.07	8.98	8.90	8.84	8.75	8.66	8.56	8.51	8.46	8.41	8.36	8.31	8.26
5	10.01	8.43	7.76	7.39	7.15	6.98	6.85	6.76	6.68	6.62	6.52	6.43	6.33	6.28	6.23	6.18	6.12	6.07	6.02
6	8.81	7.26	6.60	6.23	5.99	5.82	5.70	5.60	5.52	5.46	5.37	5.27	5.17	5.12	5.07	5.01	4.96	4.90	4.85
7	8.07	6.54	5.89	5.52	5.29	5.12	4.99	4.90	4.82	4.76	4.67	4.57	4.47	4.42	4.36	4.31	4.25	4.20	4.14
8	7.57	6.06	5.42	5.05	4.82	4.65	4.53	4.43	4.36	4.30	4.20	4.10	4.00	3.95	3.89	3.84	3.78	3.73	3.67
9	7.21	5.71	5.08	4.72	4.48	4.32	4.20	4.10	4.03	3.96	3.87	3.77	3.67	3.61	3.56	3.51	3.45	3.39	3.33
10	6.94	5.46	4.83	4.47	4.24	4.07	3.95	3.85	3.78	3.72	3.62	3.52	3.42	3.37	3.31	3.26	3.20	3.14	3.08
11	6.72	5.26	4.63	4.28	4.04	3.88	3.76	3.66	3.59	3.53	3.43	3.33	3.23	3.17	3.12	3.06	3.00	2.94	2.88
12	6.55	5.10	4.47	4.12	3.89	3.73	3.61	3.51	3.44	3.37	3.28	3.18	3.07	3.02	2.96	2.91	2.85	2.79	2.72
13	6.41	4.97	4.35	4.00	3.77	3.60	3.48	3.39	3.31	3.25	3.15	3.05	2.95	2.89	2.84	2.78	2.72	2.66	2.60
14	6.30	4.86	4.24	3.89	3.66	3.50	3.38	3.29	3.21	3.15	3.05	2.95	2.84	2.79	2.73	2.67	2.61	2.55	2.49
15	6.20	4.77	4.15	3.80	3.68	3.41	3.29	3.20	3.12	3.06	2.96	2.86	2.76	2.70	2.64	2.59	2.52	2.46	2.40
16	6.12	4.69	4.08	3.73	3.50	3.34	3.22	3.12	3.05	2.99	2.89	2.79	2.68	2.63	2.57	2.51	2.45	2.38	2.32
17	6.04	4.62	4.01	3.66	3.44	3.28	3.16	3.06	2.98	2.92	2.82	2.72	2.62	2.56	2.50	2.44	2.38	2.32	2.25
18	5.98	4.56	3.95	3.61	3.38	3.22	3.10	3.01	2.93	2.87	2.77	2.67	2.56	2.50	2.44	2.38	2.32	2.26	2.19
19	5.92	4.51	3.90	3.56	3.33	3.17	3.05	2.96	2.88	2.82	2.72	2.62	2.51	2.45	2.39	2.33	2.27	2.20	2.13

Table D4. Continued

ν_2	ν_1																		
	1	2	3	4	5	6	7	8	9	10	12	15	20	24	30	40	60	120	∞
20	5.87	4.46	3.86	3.51	3.29	3.13	3.01	2.91	2.84	2.77	2.68	2.57	2.46	2.41	2.35	2.29	2.22	2.16	2.09
21	5.83	4.42	3.82	3.48	3.25	3.09	2.97	2.87	2.80	2.73	2.64	2.53	2.42	2.37	2.31	2.25	2.18	2.11	2.04
22	5.79	4.38	3.78	3.44	3.22	3.05	2.93	2.84	2.76	2.70	2.60	2.50	2.39	2.33	2.27	2.21	2.14	2.08	2.00
23	5.75	4.35	3.75	3.41	3.18	3.02	2.90	2.81	2.73	2.67	2.57	2.47	2.36	2.30	2.24	2.18	2.11	2.04	1.97
24	5.72	4.32	3.72	3.38	3.15	2.99	2.87	2.78	2.70	2.64	2.54	2.44	2.33	2.27	2.21	2.15	2.08	2.01	1.94
25	5.69	4.29	3.69	3.35	3.13	2.97	2.85	2.75	2.68	2.61	2.51	2.41	2.30	2.24	2.18	2.12	2.05	1.98	1.91
26	5.66	4.27	3.67	3.33	3.10	2.94	2.82	2.73	2.65	2.59	2.49	2.39	2.28	2.22	2.16	2.09	2.03	1.95	1.88
27	5.63	4.24	3.65	3.31	3.08	2.92	2.80	2.71	2.63	2.57	2.47	2.36	2.25	2.19	2.13	2.07	2.00	1.93	1.85
28	5.61	4.22	3.63	3.29	3.06	2.90	2.78	2.69	2.61	2.55	2.45	2.34	2.23	2.17	2.11	2.05	1.98	1.91	1.83
29	5.59	4.20	3.61	3.27	3.04	2.88	2.76	2.67	2.59	2.53	2.43	2.32	2.21	2.15	2.09	2.03	1.96	1.89	1.81
30	5.57	4.18	3.59	3.25	3.03	2.87	2.75	2.65	2.57	2.51	2.41	2.31	2.20	2.14	2.07	2.01	1.94	1.87	1.79
40	5.42	4.05	3.46	3.13	2.90	2.74	2.62	2.53	2.45	2.39	2.29	2.18	2.07	2.01	1.94	1.88	1.80	1.72	1.64
60	5.29	3.93	3.34	3.01	2.79	2.63	2.51	2.41	2.33	2.27	2.17	2.06	1.94	1.88	1.82	1.74	1.67	1.58	1.48
120	5.15	3.80	3.23	2.89	2.67	2.52	2.39	2.30	2.22	2.16	2.05	1.94	1.82	1.76	1.69	1.61	1.53	1.43	1.31
∞	5.02	3.69	3.12	2.79	2.57	2.41	2.29	2.19	2.11	2.05	1.94	1.83	1.71	1.64	1.57	1.48	1.39	1.27	1.00

Table D5. Critical values for a two-tailed z test
(To find the critical values for a one-tailed test divide the p value in half. For example the critical values for a two-tailed test when $p = 0.1$ will be the critical values for $p = 0.05$ for a one-tailed test)

Probability value	z
0.10	1.647
0.05	1.960
0.01	2.576
0.02	2.326
0.002	3.100
0.001	3.291

Table D6. Critical values for an F_{max} test that is carried out to confirm that the variances are homogeneous before a parametric ANOVA, where $p = 0.05$, a is the number of samples or treatments and v is the degrees of freedom.

v	a										
	2	3	4	5	6	7	8	9	10	11	12
2	39.0	87.5	142	202	266	333	403	475	550	626	704.0
3	15.4	27.8	39.2	50.7	62.0	72.9	83.5	93.9	104	114	124.0
4	9.60	15.5	20.6	25.2	29.5	33.6	37.5	41.1	44.6	48.0	51.4
5	7.15	10.8	13.7	16.3	18.7	20.8	22.9	24.7	26.5	28.2	29.9
6	5.82	8.38	10.4	12.1	13.7	15.0	16.3	17.5	18.6	19.7	20.7
7	4.99	6.94	8.44	9.70	10.8	11.8	12.7	13.5	14.3	15.1	15.8
8	4.43	6.00	7.18	8.12	9.03	9.78	10.5	11.1	11.7	12.2	12.7
9	4.03	5.34	6.31	7.11	7.80	8.41	8.95	9.45	9.91	10.3	10.7
10	3.72	4.85	5.67	6.34	6.92	7.42	7.87	8.28	8.66	9.01	9.34
12	3.28	4.16	4.79	5.30	5.72	6.09	6.42	6.72	7.00	7.25	7.48
15	2.86	3.54	4.01	4.37	4.68	4.95	5.19	5.40	5.59	5.77	5.93
20	2.46	2.95	3.29	3.54	3.76	3.94	4.10	4.24	4.37	4.49	4.59
30	2.07	2.40	2.61	2.78	2.91	3.02	3.12	3.21	3.29	3.36	3.39
60	1.67	1.85	1.96	2.04	2.11	2.17	2.22	2.26	2.30	2.33	2.36

Table D7. Critical values for an *F* test for a parametric ANOVA, where $p = 0.05$, ν_1 is the degrees of freedom for numerator and ν_2 is the degrees of freedom for the denominator

ν_2	ν_1									
	1	2	3	4	5	6	8	12	24	∞
1	161.4	199.5	215.7	224.6	230.2	234.0	238.9	243.9	249.0	254.3
2	18.51	19.00	19.16	19.25	19.30	19.33	19.37	19.41	19.45	19.50
3	10.13	9.55	9.28	9.12	9.01	8.94	8.84	8.74	8.64	8.53
4	7.71	6.94	6.59	6.39	6.26	6.16	6.04	5.91	5.77	5.63
5	6.61	5.79	5.41	5.19	5.05	4.95	4.82	4.68	4.53	4.36
6	5.99	5.14	4.76	4.53	4.39	4.28	4.15	4.00	3.84	3.67
7	5.59	4.74	4.35	4.12	3.97	3.87	3.73	3.57	3.41	3.23
8	5.32	4.46	4.07	3.84	3.69	3.58	3.44	3.28	3.12	2.93
9	5.12	4.26	3.86	3.63	3.48	3.37	3.23	3.07	2.90	2.71
10	4.96	4.10	3.71	3.48	3.33	3.22	3.07	2.91	2.74	2.54
11	4.85	3.98	3.59	3.36	3.20	3.09	2.95	2.79	2.61	2.40
12	4.75	3.88	3.49	3.26	3.11	3.00	2.85	2.69	2.50	2.30
13	4.67	3.80	3.41	3.18	3.02	2.92	2.77	2.60	2.42	2.21
14	4.60	3.74	3.34	3.11	2.96	2.85	2.70	2.53	2.35	2.13
15	4.54	3.68	3.29	3.06	2.90	2.79	2.64	2.48	2.29	2.07
16	4.49	3.63	3.24	3.01	2.85	2.74	2.59	2.42	2.24	2.01
17	4.45	3.59	3.20	2.96	2.81	2.70	2.55	2.38	2.19	1.96
18	4.41	3.55	3.16	2.93	2.77	2.66	2.51	2.34	2.15	1.92
19	4.38	3.52	3.13	2.90	2.74	2.63	2.48	2.31	2.11	1.88
20	4.35	3.49	3.10	2.87	2.71	2.60	2.45	2.28	2.08	1.84
21	4.32	3.47	3.07	2.84	2.68	2.57	2.42	2.25	2.05	1.81
22	4.30	3.44	3.05	2.82	2.66	2.55	2.40	2.23	2.03	1.78
23	4.28	3.42	3.03	2.80	2.64	2.53	2.38	2.20	2.00	1.76
24	4.26	3.40	3.01	2.78	2.62	2.51	2.36	2.18	1.98	1.73
25	4.24	3.38	2.99	2.76	2.60	2.49	2.34	2.16	1.96	1.71
26	4.22	3.37	2.98	2.74	2.59	2.47	2.32	2.15	1.95	1.69
27	4.21	3.35	2.96	2.73	2.57	2.46	2.30	2.13	1.93	1.67
28	4.20	3.34	2.95	2.71	2.56	2.44	2.29	2.12	1.91	1.65
29	4.18	3.33	2.93	2.70	2.54	2.43	2.28	2.10	1.90	1.64
30	4.17	3.32	2.92	2.69	2.53	2.42	2.27	2.09	1.89	1.62
40	4.08	3.23	2.84	2.61	2.45	2.34	2.18	2.00	1.79	1.51
60	4.00	3.15	2.76	2.52	2.37	2.25	2.10	1.92	1.70	1.39
120	3.92	3.07	2.68	2.45	2.29	2.17	2.02	1.83	1.61	1.25
∞	3.84	2.99	2.60	2.37	2.21	2.10	1.94	1.75	1.52	1.00

Table D8. Critical values for an *F* test for a parametric ANOVA, where $p = 0.01$, ν_1 is the degrees of freedom for numerator and ν_2 is the degrees of freedom for the denominator

ν_2	ν_1									
	1	2	3	4	5	6	8	12	24	∞
1	4052	4999	5403	5625	5764	5859	5982	6106	6234	6366
2	98.50	99.00	99.17	99.25	99.30	99.33	99.37	99.42	99.46	99.50
3	34.12	30.82	29.46	28.71	28.24	27.91	27.49	27.05	26.60	26.12
4	21.20	18.00	16.69	15.98	15.52	15.21	14.80	14.37	13.93	13.46
5	16.26	13.27	12.06	11.39	10.97	10.67	10.29	9.89	9.47	9.02
6	13.74	10.92	9.78	9.15	8.75	8.47	8.10	7.72	7.31	6.88
7	12.25	9.55	8.45	7.85	7.46	7.19	6.84	6.47	6.07	5.65
8	11.26	8.65	7.59	7.01	6.63	6.37	6.03	5.67	5.28	4.86
9	10.56	8.02	6.99	6.42	6.06	5.80	5.47	5.11	4.73	4.31
10	10.04	7.56	6.55	5.99	5.64	5.39	5.06	4.71	4.33	3.91
11	9.65	7.20	6.22	5.67	5.32	5.07	4.74	4.40	4.02	3.60
12	9.33	6.93	5.95	5.41	5.06	4.82	4.50	4.16	3.78	3.36
13	9.07	6.70	5.74	5.20	4.86	4.62	4.30	3.96	3.59	3.16
14	8.86	6.51	5.56	5.03	4.69	4.46	4.14	3.80	3.43	3.00
15	8.68	6.36	5.42	4.89	4.56	4.32	4.00	3.67	3.29	2.87
16	8.53	6.23	5.29	4.77	4.44	4.20	3.89	3.55	3.18	2.75
17	8.40	6.11	5.18	4.67	4.34	4.10	3.79	3.45	3.08	2.65
18	8.28	6.01	5.09	4.58	4.25	4.01	3.71	3.37	3.00	2.57
19	8.18	5.93	5.01	4.50	4.17	3.94	3.63	3.30	2.92	2.49
20	8.10	5.85	4.94	4.43	4.10	3.87	3.56	3.23	2.86	2.42
21	8.02	5.78	4.87	4.37	4.04	3.81	3.51	3.17	2.80	2.36
22	7.94	5.72	4.82	4.31	3.99	3.76	3.45	3.12	2.75	2.31
23	7.88	5.66	4.76	4.26	3.94	3.71	3.41	3.07	2.70	2.26
24	7.82	5.61	4.72	4.22	3.90	3.67	3.36	3.03	2.66	2.21
25	7.77	5.57	4.68	4.18	3.86	3.63	3.32	2.99	2.62	2.17
26	7.72	5.53	4.64	4.14	3.82	3.59	3.29	2.96	2.58	2.13
27	7.68	5.49	4.60	4.11	3.78	3.56	3.26	2.93	2.55	2.10
28	7.64	5.45	4.57	4.07	3.75	3.53	3.23	2.90	2.52	2.06
29	7.60	5.42	4.54	4.04	3.73	3.50	3.20	2.87	2.49	2.03
30	7.56	5.39	4.51	4.02	3.70	3.47	3.17	2.84	2.47	2.01
40	7.31	5.18	4.31	3.83	3.51	3.29	2.99	2.66	2.29	1.80
60	7.08	4.98	4.13	3.65	3.34	3.12	2.82	2.50	2.12	1.60
120	6.85	4.79	3.95	3.48	3.17	2.96	2.66	2.34	1.95	1.38
∞	6.64	4.60	3.78	3.32	3.02	2.80	2.51	2.18	1.79	1.00

Table D9. *q* values for a Tukey's test at $p = 0.05$ and $p = 0.01$, where *a* is the number of samples and ν is the degrees of freedom for the s^2_{within} from the ANOVA calculation.

ν		a																			ν
	p	2	3	4	5	6	7	8	9	10	11	12	13	14	15	16	17	18	19	20	
5	.05	3.64	4.60	5.22	5.67	6.03	6.33	6.58	6.80	6.99	7.17	7.32	7.47	7.60	7.72	7.83	7.93	8.03	8.12	8.21	.05
	.01	5.70	6.97	7.80	8.42	8.91	9.32	9.67	9.97	10.24	10.48	10.70	10.89	11.08	11.24	11.40	11.55	11.68	11.81	11.93	.01
6	.05	3.46	4.34	4.90	5.31	5.63	5.89	6.12	6.32	6.49	6.65	6.79	6.92	7.03	7.14	7.24	7.34	7.43	7.51	7.59	.05
	.01	5.24	6.33	7.03	7.56	7.97	8.32	8.61	8.87	9.10	9.30	9.49	9.65	9.81	9.95	10.08	10.21	10.32	10.43	10.54	.01
7	.05	3.34	4.16	4.68	5.06	5.36	5.61	5.82	6.00	6.16	6.30	6.43	6.55	6.66	6.76	6.85	6.94	7.02	7.09	7.17	.05
	.01	4.95	5.92	6.54	7.01	7.37	7.68	7.94	8.17	8.37	8.55	8.71	8.86	9.00	9.12	9.24	9.35	9.46	9.55	9.65	.01
8	.05	3.26	4.04	4.53	4.89	5.17	5.40	5.60	5.77	5.92	6.05	6.18	6.29	6.39	6.48	6.57	6.65	6.73	6.80	6.87	.05
	.01	4.74	5.63	6.20	6.63	6.96	7.24	7.47	7.68	7.87	8.03	8.18	8.31	8.44	8.55	8.66	8.76	8.85	8.94	9.03	.01
9	.05	3.20	3.95	4.42	4.76	5.02	5.24	5.43	5.60	5.74	5.87	5.98	6.09	6.19	6.28	6.36	6.44	6.51	6.58	6.64	.05
	.01	4.60	5.43	5.96	6.35	6.66	6.91	7.13	7.32	7.49	7.65	7.78	7.91	8.03	8.13	8.23	8.32	8.41	8.49	8.57	.01
10	.05	3.15	3.88	4.33	4.65	4.91	5.12	5.30	5.46	5.60	5.72	5.83	5.93	6.03	6.11	6.20	6.27	6.34	6.40	6.47	.05
	.01	4.48	5.27	5.77	6.14	6.43	6.67	6.87	7.05	7.21	7.36	7.48	7.60	7.71	7.81	7.91	7.99	8.07	8.15	8.22	.01
11	.05	3.11	3.82	4.26	4.57	4.82	5.03	5.20	5.35	5.49	5.61	5.71	5.81	5.90	5.99	6.06	6.14	6.20	6.26	6.33	.05
	.01	4.39	5.14	5.62	5.97	6.25	6.48	6.67	6.84	6.99	7.13	7.25	7.36	7.46	7.56	7.65	7.73	7.81	7.88	7.95	.01
12	.05	3.08	3.77	4.20	4.51	4.75	4.95	5.12	5.27	5.40	5.51	5.62	5.71	5.80	5.88	5.95	6.03	6.09	6.15	6.21	.05
	.01	4.32	5.04	5.50	5.84	6.10	6.32	6.51	6.67	6.81	6.94	7.06	7.17	7.26	7.36	7.44	7.52	7.59	7.66	7.73	.01
13	.05	3.06	3.73	4.15	4.45	4.69	4.88	5.05	5.19	5.32	5.43	5.53	5.63	5.71	5.79	5.86	5.93	6.00	6.05	6.11	.05
	.01	4.26	4.96	5.40	5.73	5.98	6.19	6.37	6.53	6.67	6.79	6.90	7.01	7.10	7.19	7.27	7.34	7.42	7.48	7.55	.01
14	.05	3.03	3.70	4.11	4.41	4.64	4.83	4.99	5.13	5.25	5.36	5.46	5.55	5.64	5.72	5.79	5.85	5.92	5.97	6.03	.05
	.01	4.21	4.89	5.32	5.63	5.88	6.08	6.26	6.41	6.54	6.66	6.77	6.87	6.96	7.05	7.12	7.20	7.27	7.33	7.39	.01
15	.05	3.01	3.67	4.08	4.37	4.60	4.78	4.94	5.08	5.20	5.31	5.40	5.49	5.58	5.65	5.72	5.79	5.85	5.90	5.96	.05
	.01	4.17	4.83	5.25	5.56	5.80	5.99	6.16	6.31	6.44	6.55	6.66	6.76	6.84	6.93	7.00	7.07	7.14	7.20	7.26	.01
16	.05	3.00	3.65	4.05	4.33	4.56	4.74	4.90	5.03	5.15	5.26	5.35	5.44	5.52	5.59	5.66	5.72	5.79	5.84	5.90	.05
	.01	4.13	4.78	5.19	5.49	5.72	5.92	6.08	6.22	6.35	6.46	6.56	6.66	6.74	6.82	6.90	6.97	7.03	7.09	7.15	.01

Table D9. Continued

v	a																				v
	p	2	3	4	5	6	7	8	9	10	11	12	13	14	15	16	17	18	19	20	
17	.05	2.98	3.63	4.02	4.30	4.52	4.71	4.86	4.99	5.11	5.21	5.31	5.39	5.47	5.55	5.61	5.68	5.74	5.79	5.84	.05
	.01	4.10	4.74	5.14	5.43	5.66	5.85	6.01	6.15	6.27	6.38	6.48	6.57	6.66	6.73	6.80	6.87	6.94	7.00	7.05	.01
18	.05	2.97	3.61	4.00	4.28	4.49	4.67	4.82	4.96	5.07	5.17	5.27	5.35	5.43	5.50	5.57	5.63	5.69	5.74	5.79	.05
	.01	4.07	4.70	5.09	5.38	5.60	5.79	5.94	6.08	6.20	6.31	6.41	6.50	6.58	6.65	6.72	6.79	6.85	6.91	6.96	.01
19	.05	2.96	3.59	3.98	4.25	4.47	4.65	4.79	4.92	5.04	5.14	5.23	5.32	5.39	5.46	5.53	5.59	5.65	5.70	5.75	.05
	.01	4.05	4.67	5.05	5.33	5.55	5.73	5.89	6.02	6.14	6.25	6.34	6.43	6.51	6.58	6.65	6.72	6.78	6.84	6.89	.01
20	.05	2.95	3.58	3.96	4.23	4.45	4.62	4.77	4.90	5.01	5.11	5.20	5.28	5.36	5.43	5.49	5.55	5.61	5.66	5.71	.05
	.01	4.02	4.64	5.02	5.29	5.51	5.69	5.84	5.97	6.09	6.19	6.29	6.37	6.45	6.52	6.59	6.65	6.71	6.76	6.82	.01
24	.05	2.92	3.53	3.90	4.17	4.37	4.54	4.68	4.81	4.92	5.01	5.10	5.18	5.25	5.32	5.38	5.44	5.50	5.54	5.59	.05
	.01	3.96	4.54	4.91	5.17	5.37	5.54	5.69	5.81	5.92	6.02	6.11	6.19	6.26	6.33	6.39	6.45	6.51	6.56	6.61	.01
30	.05	2.89	3.49	3.84	4.10	4.30	4.46	4.60	4.72	4.83	4.92	5.00	5.08	5.15	5.21	5.27	5.33	5.38	5.43	5.48	.05
	.01	3.89	4.45	4.80	5.05	5.24	5.40	5.54	5.65	5.76	5.85	5.93	6.01	6.08	6.14	6.20	6.26	6.31	6.36	6.41	.01
40	.05	2.86	3.44	3.79	4.04	4.23	4.39	4.52	4.63	4.74	4.82	4.91	4.98	5.05	5.11	5.16	5.22	5.27	5.31	5.36	.05
	.01	3.82	4.37	4.70	4.93	5.11	5.27	5.39	5.50	5.60	5.69	5.77	5.84	5.90	5.96	6.02	6.07	6.12	6.17	6.21	.01
60	.05	2.83	3.40	3.74	3.98	4.16	4.31	4.44	4.55	4.65	4.73	4.81	4.88	4.94	5.00	5.06	5.11	5.16	5.20	5.24	.05
	.01	3.76	4.28	4.60	4.82	4.99	5.13	5.25	5.36	5.45	5.53	5.60	5.67	5.73	5.79	5.84	5.89	5.93	5.98	6.02	.01
120	.05	2.80	3.36	3.69	3.92	4.10	4.24	4.36	4.48	4.56	4.64	4.72	4.78	4.84	4.90	4.95	5.00	5.05	5.09	5.13	.05
	.01	3.70	4.20	4.50	4.71	4.87	5.01	5.12	5.21	5.30	5.38	5.44	5.51	5.56	5.61	5.66	5.71	5.75	5.79	5.83	.01
∞	.05	2.77	3.31	3.63	3.86	4.03	4.17	4.29	4.39	4.47	4.55	4.62	4.68	4.74	4.80	4.85	4.89	4.93	4.97	5.01	.05
	.01	3.64	4.12	4.40	4.60	4.76	4.88	4.99	5.08	5.16	5.23	5.29	5.35	5.40	5.45	5.49	5.54	5.57	5.61	5.65	.01

Table D10. Critical values for the Mann–Whitney U test (*U*) at $p = 0.05$ for a two-tailed test, where n_1 is the number of observations in sample 1 and n_2 is the number of observations in sample 2

(To find the critical values for a one-tailed test divide the p value in half. For example the critical values for a two-tailed test when $p = 0.1$ will be the critical values for $p = 0.05$ for a one-tailed test)

n_1	n_2																		
	2	3	4	5	6	7	8	9	10	11	12	13	14	15	16	17	18	19	20
2							0	0	0	0	1	1	1	1	1	2	2	2	2
3				0	1	1	2	2	3	3	4	4	5	5	6	6	7	7	8
4			0	1	2	3	4	4	5	6	7	8	9	10	11	11	12	13	13
5		0	1	2	3	5	6	7	8	9	11	12	13	14	15	17	18	19	20
6		1	2	3	5	6	8	10	11	13	14	16	17	19	21	22	24	25	27
7		1	3	5	6	8	10	12	14	16	18	20	22	24	26	28	30	32	34
8	0	2	4	6	8	10	13	15	17	19	22	24	26	29	31	34	36	38	41
9	0	2	4	7	10	12	15	17	20	23	26	28	31	34	37	39	42	45	48
10	0	3	5	8	11	14	17	20	23	26	29	33	36	39	42	45	48	52	55
11	0	3	6	9	13	16	19	23	26	30	33	37	40	44	47	51	55	58	62
12	1	4	7	11	14	18	22	26	29	33	37	41	45	49	53	57	61	65	69
13	1	4	8	12	16	20	24	28	33	37	41	45	50	54	59	63	67	72	76
14	1	5	9	13	17	22	26	31	36	40	45	50	55	59	64	67	74	78	83
15	1	5	10	14	19	24	29	34	39	44	49	54	59	64	70	75	80	85	90
16	1	6	11	15	21	26	31	37	42	47	53	59	64	70	75	81	86	92	98
17	2	6	11	17	22	28	34	39	45	51	57	63	67	75	81	87	93	99	105
18	2	7	12	18	24	30	36	42	48	55	61	67	74	80	86	93	99	106	112
19	2	7	13	19	25	32	38	45	52	58	65	72	78	85	92	99	106	113	119
20	2	8	13	20	27	34	41	48	55	62	69	76	83	90	98	105	112	119	127

Table D11. Critical values for the Wilcoxon's matched pairs test (T) between $p = 0.1$ and $p = 0.002$ for a two-tailed test, where N is the number of pairs of observations used to provide ranks for the calculation (i.e. not those where $d = 0$)
(To find the critical values for a one-tailed test divide the p value in half. For example the critical values for a two-tailed test when $p = 0.1$ will be the critical values for $p = 0.05$ for a one-tailed test)

N	p			
	0.1	0.05	0.02	0.002
5	0			
6	2	0		
7	3	2	0	
8	5	3	1	
9	8	5	3	
10	10	8	5	0
11	13	10	7	1
12	17	13	9	2
13	21	17	12	4
14	25	21	15	6
15	30	25	19	8
16	35	29	23	11
17	41	34	27	14
18	47	40	32	18
19	53	46	37	21
20	60	52	43	26
21	67	58	49	30
22	75	65	55	35
23	83	73	62	40
24	91	81	69	45
25	100	89	76	51
26	110	98	84	58
27	119	107	92	64
28	130	116	101	71
30	151	137	120	86
31	163	147	130	94
32	175	159	140	103
33	187	170	151	112

Glossary

Absolute You may find that having calculated the difference between two values you have a negative value. For some parts of some calculations any negative sign must be removed. In these instances the number without its sign is called the absolute value and is indicated as |number|. For example |−10| would be used in the calculation as +10.

Aim A generalized statement about the topic you are investigating.

Association Where one variable changes in a similar manner to another variable for reasons unknown (see Relationship).

a priori Before the event. You need to be aware when you are designing your experiments or using some statistical tests, such as a chi-squared test, whether you have a reason to expect a specific outcome. For example one *a priori* expectation arises when you are carrying out genetic crosses and wish to compare your observed data with an expected segregation ratio (Chapter 5). This notion of 'expectation' must not be confused with a personal preference or educated hunch.

Bar chart A figure for observations from an investigation with one treatment variable. The data are measured on a nominal or ordinal scale. The *x*-axis is the treatment variable. The *y*-axis is the number of observations (or frequency or percentage). Each category from the nominal or ordinal scale is represented as a discrete bar on the figure. There must be a space between each bar to indicate that the data are not measured on a continuous scale. The tops of these bars must not be joined together by a trend line (Fig. 10.3.). If many observations are recorded within each category the variation in each category may be represented by confidence intervals.

Bimodal When a distribution has two high points (peaks).

Categorical When each observation may fall into only one of two or more mutually exclusive categories (e.g. blood groups A, B, AB, or O). The categories may be either qualitative (e.g. flower colours described as red, pink, blue, green, yellow, purple) or quantitative (e.g. the number of piglets in a litter 0, 1, 2, 3, 4, 5, 6, etc.).

Confidence limits The values derived from a sample between which the population mean probably falls.

Control An experimental baseline against which any effects of the treatment(s) may be compared. It must be an integrated part of an experiment.

Continuous Observations that do not fall into a series of distinct categories and may take any value in the scale of measurement e.g. height (cm).

Curvilinear When data are plotted and where part or all of the distribution is found to curve rather than follow a straight line.

Data A number of observations or measurements on the subject you are investigating.

Degrees of freedom (ν) You may have a sample of three observations and a known total of 25. If two of the three observations are 7 and 15 the third observation has to be 3 so that the total is 25. There is no freedom over what the value of the last observation must be. This limit to choice is described by the degrees of freedom. The level of constraint results from the experimental design and type of data. In statistical tests these 'limits' are often allowed for in the calculation process. For many tests the degrees of freedom is $n-1$, as in our example. For some tests there are more constraints so there are only $n-2$ degrees of freedom.

Denominator In an equation where one term is divided by another, the value that is being used to divide is the denominator. For example, in the fraction 2/4, 4 is the denominator.

Dependent and independent variables These are terms used in investigations with two or more variables. They are used to imply several different meanings and as such the context of their use has to be considered carefully to appreciate which meaning is intended. These terms can mean:

a. that the independent variable causes the change in the second dependent variable (i.e. a relationship)
b. that the independent variable is taken without sampling error, is usually under the investigator's control and that there is an association (or relationship) between the two variables
c. that the independent variable is taken without sampling error and is usually under the investigator's control.

We tend to use the second of these definitions. The independent variable can be ordinal or interval and is plotted on the *x*-axis. The dependent variable is usually measured on an interval scale or is a derived variable such as frequency, proportion, percentage or rate.

Derived variables The unit of measure for the observations is the result of a calculation, e.g. ratio, proportion, percentage, rate.

Discrete Observations fall into a series of distinct categories. The number of categories is limited, e.g. the number of eggs in a clutch; colours (red, green, blue).

Expectation A mathematical (e.g. Gaussian equation) or biological (e.g. genetic segregation ratio) predicator of the outcome of an experiment. The predictor can be consistently applied. Your personal prediction is not a statistical expectation.

Expected The values generated in response to a mathematical or biological predictor such as the Gaussian equation or a genetic segregation ratio.

Experimental investigation An investigation that sets out to test a hypothesis.

General hypotheses The hypotheses are phrased in a general way. For example, there is no difference between the samples. See 'specific hypotheses'.

Histogram A type of figure for observations from an investigation with one treatment variable. The data are measured on a continuous (interval) scale. There are many data points along the *x*-axis usually because the treatment variable is not under the investigator's control. Therefore, these observations are organized into classes in a frequency table (10.4.) and these classes and the number of observations in each class are used for the *x*- and *y*-axes respectively. Each class is represented as a bar and each bar abuts the next, indicating that the scale of measurement is continuous not discrete. The tops of these bars should not be joined together with a trend line (Fig. 3.4.). If many observations are recorded within each class the variation in each class may be represented by confidence intervals.

Hypothesis The formal phrasing of each objective; includes details relating to the experiment and the way in which the data will be tested statistically.

Independent observations All observations are collected in a consistent manner so that no one observation is more similar to any other observation. For example, in a study comparing the diets of undergraduates during 1 week from two departments, in one department the investigators recorded the diets for 100 students but in the other smaller department they were only able to sample 5 students. To increase the data for this smaller department they therefore recorded the diet for each student on 3 separate occasions. The diet information from the small department is not independent. Any one student is contributing 3 results to the study and therefore these results are not independent of each other.

Independent variable see Dependent variable

Interaction The response to a treatment shown by one group of items is reversed in a second sample in such a way that it is clear that the two treatment variables are not having the same effect in the two groups. For example, the number of rabbits was recorded in two locations A and B at three points in the year. In location 1 the number of rabbits increased during the year, but in location 2 the number of rabbits decreased during the year. The effect of 'time' on the number of rabbits is different for the two samples, so the effect of time is dependent on location. Time and location would be said to be interacting.

Interpolation The estimation of a value when only nearby values are known, one greater than the required value, and the other less than the required value. For a worked example see 4.3.6.

Interval These data are measured on a continuous scale; the data are rankable and it is possible to measure the difference between each observation. Interval scales include temperature (°C), distance (cm), and date.

Item One representative of the population you wish to measure. In other texts an item may be referred to as an 'experimental unit', subject, or case.

Line graph A figure for observations from an investigation with one treatment variable. The data are measured on a continuous (interval) scale. There are few data points along the *x*-axis (treatment variable) usually because the treatment variable is under the investigator's control. Each observation or mean from a group of observations for a given *x* value is plotted as a point on the figure. Variation around a mean can be indicted using confidence intervals (Fig. 10.4.). These points may be joined together with a trend line.

Matched When two observations on the same scale are made for each item, then these two sets of observations are said to be matched. For example, if in a study of heart rate a resting rate was taken for 10 people first and then following exercise another heart-rate reading was taken for the same 10 people. For each person there is a before and after reading, so these data are matched. An alternative approach would be to assign 10 people at random and record a resting heart rate and then assign a different group of 10 people to the exercise treatment and take an 'after exercise' heart rate from this second, independent group. These data would not be matched (unmatched).

Median One measure of central tendency, calculated as the middle value of an ordered data set.

Mean One measure of central tendency for interval data where the distribution is known. For any one distribution the sample mean ($\bar{x}$) and **population mean** (μ) are calculated in the same way. The sample mean is used as an estimate of the population mean and a value called the confidence interval (3.7.) can be calculated to demonstrate the area around a sample mean in which the population mean will probably fall. The sample mean for normally distributed data is calculated as the sum (Σ) of all observations (x) divided by the number of observations in the data set (n).

Mode One measure of central tendency. When the data are placed in categories (which may be natural, (for example the number of blossoms on a rose plant) or imposed on continuous data, (for example the heights of trees placed in ranges of 0.00–0.99 m; 1.00–1.99 m; etc.)), the mode is the category that contains the greatest number of observations.

Nominal Observations fall into discrete categories. The categories have no specified order, e.g. colours (red, green, blue).

Non-parametric data Data that, when plotted, have a non-normal distribution or more usually where the distribution is unknown. When testing hypotheses and you have non-parametric data you may either transform the data (3.9.) or use non-parametric statistics (Chapters 5, 6, and 8). Parametric data may also be analysed using non-parametric statistics but not the other way round.

Non-treatment variable Any factor(s) that may influence your observations but is not the factor(s) being investigated in the experiment.

Numerator In an equation where one term, is divided by another term, the value that is being divided into is the numerator. For example, in the equation 2/4, 2 is the numerator.

Objective Provides succinct and specific details about an experiment that is being carried out in order to examine all or part of an aim.

Observation A single measurement taken from an item.

Observational investigation An investigation which does not usually set out to test a hypothesis (for example, a general survey of a woodland).

One-tailed test When the hypothesis testing is such that only one end of a distribution is being considered. One-tailed tests are most often encountered when testing specific hypotheses.

Ordinal scale Observations that fall into discrete categories and the categories have an order (they can be ranked), e.g. the number of eggs in a clutch; the ACFOR scale.

Orthogonal A particular experimental design where every category for one treatment is found in combination with every category for every other treatment.

Outlier An observation that is noticeably different from all other observations, one that does not follow the apparent trend.

Parametric data Measured on an interval scale and so quantitative, continuous, and rankable. The data have to have a normal distribution, e.g. height (cm) (Table 3.1.). When testing hypotheses and you have parametric data you should use parametric statistics.

(Statistical) Population All the individual items that are the subject of your research.

Pie chart A figure for observations from an investigation with one treatment variable. The data are measured on a nominal or ordinal scale. A circle is divided up proportionately to reflect the relative proportions (or percentages) of the observations in each category or sample (Fig. 10.2.). If many observations are recorded within each category you should consider using a bar chart so that the variation in each category may be represented by confidence intervals.

Power of a test Table G1. illustrates the four possible outcomes from hypothesis testing. Two of these are where the actual biology accords with the outcome from the hypothesis testing – either there really is a difference and the statistical test detects this or there is not a difference and the statistical test confirms this. The probability of the first of these happening is called the power of a test.

There are two other possible outcomes from hypothesis testing neither of which are desirable. The first is that there is not a biological difference but the statistical test indicates that there is, or that there really is a difference biologically but the statistical test does not pick this up. These potential errors from hypothesis testing are called Type I and Type II errors respectively.

Non-parametric tests are generally considered to be less powerful than parametric tests because they are less vigorous in their conditions of use. This is only true if the assumptions behind the use of a statistical test are met. If not, then the value of using any test is undermined. Other factors that affect the likelihood of detecting a real difference include the use of a larger sample size, which is more likely to detect a real trend even when the difference is small, and a small *p* value, which decreases the probability of a Type 1 error occurring.

(The term power used here is not the same as that used in maths (Appendix c).)

Qualitative Observations are assigned to named, descriptive categories that are mutually exclusive and non-numerical, e.g. colours (red, green blue). The data are always discrete. Qualitative answers in questionnaires are descriptive answers.

Quantitative Observations are numerical and consist of named, ordered categories, e.g. number of eggs in a clutch or the height (cm). Data may be either discrete or continuous.

Random sampling Each item must have an equal chance of being sampled each time.

Range The range is the difference between the highest and lowest observations.

Rankable Observations consist of named categories or values that have an order to them. For example, finishing positions in a race or the height (cm).

Relationship Where one variable is directly responsible for causing the change in another variable.

Replication Where more than one item is exposed to a treatment; or more than one group of items are exposed to one treatment; or more than one item is included in a sample.

Research The process by which you try to find out more about a particular topic.

Sample A number of observations recorded from a subset of items in the population.

Sampling error The variation in a set of data that is due to the effects of non-treatment variables. Other authors may

Table G1. The power of a test, Type I and Type II errors

	The analysis does:	
Experimentally there really:	detect a difference. (The null hypothesis is rejected.)	not detect a difference. (The null hypothesis is not rejected.)
is a difference. (The null hypothesis should be rejected.)	Correct outcome The power of a test	Incorrect outcome Type II error
is not a difference (The null hypothesis should not be rejected.)	Incorrect outcome Type I error	Correct outcome

use the terms 'noise' or 'within treatment variation' when referring to sampling error.

Scatter plot A figure that can be drawn when two or more observations are made for each item. Each of these observations relates to one treatment variable (e.g. Fig. 6.12.). The data can be measured on any scale, although you must be very careful if you infer any trends in the data for nominal scales. If you have an independent variable this is plotted on the *x*-axis. The second variable, which may be a dependent variable, is plotted on the *y*-axis. If more than two variables are recorded, the plot will be three (or more) dimensions. The two observations from each item are marked as a point on the figure. If both measurements are interval, these points may be joined together with a trend line. You may carry out regression analysis and if significant you may draw a regression line (e.g. Figs 6.14., 6.15.). If more than one *y* value has been recorded for each *x* value you may use confidence intervals to reflect the variation around the mean *y* value.

Skew When observations are plotted the distribution may not be symmetrical. Where there is a tail of observations to the right the distribution is said to have a positive skew (Fig. 3.10.). Where there is a tail of observations to the left the distribution is said to have a negative skew (Fig. 3.11.).

Specific hypotheses Statistical tests of hypotheses may ask the question 'is there a difference?. This is an example of a general hypothesis. Some statistical tests allow you to ask more discerning (specific) questions. In the Tukey's test, where you are comparing many samples, you are able to compare all the means of these samples in all possible pairwise combinations. In a non-parametric ANOVA you examine the medians for each sample and devise your own specific arrangements of the medians as the basis for your hypothesis testing. For example is median A is greater than median B which is greater than median C? Specific tests of hypotheses are usually one-tailed.

Standard deviation Is a measure of the variation in data where the distribution is known. It is calculated as the square root of the variance and is a value with the same units as the original observations.

Systematic/Periodic sampling A regularized sampling method where an item at regular intervals is chosen for observation.

Transforming data A mathematical calculation that has the effect of squeezing and/or stretching the scale you used when making your measurements so that the distribution takes on the approximate shape of a bell-shaped curve. The process of transforming data can be carried out when you wish to normalize non-parametric data.

Treatment (definition 1) When carrying out an experiment your items are exposed to a particular environment that is manipulated by the investigator.

Treatment (definition 2) A statistical term for any samples that are being compared.

Treatment variables The factor(s) under investigation. For example, in an investigation into the effect of pH on plant growth, pH is the treatment variable.

Two-tailed test Where the hypothesis testing is against both ends of a distribution. Most general tests of hypotheses would usually be two-tailed.

Type I and Type II errors If, on the basis of using a particular *p* value, we reject the null hypothesis even when it is really true, this is called a Type I error. If we 'accept' the null hypothesis even when it is false this is called a Type II error. (See Power of a test.)

Unmatched See Matched.

Unimodal. The description of a distribution, that has a single high point or peak (see Bimodal).

Variance A measure of the variation in the data which for normally distributed data is calculated as:

$$s^2 = \frac{\sum (x - \bar{x})^2}{n - 1}$$

Variation When the observations within your data set do not all have the same value.

Variable The factor(s) that bring about variation between observations so that not all observations on items, a sample, or in a population, are the same.

References

BARNARD, C., GILBERT, F. & McGREGOR, P. (2001). *Asking questions in biology.* Longman Scientific and Technical, Harlow.

CROTHERS, J.H. (1981). On the graphical presentation of quantitative data. *Field Studies* **5**: 487–511.

GRAFEN, A. & HAILS, R. (2002). *Modern statistics for the life sciences.* Oxford University Press, Oxford.

HAWKINS, D. (2005). *Biomeasurement: understanding, analyzing and communicating data in the biosciences.* Oxford University Press, Oxford.

HEATH, D. (1995). *An introduction to experimental design and statistics for biology.* UCL Press, London.

FOWLER, J., COHEN, L. & JARVIS, P. (1998). *Practical statistics for field biologists.* John Wiley and Sons, Chichester.

LEGENDRE, P. & LEGENDRE, L. (1998). *Numerical ecology.* 2nd English edn. Elsevier Science, Amsterdam.

MEDDIS, R. (1984). *Statistics using ranks: a unified approach.* Blackwell Publishers, Oxford.

RADFORD, M. (2001). *Animal welfare law in Britain.* Oxford University Press, Oxford.

ROBSON, C. (2002). *Real world research: A resource for social scientists and practitioner researchers.* 2nd ed. Blackwell Publishers, Oxford.

RODWELL, J.S. (ed.) (1991). *British plant communities vol 1. Woodlands and Scrub.* Cambridge University Press, Cambridge.

SCHEIRER, C.J., RAY, W.S. & HARE, N. (1976). The analysis of ranked data derived from completely randomised factorial designs. *Biometrics* **32**: 429–434.

SOKAL, R.R. & ROHLF, F.J. (1981). *Biometry.* 2nd ed. Freeman, New York.

WILLIAMS, D.A. (1976). Improved likelihood ratio tests for complete contingency tables. *Biometrica* **63**: 33–37

WRIGHT, N.R. (1997). Breath alcohol concentrations in men 7–8 hours after prolonged, heavy drinking: influence of habitual alcohol intake. *Lancet* **349**:182

Index

S

T

U

V

W

Y

Z